HATE CRIMES

To Dennis, for his limitless support; to Allison, for her patience while I worked; and to Quinn, for waiting until the book was done.

Causes, Controls, and Controversies

Phyllis B. Gerstenfeld
California State University, Stanislaus

SAGE Publications
International Educational and Professional Publisher
Thousand Oaks ▪ London ▪ New Delhi

For information:

Sage Publications, Inc.
2455 Teller Road
Thousand Oaks, California 91320
E-mail: order@sagepub.com

Sage Publications Ltd.
6 Bonhill Street
London EC2A 4PU
United Kingdom

Sage Publications India Pvt. Ltd.
B-42, Panchsheel Enclave
Post Box 4109
New Delhi 110 017 India

Printed in the United States of America

Library of Congress Cataloging-in-Publication Data

Library of Congress Cataloging-in-Publication Data

Gerstenfeld, Phyllis B.
Hate crimes: Causes, controls, and controversies / Phyllis B. Gerstenfeld.
    p. cm.
Includes bibliographical references and index.
ISBN 0-7619-2813-8 (cloth: alk. paper)
ISBN 0-7619-2814-6 (pbk.: alk. paper)
  1. Hate crimes—United States. 2. Hate crimes—United States—Prevention. 1. Title.
HV6773.52 G47 2003
364.15′0973—dc21

2003006147

This book is printed on acid-free paper.

03   04   05   06   10   9   8   7   6   5   4   3   2   1

Acquisitions Editor:	Jerry Westby
Editorial Assistant:	Vonessa Vondera
Production Editor:	Denise Santoyo
Copy Editor:	Meredith Brittain
Typesetter:	C&M Digitals (P) Ltd.
Indexer:	Nara Wood
Cover Designer:	Michelle Lee

CONTENTS

FOREWORD

At the beginning of the 21st century, an amazing thing happened: The Attorney General for the United States of America filed federal hate crime charges against an alleged murderer not only for what he is accused of doing, but for who he chose to do it to and why as well.

What he is accused of doing is murder. *Who* he is accused of doing it to is two women who happen to be lesbians. *Why* he allegedly did it is because of hatred of women and hatred of lesbians.

Nearly 6 years after two women were bound and gagged and had their throats slit while camping and hiking in the Shenandoah National Park, U.S. Attorney General John D. Ashcroft held a nationally televised press conference on April 10, 2002, to announce that the U.S. Justice Department invoked the federal hate crimes statute for the first time to charge the alleged murderer with hate crime. In announcing the indictment, Ashcroft spoke at length about his meeting with the parents of the victims and about the lives and character of the young women: two Midwesterners who migrated to New England, met and became lovers, and shared a love of science and the outdoors. Justifying the invocation of federal hate crime law, which carries with it enhanced penalties, Ashcroft said that "criminal acts of hate run counter to what is best in America—our belief in equality and freedom. The Department of Justice will aggressively investigate, prosecute and punish criminal acts of violence and vigilantism motivated by hate and intolerance." Moreover, he said that "we will pursue, prosecute, and punish those who attack law-abiding Americans out of hatred for who they are" and "hatred is the enemy of justice, regardless of its source" (Department of Justice, 2002b).

According to prosecutors, Darrell David Rice, a computer programmer from Columbia, Maryland, is, by his own account, a man who hates lesbians and enjoys intimidating and assaulting women. According to law enforcement, sometime after being arrested, Rice told law enforcement officials that he intentionally selected women to assault "because they are more vulnerable than men," that he "hates gays," and that the victims in this case "deserved to die because they were lesbian whores" (Department of Justice, 2002b, p. A1). Accordingly, lead Assistant U.S. Attorney Tom Bondurant, Jr., plans to argue that Rice chose to slit the throats of the two young women because of their gender and because of their "actual or perceived sexual orientation" (Department of Justice, 2002b). According to court documents, Bondurant will introduce evidence of the defendant's numerous physical and verbal assaults upon randomly selected women, including acts of road rage, physical assaults, demeaning sexual comments, and threats of injury or death. The U.S. Assistant Attorney plans to argue in court that "the defendant's killing of the two women was part of an ongoing plan, scheme or modus operandi to assault, intimidate, injure and kill women because of their gender" (Kellman, 2002).

Regardless of how this historic case concludes, it serves to highlight that, by the beginning of the 21st century, "hate crime" found a home in social, political, and legal discourse in the United States. Although it remains an empirical question whether the United States is experiencing greater levels of hate- or bias-motivated violence than in the past, it is beyond dispute that the term *hate crime* has found a home in various institutional spheres of modern life. From the introduction and politicization of the term in the late 1970s to the continued enforcement of hate crime law at the beginning of the 21st century, modern civil rights movements constructed the problem of bias-motivated violence in ways that distinguish it from other forms of violent crime; state and federal politicians made legislation that defines the parameters of hate crime in ways that distinguish it from other types of violent crime; judicial decision makers elaborated and enriched the meaning of hate crime as they determined the constitutionality of "hate crime" as a legal concept that distinguishes types of violence based on the motivation of the perpetrator; and law enforcement officials continue to investigate and prosecute bias-motivated incidents as a special type of crime that warrants enhanced penalties.

The result of these changes is significant. In the United States, the use of the term *hate crime* is now commonplace. In the latter part of the 20th century it became increasingly understood that criminal conduct takes on a new meaning when it involves an act motivated by bigotry. Violence born of bigotry and manifested as discrimination has been resignified and reacted to in ways that result in the reconfiguration of violence directed toward minorities. In particular, people of color, Jews, gays and lesbians, women, and those with disabilities increasingly have been recognized as victims of hate crime, whereas other vulnerable victims—for example, union members, octogenarians, the elderly, children, and police officers—have not. This classification reflects the politics of hate crime in the United States. More recently, the U.S.-born concept of hate crime has diffused across international borders as various Western countries, especially those sharing a predominantly English-speaking culture, appropriate and deploy the concept to reference bias-motivated conduct in their respective legal and cultural milieus. Australia, for example, has outlawed at the federal, state, and territory level words and images that incite hatred toward particular groups of people. Relying on discrimination law, Australian legislators have outlawed conduct that constitutes "vilification" or "racial hatred." Britain and Canada have also passed a series of laws designed to curb racial-ethnic violence. Finally, Germany has passed laws that forbid "public incitement" and "instigation of racial hatred," including prohibiting the distribution of Nazi propaganda or literature liable to corrupt youth. Unlike the United States, other countries have adopted a fairly limited view of hate crime, focusing primarily on racial, ethnic, and religious violence, and still other countries—mostly in the non-Western world—have not adopted the term to reference racial, ethnic, religious, and other forms of intergroup conflict.

This domestic and international social change brings with it a plethora of questions about the causes, manifestations, and consequences of hate crimes, as well as the larger social context within which they occur, take shape, get reacted to, and (arguably) seem to be proliferating. For example, why has bias-motivated violence and its attendant categories of victimization only recently come to the fore and become recognized as a serious social problem in the United States, especially because violence directed at people because of their real or imagined characteristics is as old as humankind? And on a related note, why is it that injuries against some people—Jews, people of color, gays and lesbians, and, on occasion, women and those with disabilities—are increasingly recognized by the law and in the public's mind as a "hate crime," whereas other types of bias-motivated violence continue to go unnoticed? What is the nature of the acts that constitute hate crimes? Who commits hate crimes and why? Who is most likely to be victimized by hate crimes and why? In what ways are hate crimes and efforts to curb them connected to larger social movements? Under what conditions and how do communities in which hate crimes occur respond to such acts? What types of behaviors seem to be getting center stage in both public and policy discussions of hate crimes? Conversely, what types of behaviors evoke the attention of those charged with controlling hate crimes and/or protecting civil liberties? Who are the relevant political players and what organizations, institutions, and constituencies are associated with both the proliferation and the social control of hate crimes? Finally, how have social control efforts been undertaken, and to what degree have they been effective? With these questions in mind, this book constitutes the most comprehensive treatment of the topic of hate crime (broadly construed).

This book deploys an exceptionally accessible language to expose students, activists, policymakers, and scholars alike to up-to-date information that they can digest and use to make sense of the causes, manifestations, and consequences of the types of violence that might fall under the rubric of hate crime. As such, it does more than simply provide "the facts" of hate crime—who is doing what to whom, how often, and under what conditions. It goes further. It takes an interdisciplinary perspective to make sense of the way groups do and do not get along; the way race, ethnicity, religion, gender, sexuality, (dis)abilities, and so on are organized; and the way systems of social control, especially law, work. In the process, it helps readers understand why the murder of Julianne Marie Williams and Laura Winnans, the two women described in the case presented in the opening paragraph of this foreword, is historic insofar as it is being treated significantly differently than most murders that have occurred in U.S. history, including other murders that occurred the same year (1996). However, this is more than a book on one specific type of violent conduct and criminal activity (i.e., hate crime) and the social control efforts designed to curb such violence (bias-motivated violence). To its credit, the book treats the study of hate crimes as a window through which a variety of social structures and processes can be rendered visible and amenable to examination, especially those related to social stratification, social stability, social change, and social control.

—Valerie Jenness
Department of Criminology, Law,
and Society
Department of Sociology,
University of California, Irvine, California

PREFACE

New acquaintances often ask me what I study. When I tell them hate crimes, I am usually met with a few moments of surprised silence. Soon, though, they begin to ask me questions about incidents they have heard about, such as the murders of Matthew Shepard and James Byrd, Jr.; or they talk about skinheads or the Klan; or, if they follow current legal developments, they voice opinions about recent laws or Supreme Court opinions. Hate crimes seem to be a topic of some interest to nearly everybody, and yet few people really know much about them.

Hate crimes are a new area of academic study. Little more than a decade ago, few people had even heard the term. In the past few years, however, there has been a small explosion of research on hate crimes, with new articles and books appearing at an increased rate. In this book, I have attempted to present as complete a picture as possible of the current state of knowledge about hate crimes while at the same time acknowledging the significant gaps that remain.

Researching and writing this book was a challenge for many reasons, one of which is the quick rate at which new information on hate crimes is appearing. As I was writing, new events transpired and new studies were published. To the extent possible, I have incorporated these throughout the book.

A second challenge was the diversity of disciplinary approaches that have been taken to study hate crimes. Research on hate crimes has been undertaken by scholars in the fields of sociology, criminology, criminal justice, psychology, economics, political science, public policy, and law, among others. Each of these perspectives has provided useful insight about a complex topic. Therefore, I have taken a multidisciplinary approach in this book.

A third challenge was the large scope of issues related to hate crimes. A thorough study of the topic includes an examination of, for example, the causes of prejudice, the history and operation of hate crime legislation, the activities of organized extremist groups, international manifestations and solutions to hate crimes, and the consequences of hate crimes upon victims and communities, as well as many other subjects. This book addresses the many facets of hate crimes, providing a complete picture of an intricate problem.

Hate crimes are of great concern to us as a nation and have the potential to affect the lives of many of us as individuals—that is why I chose to write this book. I hope, of course, that this book will help others learn a great deal about the subject. But increased knowledge is not my only goal. There exist many controversies concerning this subject, and I hope that this book will spark discussion and debate. In addition, many important questions about hate crimes remain unanswered. I hope that this book will inspire others to make their own attempts to address those questions. Ultimately, I hope that this book will help to reduce the numbers of hate crimes that occur.

I have been writing this book in my head for a long time, but it was only with the assistance of many others that it finally made it onto paper. I would like to extend special thanks to Gabi Clayton, Michael Lieberman, Michael Novick, Wendy Patrick, and Karen Zatz, all of whom very kindly contributed original pieces for the Narrative Portraits contained within this book. Their contributions have made the book richer and more interesting. I would also like to thank Valerie Jenness, who took time out of a very busy schedule to write the foreword to this book.

I appreciate the support for this project that I have received from my colleagues. I would especially like to acknowledge the endless support I received from Diana Grant. Dr. Grant and Chau-Pu Chiang were my valuable partners for some of the research discussed in this book. Alan Tomkins provided the mentoring for the dissertation research that was the genesis of my work on hate crimes. Thanks also to my students, whose questions and comments over the years helped shape this book and whose enthusiasm helped give me the impetus to write it.

The people at Sage Publications have made my first experience writing a book an enjoyable one, and they have contributed a great deal to the quality of this book. Special thanks to Jerry Westby, Denise Santoyo, Meredith Brittain, and Vonessa Vondera for all of their able advice and assistance, and for being able to manage a very tight schedule of deadlines.

I would also like to thank all of the reviewers for providing valuable critiques: Paul Becker, University of Dayton; Randy Blazak, Portland State University; Tim Buzzell, Baker University; Katherine A. Culotta, Indiana State University; Art Jipson, University of Dayton; Rebecca Katz, Morehead State; Victoria Munoz, Wells College; Barbara Perry, Northern Arizona University; Carolyn Petrosino, Bridgewater State College; Phillip Neil Quisenberry, University of Kentucky; Sarah A. Soule, University of Arizona; and Anne Sullivan, Salem State College.

Finally, I would like to thank my friends and family for encouraging me in my work, providing emotional support when I needed it, and generally helping to keep me going. Dennis Behrens deserves special appreciation for being a sounding board for my ideas, urging me onward when I needed encouragement, cheering for me when I reached my goals, and holding down the fort while I worked. And, of course, thank you to Allison and Quinn for giving me what I needed most: the time to work and the frequent reminders of what really matters.

ACKNOWLEDGMENTS

For permission to reprint from the following, grateful acknowledgment is made to the publishers and copyright holders:

Chapter 2:

Excerpt from D. Luzadder (1999) reproduced with permission of the *Rocky Mountain News,* copyright 1998.

Excerpt from G. Freeman (1998) reproduced with permission of the *St. Louis Post-Dispatch,* copyright 1998.

Pieces by K. Zatz and M. Lieberman printed courtesy of the authors.

Chapter 3:

Excerpt from J. J. Nolan & Y. Akiyama (1999) reproduced with permission of Sage Publications.

Piece by W. Patrick printed courtesy of the author.

Chapter 4:

Excerpt from L. L. King (1969) reproduced with permission of the author. Originally published in King, L. L. (1969). *Confessions of a white racist*. New York: Viking Press.

Excerpt from O. G. Davidson (1996) reproduced with permission of the author. Originally published in Davidson, O. G. (1996). *The best of enemies: Race and redemption in the New South*. New York: Scribner.

Chapter 5:

Excerpt from P. Simms (1996) reproduced with permission of the University Press of Kentucky. Originally published in Simms, P. (1996). *The Klan*. Lexington, KY: University Press of Kentucky.

Excerpt from J. Ronson (2002) reproduced with permission of Macmillan, London, UK, and Simon & Schuster, New York. Originally published in Ronson, J. (2002). *Them: Adventures with extremists*. New York: Simon & Schuster.

Excerpt from "The Death of Emmett Till" copyright © 1963 by Warner Bros. Inc. Copyright renewed 1991 by Special Rider Music. All rights reserved. International copyright secured. Reprinted by permission.

Chapter 6:

Strange Fruit—Lewis Allen. © 1939 (Renewed) by Music Sales Corporation. All rights outside the United States controlled by Edward B. Marks Music Company. Used by permission. All rights reserved. © 1939 (Renewed) by Music Sales Corporation (ASCAP). International Copyright Secured. All Rights Reserved. Reprinted by permission.

Excerpt from C. Brenner (1992) reproduced with permission of Sage Publications. Originally published in Brenner, C. (1992). Survivor's story: Eight bullets. In G. M. Herek and K. T. Berrill (Eds.), *Hate crimes: Confronting violence against lesbians and gay men*. Newbury Park, CA: Sage.

Excerpt from V. M. Hwang (2001) reproduced with permission of Rowman & Littlefield Publishers. Originally published in Hwang, V. M. (2001). The interrelationship between anti-Asian violence and Asian America. In P. W. Hall and V. M. Hwang (Eds.), *Anti-Asian violence in North America: Asian American and Asian Canadian reflections on hate, healing, and resistance*. Walnut Creek, CA: Rowman & Littlefield.

Chapter 7:

"The Jigsaw Classroom" by E. Aronson reproduced with permission of the author.

Piece by G. Clayton printed courtesy of the author.

Piece by M. Novick reprinted by permission from Michael Novick, editor of *Turning the Tide*. People Against Racist Terror (310) 288-5003.

Chapter 8:

Excerpt from S. M. Weine (1999) reproduced with permission of Rutgers University Press. Originally published in Weine, S. M. (1999). *When history is a nightmare: Lives and memories of ethnic cleansing in Bosnia-Herzegovina.* New Brunswick, NJ: Rutgers University Press.

Excerpt from F. Keane (1995) reproduced by permission of Penguin Books Ltd. Originally published in Keane, F. (1995). *A season of blood: A Rwandan journey.* New York: Penguin.

Chapter 1

INTRODUCTION

Shanika Williams was happy to have found a nice house to rent in Holiday, Florida. It was in a good family neighborhood, perfect for her children, who were 9, 6, and 4 years old. Shortly before Shanika and her children were to move into the house, however, someone broke in and scrawled messages on the freshly painted walls. Among the messages were "KKK" and "White Power Rules." Although she was upset, Williams, who is African American, declared that she was determined that these acts would not send her running away (Samolinksi, 2001).

Balbir Singh Sodhi immigrated from India to the United States. He and his family worked long hours for years as taxi drivers, finally saving enough money to buy a gas station and convenience store in Mesa, Arizona. On the afternoon of September 15, 2001, Sodhi was planting flowers in front of his new store when Frank Roque drove by and shot him to death, mistaking Sodhi in his Sikh turban and beard for a Muslim. Later that same day, Roque shot into the home of an Afghan American family and fired at a Lebanese American gas station attendant. When he was arrested, he told the police, "I am a patriot" (Cooperman, 2002).

On September 2, 2002, Trev Broudy was hugging a friend goodbye on a quiet street in West Hollywood. Suddenly, two men began attacking them with a bat and a metal pipe. Broudy was beaten severely enough that he remained in a coma for 9 days; his friend, who was able to escape to his parked car, was injured as well. An hour later in the same neighborhood, another man was beaten by two people in the same manner and suffered severe bruising. And a few weeks later, a fourth man was also attacked by two men wielding a bat and a pipe. All the victims were gay (Edds, 2002). Two men were later arrested for the beatings.

In January 2001, Kenneth Luker and two friends drove a backhoe and lawnmower over the grounds of two Jewish cemeteries in New Jersey, toppling and damaging numerous headstones. Luker was subsequently caught and sentenced to prison for 9 months. Four months after being paroled, he again vandalized the same cemeteries, toppling 40 gravestones and causing tens of thousands of dollars in damage. This time, he was sentenced for up to 5 years ("Repeat Cemetery Vandal," 2002).

Fred Martinez, Jr., a 16-year-old Navajo youth, lived in a small Colorado town. He was gay and transgendered, often preferring to wear makeup and women's clothing. He suffered such frequent taunts from classmates that he eventually dropped out of school. On June 16, 2001, he was bludgeoned to death with a rock by 18-year-old Shaun Murphy, who later was allegedly heard bragging about beating up a "fag." Murphy eventually pled guilty to second-degree murder and was sentenced to 40 years in prison (Frazier, 2002).

Books and articles about hate crimes often start with the stories of Matthew Shepard, the gay University of Wyoming student who was murdered in 1998, and James Byrd, Jr., the black father of three who was dragged to death behind a pickup truck in Texas that same year. Horrible as those two cases were, they represent only a tiny fraction of the hate crimes that occurred in 1998; the stories of the other victims are rarely heard. Unfortunately, each year at least thousands, and more likely tens of thousands, of Americans have stories to tell like those of Shanika Williams, Balbir Singh Sodhi, Trev Broudy, Fred Martinez, Jr., and the families whose loved ones are buried in the cemeteries Kenneth Luker damaged.

Hate crimes happen in small towns and large cities. They happen in every state: north, south, east, and west. They involve everything from simple graffiti to brutal murders. They may be called hate crimes, bias crimes, civil rights crimes, or ethnic intimidation. All are crimes committed because of the race, religion, ethnicity, sexual orientation, or other group membership of the victim. The precise groups that are included in the definition of hate crime vary from state to state. Race and religion are always included in the definition, but it may or may not include sexual orientation, gender, national origin, physical or mental disability, age, or political affiliation (Gerstenfeld, 1992).

Although bigotry is probably as old as humanity, the term *hate crime* is a new one, as is the idea of special treatment of these offenses. The first hate crime law was passed around 1980, and two decades later, 43 states, the District of Columbia, and the federal government had all enacted some form of hate crime legislation (Anti-Defamation League [ADL], 2001). The phrase has entered the popular vernacular, and the media frequently use it. For example, in the first 6 months of 2002, it appeared 940 times in major newspapers alone.

In spite of the rather swift acceptance of the concept of hate crimes (or maybe because of it), the pathway for contending with these offenses has been rough and winding. Relatively little research has been conducted on who commits hate crimes, against whom, and why. Although a variety of legal and other approaches have been implemented for hate crime prevention, virtually no studies have assessed the effectiveness of these approaches. There have also been a number of powerful controversies related to hate crimes— sometimes on constitutional grounds, sometimes on policy grounds, and sometimes on moral grounds. Misconceptions and premature conclusions about hate crimes abound.

This book explores what is known, and what is not known, about hate crimes. It addresses a broad range of issues from a variety of perspectives. As you read, you will be asked to assess the existing data and arguments, and to develop your own viewpoints. This is not an easy task; even many of those who have studied and thought about hate crimes for years sometimes have difficulty reaching firm conclusions. The one thing that can be said with certainty about hate crimes is that there is very little that can be said with certainty.

OVERVIEW OF THIS BOOK

Chapter 2

What, exactly, are hate crimes, and how are they differentiated from other crimes? How was bigotry treated by criminal laws prior to the recent passage of hate crime laws? What forces led to those laws being enacted? Why, in fact, do so many people believe that special laws are needed for bias-motivated offenses? These are some of the questions addressed in Chapter 2.

As you will see, special hate crime laws were not the first legal remedy for prejudice-inspired offenses. In fact, some federal civil rights legislation dates to the end of the

Civil War. However, before hate crime laws were enacted, existing ways of handling these offenses were severely limited in scope and application, which led some people to argue that new laws were needed.

Several claims have been made as to why special hate crime laws are necessary, or at least desirable. These claims boil down to three basic assertions: first, that hate crime offenders deserve increased punishments because hate crimes are worse than ordinary crimes; second, that hate crime laws will deter these offenses; and third, that the laws will have the symbolic effect of stating that this kind of behavior is unacceptable. Chapter 2 discusses each of these arguments in detail. It also summarizes the small but growing body of literature on the impact of hate crimes on victims and the community.

Chapter 2 also explores the many forms that hate crime laws take in different jurisdictions. Depending on the state, bias-motivated crime might result in a "penalty enhancer" or a separate substantive offense, or be considered a special circumstance. In some states, hate crime might also give rise to civil lawsuits. Some jurisdictions have enacted additional related laws as well, such as laws requiring that law enforcement agencies report hate crimes, or laws requiring specialized training for police officers.

Seventeen-year-old Robert Anthony Viktora, who was white, burned a cross on a black family's front lawn. He was charged with a variety of offenses, but in 1992 (in *R. A. V. v. City of St. Paul*) the Supreme Court found one of his convictions to be a violation of the First Amendment. Nineteen-year-old Todd Mitchell, who was black, urged his friends to "go get" a 14-year-old white boy who happened to be walking by. The friends beat the boy badly enough to put him in a coma. Mitchell also faced a variety of charges, but this time, only a year after Viktora's case, the Supreme Court upheld those convictions (*Wisconsin v. Mitchell,* 1993).

What distinguished Viktora's case from Mitchell's? The former involved hate *speech,* whereas the latter involved hate *crime.* These terms are often used interchangeably by laypeople and the media, and the distinction may seem unimportant. In reality, however, hate crime and hate speech are very different from legal and policy standpoints and in the way they are manifested and experienced. Certainly, the outcomes were very different for these two young men. The line between hate speech and hate crime may sometimes be very thin, however, such as when people make general threats against others. Chapter 2 describes the distinctions between these two closely related ideas and also summarizes some of the arguments that have been made for and against hate speech restrictions.

Chapter 3

Chapter 3 considers the major constitutional and policy controversies surrounding hate crime legislation. At times, these controversies have become quite heated, inspiring debate among lawyers, legislators, advocates, and scholars. Constitutional issues have received perhaps the most attention and commentary, and they have generated many court cases. Specifically, it has been argued that hate crime laws infringe on a number of rights, including those involving equal protection and freedoms of association, expression, and thought. We consider these arguments and discuss how they have been treated by the courts.

One of the most problematic aspects of hate crimes is that they are the only crimes that require proof of the offender's motive. This requirement creates difficulties of a practical nature; because the legal system does not employ psychics, it is hard to determine what was going through a defendant's mind when he or she was committing a hate crime. This dilemma also creates legal difficulties. One of the strongest arguments against hate crime laws is that they punish thoughts. I admit to still feeling ambivalent on this matter myself. After reading Chapter 3, see if you can reach a conclusion of your own.

Chapter 3 also deals with some of the other particularly thorny dilemmas concerning hate crimes. One of these issues is deciding which groups to protect under the laws. All laws protect at least religion, race, and ethnicity, but many protect numerous other categories as well. Some of the fiercest battles surrounding hate crimes have erupted over whether gender and sexual orientation should be included. Jurisdictions are split on these issues.

The gender question is interesting because women's advocates have, at times, found themselves opposing one another. A number of policy-based arguments have been made for and against including gender. Debatable as this issue is, however, it has been surpassed in contentiousness by the question of whether sexual orientation should be protected. The problem here is the symbolic nature of hate crimes: Opponents claim that including sexual orientation would put an official governmental seal of approval on homosexuality, whereas proponents argue that excluding it sends the message that gay-bashing is socially acceptable. This issue has resulted in the repeated derailment of federal hate crime legislation. It is ironic that the debate over hate crimes has, at times, become an arena for the display of biases.

Once the complexities of the laws themselves are resolved, however, the problems do not end. As Chapter 3 discusses, there are a variety of major difficulties associated with the law enforcement and prosecutorial responses to hate crimes. If you were the victim of one of these crimes, would you report it to the police? If so, you are very much in the minority; the research indicates that very few victims report the crimes. If you did report it, how would the police respond? If the police did consider it a hate crime, would it get recorded as such? Would a prosecutor bring hate crime charges against the offender? Would a conviction result? Again, the answer to these last three questions is "Probably not." Chapter 3 examines the reasons why. This is a particularly important issue because hate crime laws are meaningless if they sit on the shelf unused.

A final issue considered in Chapter 3 is one that seems, at first anyway, paradoxical: Some people have argued that hate crime laws may actually harm minorities. How so? They might inspire complacency among policymakers, who will fail to pursue meaningful solutions to bigotry. The laws might also result in resentment toward minorities, who could be perceived as receiving "special treatment." And, because the institutions that are charged with implementing hate crime laws are themselves frequently afflicted with prejudices and bias, the laws could be used as a means of disempowering members of minority groups.

Chapter 4

In the next two chapters, the focus turns from the laws concerning hate crimes to the people who commit them. Chapter 4 explores the research on who perpetrates hate crimes and why. Chapter 5 looks at white supremacist and other organized hate groups.

When you think about people who commit hate crimes, what images come to mind? If you are like most people, you probably think of white-robed members of the Ku Klux Klan, or perhaps neo-Nazi skinheads with their red suspenders and shaved heads. In truth, however, only a small proportion of hate crimes are committed by people like these. In fact, most offenders belong to no organized hate groups at all. Instead, most appear to be young people out looking for thrills or peer acceptance.

It is tempting for many of us to feely smugly superior to those who perpetrate hate crimes and to think of them as deviants and fanatics. As Chapter 4 discusses, however, most hate crime offenders do not fit this profile. In fact, for the most part, *they* are *us*. In addition to presenting recent research on typologies of offenders, Chapter 4 also explores psychological studies on how prejudice develops and how stereotypes affect our perceptions of other people.

Are hate crimes the result of bigoted personalities or of a biased environment? This question is addressed in Chapter 4 as well. Traditionally, much of the emphasis was on personal factors, such as parenting style and the authoritarian personality. More recently, however, the focus has shifted to situational factors, such as groups pressures, economics, the social surroundings, and the culture in general. Given what we are learning about hate crime offenders, these latter factors seem to bring more promising explanations. I discuss the research relevant to all these theories, including some surprising recent studies that indicate that, contrary to previous expectations, hate crimes may not be strongly linked to adverse economic conditions. As you read Chapter 4, you might ask yourself what differentiates you from the individuals who attacked Shanika Williams, Balbir Singh Sodhi, Trev Broudy, Matthew Shepard, James Byrd, Jr., Fred Martinez, Jr., and others.

Consider the clubs or other social organizations to which you belong. What attracted you to them and why did you join? Chapter 4 asks the same questions about extremist groups: What leads someone to become a member of the Klan or the neo-Nazis or other hate groups?

Chapter 5

In Chapter 5, the history and activities of extremist groups are explored in detail. I begin with the somewhat difficult problem of defining a hate group. There are currently hundreds of these groups in the United States, and their relationships to one another can be confusing at best. In this chapter, I discuss what these groups have in common as well as attempts that have been made to divide them into types or categories.

Chapter 5 then discusses in detail three major parts of the American hate group scene: the Ku Klux Klan, the racist skinheads, and the racist militias. The histories of each are described. You will probably find that some of your preconceptions about these groups are inaccurate. For example, did you know that there is not one single Klan, but rather several separate groups, many of whom often actually oppose one another? Did you know that not all skinheads or militia members are racist?

There is much variety in the names of hate groups, in their origins, and in their costumes and symbols. However, most of these groups share essentially the same beliefs. Chapter 5 describes the ideologies of hate groups. What themes do they have in common? Whom do they hate?

This chapter also explores in more detail a question raised in Chapter 4: How do hate groups recruit new members? As you will see, it appears that their methods are not very different from those of mainstream organizations such as churches, fraternities, and social clubs. Furthermore, you may be surprised to learn that most people do not join hate groups because they are bigoted; instead, they become bigoted because they join hate groups. Recruitment is an important issue among hate groups because, as it turns out, membership turnover is high. Chapter 5 examines the small amount of literature on why people leave these groups.

Another topic addressed in Chapter 5 is the activities of hate groups. Obviously, members do not spend every moment of their waking lives committing hate crimes. Interestingly, little research has focused on the specific actions of these groups, but this chapter does discuss what is known. Again, you might ask yourself how these groups differ from those you belong to.

Finally, Chapter 5 turns to the role of women in hate groups. This topic has been ignored completely by most of those who study extremists. Others assume that extremist groups are always antifemale and that women are always restricted to secondary and subsidiary roles. The truth is more complex. To a great extent, hate group rhetoric is anti-female and antifeminist. At the same time, however, many groups specifically try to reach out to women because male members tend to defect if their wives and girlfriends are not supportive.

Chapter 6

In Chapter 6, the attention shifts from those who commit or encourage hate crimes to those who are victimized by them. I begin by examining the data on hate crime victims. These data have a number of significant limitations that must be kept in mind, so the real picture is incomplete.

We then look at three of the most common groups of hate crime victims: African Americans, Jews, and gays and lesbians. In each case, we examine the history of offenses against the group as well as their causes and manifestations today. In doing so, we are faced with an immediate problem: Even though blacks may be the most common victims of hate crimes, virtually no research has focused on this topic. We explore the reasons for this deficiency and the results of the few studies that exist. We also look at the considerably larger literature on anti-Semitic and homophobic crimes.

Three categories of "other" victims are also discussed. First, we look at crimes committed because of the victims' ethnicity. One thing that distinguishes these crimes from those based on race is that Americans of Asian, Latino, and Middle Eastern descent (regardless of their race) tend to be viewed as perpetual foreigners no matter how long they and their families have lived in this country. Next, Chapter 6 examines those who are targeted because of their disabilities. About half of all states include this category in their legislation, but again, there has been almost no relevant research. Third, we briefly consider crimes against women and revisit the question (raised in Chapter 3) of whether these ought to be classified as hate crimes.

Chapter 7

Aside from hate crime laws, which are discussed in detail in Chapters 2 and 3, what other methods are used to prevent hate crimes? Chapter 7 begins with a discussion of the psychological literature on prejudice reduction. Although this literature is rather extensive and has been developed for several decades, it has, for the most part, not been used to inform or shape hate crime prevention policies.

The chapter then continues with a discussion of the history, goals, and activities of four groups that have played a major part in antihate advocacy efforts. The Anti-Defamation League (ADL) and the Simon Wiesenthal Center (SWC) are both Jewish organizations that have fought anti-Semitism and intolerance in general. The Southern Poverty Law Center (SPLC), which began as a civil rights law firm, has evolved into watchdog and educational roles and has continued its legal efforts on behalf of victims. And the National Gay and Lesbian Task Force (NGLTF) has played a strong part in lobbying and research efforts. Each of these groups takes a somewhat different approach to the complex problem of hate.

There have also been a variety of government initiatives aimed at fighting hate that have been enacted at the local, state, and federal levels. This legislation has tended to focus on law enforcement or educational efforts. Chapter 7 critically examines these initiatives.

It does not take a huge organization or a government agency to attempt to prevent hate crimes. In addition, there have been many grassroots-level antihate campaigns. Some of these organizations have been started by victims and their families. Just as the Internet allows hate groups to spread their messages, it has also made it possible for even small antihate groups to establish an international presence. We look at some of these groups, including such innovative ideas as Project Lemonade, in which hate group activity is used as a way to earn money for fighting prejudice.

Chapter 7 also takes a look at several nonpunitive methods of dealing with hate crime offenders. One method that appears to show promise is the use of victim-offender mediation. In some cases, special counseling or educational programs have been tried. Another tactic has been lawsuits against perpetrators of hate crimes, the groups with which they are affiliated, or even their employers or families.

The last section of Chapter 7 critically evaluates hate crime prevention programs in light of what is known about prejudice and hate crimes. The chapter makes a number of recommendations; as you read, consider the extent to which you agree with those recommendations and whether you have some of your own.

Chapter 8

Most of this book looks at hate crimes in the United States. However, there is nothing uniquely American about these offenses, and perhaps something can be learned by examining other countries' problems and solutions. This is the intent of Chapter 8, which investigates the international aspects of hate crimes.

I faced two major problems when I wrote Chapter 8. First, how can hate crimes be distinguished from such close cousins as terrorism, genocide, and civil war? I have attempted to deal with this issue; however, of necessity, the discussion occasionally must exceed the (rather arbitrary) boundaries that I set. Second, it turns out that almost all of the research on hate crimes abroad has focused on European and English-speaking countries, mostly ignoring the significant problems faced by nations in Asia, Africa, and Latin America. Unfortunately, then, the discussion in Chapter 8 is more limited than it should be.

In recent years, there has been a marked rise in the prominence of nationalism and the radical right in Europe. We look at the potential causes of this as well as its effects on people who live in these countries. One result has been an apparent sharp increase in xenophobia and in hate crimes.

Chapter 8 takes a close look at the problems and legal solutions to hate crimes in four countries: Germany, the United Kingdom, Australia, and Canada. For each country, I consider the cultural and societal factors that have contributed to racial and ethnic conflict and the manifestations of that conflict. I also examine the ways in which each country has chosen to approach the problem of hate crimes and compare those methods to those adopted in the United States.

Chapter 8 then transcends borders to examine the ways in which hate groups have established international networks. These networks have been facilitated greatly by the Internet and, although their scope is unknown, there is increasing concern about their repercussions. Is this truly a growing problem? This chapter explores the existing research.

As hate exceeds borders, controlling it becomes no longer simply a domestic problem. But policing internationally, especially when countries have different and often conflicting laws, is problematic at best. Currently, for instance, French and U.S. courts are at odds over the legal liability of the American Internet site Yahoo! for Nazi paraphernalia offered on the auctions the site sponsors. Canada and Germany have both attempted prosecutions of citizens of other countries for material those people placed on Web sites. This issue of the globalization of hate is one that is evolving very rapidly. Chapter 8 introduces some of the problems and disputes involved.

Chapter 9

Chapter 9 is the final chapter of this book. Its purpose is to summarize and draw together the themes of the previous chapters. It also takes a bit of a chance and makes predictions and recommendations about the future of hate crimes. Specifically, I highlight a variety of topics that are particularly in need of more research. I also make suggestions for ways to improve data collection, law enforcement approaches, and legislation. The issue of globalized hate is emphasized as well, and ideas for how to better fight hate are discussed.

GOALS OF THIS BOOK

I hope to accomplish several goals with this book. One major goal, of course, is to educate readers about many of the issues and aspects of hate crimes. Not only do I hope to introduce you to new information, but I also want to dispel some of the misinformation that you might have previously gathered.

Second, I want to make you aware of the many controversies surrounding hate crimes. I approach those controversies from a neutral standpoint, and you will be encouraged to critically evaluate the relevant research and data. In the end, you may reach your own conclusions or at least identify the subjects about which you need to know more. You may even be encouraged to do research of your own.

Finally, although prejudice and its effects will never be completely eliminated, they can certainly be reduced. The most fruitful methods of achieving this reduction will likely come not from more punitive approaches but rather through the personal efforts of many individuals. I hope that this book will inspire you to look for ways that you can contribute to fighting bigotry and hate crimes.

DISCUSSION QUESTIONS

1. Where and when have you heard about hate crimes? Do you think what you have heard is accurate and representative?

2. Can you give a precise definition of a hate crime? What conceptions do you have about hate crimes?

3. Why are hate crimes worth studying? Are there things that differentiate them from "ordinary" crimes?

Chapter 2

THE WHYS AND HOWS OF HATE CRIME LAWS

When did the first hate crime occur? If we narrowly define the term *hate crime* to include only acts that are formally and legally defined as hate crimes, then the first one likely occurred very soon after the earliest hate crime laws were enacted around 1980. But if we define the term more broadly, to include any violent or invasive act committed because of the victim's group, then the first hate crime probably happened a million years ago, when one *homo erectus* bashed another over the head with his club. As we will see in Chapter 4, bias seems to be an innate human trait.

Throughout recorded history, it is easy to find clear examples of acts that would be labeled hate crimes today. In America, people have frequently been persecuted (and sometimes prosecuted) because they were Native American or African American or Japanese American, or because they were Mormon or Jewish or Muslim, or because they were female or gay or disabled—to name only a few examples (see, e.g., Jacobs & Potter, 1998, pp. 59-63; Newton & Newton, 1991; Petrosino, 1999). In the South between the years 1882 and 1930, at least 2,500 African Americans were lynched. This works out to an average of nearly one lynching per week (Tolnay & Beck, 1995).

What is a hate crime? Although the answer may seem obvious, it is not. To begin with, it is not simply a crime in which the offender hates the victim. In fact, most crimes involving hatred between the parties would not fall under the legal definition of hate crimes. Moreover, for an act to be classified as a hate crime, technically the act does not need to be motivated by hatred at all.

The simplest definition of *hate crime* is this: a criminal act which is motivated, at least in part, by the group affiliation of the victim. As Chapter 3 discusses, only certain types of group affiliations are protected by hate crime laws; typically, these include race, religion, and ethnicity. They may also include sexual orientation, gender, disability, and other categories as well.

Although this definition is legally correct, it is important to note that it is not always as precise as we might want. For example, there may often be significant overlap between the definition of hate crimes and those of gang violence, terrorism, political violence, and even war. Was the bombing of the federal building in Oklahoma City a hate crime? The September 11 attacks in America? The Holocaust? What about violence between the Palestinians and the Israelis, between Northern Ireland's Catholics and Protestants, between the Serbs and the Croats, or between the Tutsis and the Hutus? Some of these difficult questions will be discussed in more detail in Chapter 8.

It is also sometimes difficult to distinguish between criminal and noncriminal acts of bias. If a white supremacist prints a magazine urging white people to unite and form a revolution to evict all nonwhites from the country, is this a crime, or the exercise of freedom of speech? This is an important question because, as discussed in this chapter and the next, the First Amendment limits the government's power to restrict freedom of expression.

Finally, it should also be noted that, although most jurisdictions have some sort of law permitting the government to prosecute those who commit hate crimes, some also allow civil suits under these circumstances. In a civil case, it is the aggrieved person, rather than the government, who brings the wrongdoer to trial and, if the plaintiff wins, the defendant pays damages rather than going to jail. As discussed later in this chapter, civil bias suits have, at times, been a useful tool for compensating certain victims.

Most of the discussion in this book will focus on hate crimes as criminal acts, committed by individuals or small groups (as opposed to governments or large political factions), in which the victim was targeted primarily because of his or her group. These are the kinds of acts that hate crime laws were intended to address.

PUNISHING HATE
BEFORE HATE CRIME LAWS

Prior to the enactment of special hate crime laws, there were several ways in which the legal system dealt with acts motivated by bias. Of course, the perpetrator could always be prosecuted for the underlying crime. After the Civil War, violence against black Southerners became a serious problem (this was the era in which the Ku Klux Klan was born), and many Northerners remained skeptical that local authorities in the South would prosecute whites for attacking blacks. In fact, many feared that the local authorities frequently participated in these attacks themselves. Consequently, Congress passed the Civil Rights Acts in 1871. These laws permit the federal government to prosecute people who have deprived others of their civil rights, either in concert with other perpetrators (18 U.S.C. § 241) or in their capacity as government employees (18 U.S.C. § 242). These two laws, especially section 242, have been used to punish people in a wide variety of situations, many of which have been unrelated to race or bias (Jacobs & Potter, 1998). A well-known example of the usage of these laws was the prosecution of four Los Angeles police officers for the 1992 attack on Rodney King (Levenson, 1994). A third law (42 U.S.C. § 1983) allows a person to sue a state or local government employee in federal court when that employee interferes with his or her constitutional or civil rights.

Nearly 100 years after passing the Civil Rights Acts, Congress passed additional civil rights legislation. Among the new laws was 18 U.S.C. § 245, which criminalizes interfering with another person's enjoyment of certain activities or benefits when the interference is due to the person's race, color, religion, or national origin. Among the listed protected activities are attending school, being employed, traveling, and eating in restaurants. This statute, which was meant to protect those who were engaged in the civil rights movement, is rarely used (Gilbert & Marchand, 1999; Jacobs & Potter, 1998). See Box 2.1 for the language of the current federal Civil Rights Acts that are most relevant to hate crimes.

Box 2.1   The Federal Civil Rights Acts

18 UNITED STATES CODE § 241

If two or more persons conspire to injure, oppress, threaten, or intimidate any person in any State, Territory, Commonwealth, Possession, or District in the free exercise or enjoyment of any right or privilege secured to him by the Constitution or laws of the United States, or because of his having exercised the same; or

If two or more persons go in disguise on the highway, or on the premises of another, with intent to prevent or hinder his free exercise or enjoyment of any right or privilege so secured—

They shall be fined under this title or imprisoned not more than ten years, or both; and if death results from the acts committed in this section or if such acts include kidnapping or an attempt to kidnap, aggravated sexual abuse or an attempt to commit aggravated sexual abuse, or an attempt to kill, they shall be fined under this title or imprisoned for any term of years or for life, or both, or may be sentenced to death.

18 UNITED STATES CODE § 242

Whoever, under color of any law, statute, ordinance, regulation, or custom, willfully subjects any person in any State, Territory, Commonwealth, Possession, or District to the deprivation of any rights, privileges, or immunities secured or protected by the Constitution or laws of the United States, or to different punishments, pains, or penalties, on account of such person being an alien, or by reason of his color, or race, than are prescribed for the punishment of citizens, shall be fined under this title or imprisoned for not more than a year, or both; and if bodily injury results from the acts committed in violation of this section or if such acts include the use, attempted use, or threatened use of a dangerous weapon, explosives, or fire, shall be fined under this title or imprisoned not more than ten years, or both; and if death results from the acts committed in violation of this section or if such acts include kidnapping or an attempt to kidnap, aggravated sexual abuse or an attempt to commit aggravated sexual abuse, or an attempt to kill, they shall be fined under this title or imprisoned for any term of years or for life, or both, or may be sentenced to death.

18 UNITED STATES CODE § 245

. . .

(b) Whoever, whether or not acting under color of law, by force or threat of force willfully injures, intimidates or interferes with, or attempts to injure, intimidate or interfere with—

. . . (2) any person because of his race, color, religion or national origin and because he is or has been—

(A) enrolling in or attending any public school or college;

(B) participating in or enjoying any benefit, service, privilege, program, facility or activity provided or administered by any State or subdivision thereof;

(Continued)

Box 2.1 (Continued)

(C) applying or enjoying employment, or any perquisite thereof, by any private employer or any agency of any State or subdivision thereof, or joining or using the services or advantages of any labor organization, hiring hall, or employment agency;

(D) serving, or attending upon any court of any State in connection with possible service, as a grand or petit juror;

(E) traveling in or using any facility of interstate commerce, or using any vehicle, terminal, or facility of any common carrier by motor, rail, water, or air;

(F) enjoying the goods, services, facilities, privileges, advantages, or accommodations of any inn, hotel, motel, or other establishment which provides lodging to transient guests, or of any restaurant, cafeteria, lunchroom, lunch counter, soda fountain, or other facility which serves the public and which is principally engaged in selling food or beverages for consumption on the premises, or of any gasoline station, or of any motion picture house, theater, concert hall, sports arena, stadium, or any other place of exhibition or entertainment which serves the public. . . .

42 UNITED STATES CODE § 1983

Every person who, under color of any statute, ordinance, regulation, custom, or usage of any State or Territory or the District of Columbia, subjects, or causes to be subjected, any citizen of the United States or other person within the jurisdiction thereof to the deprivation of any rights, privileges, or immunities secured by the Constitution and laws, shall be liable to the party injured in an action at law, suit in equity, or other proper proceeding for redress, except that in any action brought against a judicial officer for an act or omission taken in such officer's judicial capacity, injunctive relief shall not be granted unless a declaratory degree was violated or declaratory relief was unavailable. For the purposes of this section, any Act of Congress applicable exclusively to the District of Columbia shall be considered to be a statute of the District of Columbia.

By the 1970s, then, a number of legal avenues existed for prosecuting certain people who committed crimes because of their victims' groups. However, the federal statutes suffer from several important limitations. They are *federal* laws, and federal jurisdiction is limited, both legally and practically. From a legal standpoint, according to the Constitution, the federal government may criminalize primarily those acts that affect interstate commerce. In the 1960s, the United States Supreme Court interpreted the phrase "interstate commerce" very broadly. For example, in *Heart of Atlanta v. United States* (1964) and *Katzenbach v. McClung* (1964), in which the Supreme Court upheld the Civil Rights Act of 1964, the Court found that the activities of restaurants and hotels could be federally regulated, even if most of the clientele were local residents. In 1985, the Court held that arson of an apartment building affected interested commerce as well (*United States v. Russell,* 1985).

However, in more recent years, the Court has taken a much narrower view of the Commerce Clause. In 1995, in *United States v. Lopez,* the Court struck down the Gun-Free School Zones Act, which made it a federal crime to possess a gun within a school zone. According to the Court, there was not a sufficient nexus between gun ownership and interstate commerce (Dral & Phillips, 2001).

Of particular relevance to hate crimes is *United States v. Morrison* (2000). In this case, the Supreme Court declared unconstitutional a portion of the Violence Against Women Act that allowed rape victims to sue rapists in federal court. Among other things, the Court found that the law exceeded Congress's authority under the Commerce Clause. As the law currently stands, then, the federal government's power to prohibit bias-motivated crime is, at the very least, severely in question (Hasenstab, 2001; Peek, 2001).

Another legal limitation lies within the language of the Civil Rights Acts themselves. These laws were enacted under specific circumstances and with the purpose of addressing specific problems (e.g., the activities of the Klan during Reconstruction). The language of the laws does not encompass many potential bias-related crimes. Section 241 does not cover crimes committed by individual actors; sections 242 and 1983 apply only to acts committed by government employees or agents; and section 245 protects only certain acts on the part of the victims. Furthermore, section 242 only protects a victim who was selected because of "being an alien, or by reason of his color, or race," and 245 mentions only "race, color, religion or national origin." Therefore, neither law would apply if a crime were committed because of the victim's sexual orientation, disability, age, or gender.

Even if the federal government *could* use the laws in the Civil Rights Acts to punish hate crimes, there is a serious question of whether it *would.* Due to issues concerning the role of federalism, as well as limited resources, the federal government generally plays a minor role in the prosecution of crimes; most criminal prosecutions are conducted by the states. For example, in 2000 the federal government made 115,589 arrests (Department of Justice, 2001a). In that same time period, there were 1,385,361 arrests in the state of California alone (California Department of Justice, 2001a); state prosecutors in districts with populations over 500,000 (this encompasses about 45% of the U.S. population) reported more than 4 million arrests (Department of Justice, 2001b). Not only are federal prosecutions a tiny fraction of all prosecutions in the United States, but they also tend to concentrate on only a few types of offenses. Of the federal arrests in 2000, about 30% were for drug-related offenses, and about 22% were related to immigration matters (Department of Justice, 2001a).

If the federal government does not have the power, the resources, or the inclination to address most bias-motivated crime, then it up to the states. Without specific hate crime laws, there are two potential ways in which these offenders can be prosecuted. The first route, which is by far the more limited, is to prosecute them via laws prohibiting very specific activities. For example, several states have laws against wearing masks in public places (see Box 2.2). Many of these laws were enacted in direct response to a resurgence of Ku Klux Klan violence in the 1920s (Allen, 1991). Other examples include laws against desecrating cemeteries or places of worship. In addition, in some states bias motivation is considered a special circumstance in capital cases, which makes the defendant eligible for the death penalty (Gerstenfeld, 1992). Of course, the problem with these laws is that they encompass only a small percentage of bias-motivated crimes. Of the 7,876 hate crimes reported to the Federal Bureau of Investigation (FBI) in 1999, only 274 (less than 4%) occurred at houses of worship, and only 17 were homicides (2001b).

Box 2.2   Specialized State Laws

OFFICIAL GEORGIA CODE ANNOTATED § 16-11-38

(a) A person is guilty of a misdemeanor when he wears a mask, hood, or device by which any portion of the face is so hidden, concealed, or covered as to conceal the identity of the wearer and is upon any public way or public property or upon the private property of another without the written permission of the owner or occupier of the property to do so.

(b) This Code section shall not apply to:

(1) A person wearing a traditional holiday costume on the occasion of the holiday;

(2) A person lawfully engaged in trade and employment or in a sporting activity where a mask is worn for the purpose of ensuring the physical safety of the wearer, or because of the nature of the occupation, trade, or profession, or sporting activity;

(3) A person using a mask in a theatrical production including use in Mardi Gras celebrations and masquerade balls; or

(4) A person wearing a mask prescribed in emergency management drills or emergencies.

TITLE 13 VERMONT STATUTES ANNOTATED § 1456

Any person who intentionally and maliciously sets fire to, burns, causes to be burned, or aids or procures the burning of a cross or a religious symbol, with the intention of terrorizing or harassing a particular person or persons, shall be subject to a term of imprisonment of not more than two years or a fine of not more than $5,000.00, or both.

WISCONSIN STATUTES § 943.012

Whoever intentionally causes damage to, intentionally marks, draws or writes with ink or another substance on or intentionally etches into any physical property of another, without the person's consent and with knowledge of the character of the building, is guilty of a class E felony if the property consists of one or more of the following:

(1) Any church, synagogue or other building, structure or place primarily used for religious worship or another religious purpose.

(2) Any cemetery, mortuary or other facility used for burial or memorializing the dead.

(3) Any school, educational facility or community center publicly identified as associated with a group of persons of a particular race, religion, color, disability, sexual orientation, national origin or ancestry or by an institution of any such group.

(4) Any personal property contained in any property under subs. (1) to (3) if the personal property has particular significance or value to any group of persons of a particular race, color, disability, sexual orientation, national origin or ancestry and the actor knows the personal property has particular significance or value to that group.

CALIFORNIA PENAL CODE § 190.2

(a) The penalty for a defendant who is found guilty of murder in the first degree is death or imprisonment for life without the possibility of parole if one or more of the following special circumstances has been found under Section 190.4 to be true:

· · ·

(6) The victim was intentionally killed because of his or her race, color, religion, nationality, or country of origin.

· · ·

The second, more general way to deal with people who commit bias-motivated crimes is simply to prosecute them for the underlying crime. If they burn a cross in someone's yard, prosecute them for trespassing or arson; if they deface a synagogue, prosecute for vandalism; if they attack someone they think is gay, prosecute for assault and battery. The advantage of this plan is its simplicity: No special laws need to be created, police and prosecutors need no special training, and nobody needs to question the laws' constitutionality. Beginning in the late 1970s, however, some people and advocacy groups began to argue that this was not sufficient, and that special hate crime laws should be passed.

ARGUMENTS FOR HAVING HATE CRIME LAWS

A variety of arguments have been made as to why bias-motivated acts merit their own special laws.[1] Broadly speaking, these arguments fall into three (interrelated) categories: retribution, deterrence, and symbolic effects.

Retribution: Hate Crimes Are Different From Ordinary Crimes

This category of arguments asserts that hate crime laws should be enacted because this type of offender deserves to be punished more than "ordinary" offenders. According to this argument, when it comes to crime, hate is different in several ways.

Victims May Suffer More Psychological Trauma

To begin with, many people have argued that hate crime victims suffer more, psychologically and emotionally, than victims of similar crimes that are not bias motivated (see, e.g., Gellman, 1991; Lawrence, 1999; B. Levin, 1999; Weisburd & B. Levin, 1994). In *Wisconsin v. Mitchell* (1993), this claim was a central focus of many of the briefs that were submitted in support of Wisconsin's hate crime law (see Box 2.3) and was accepted by Chief Justice Rehnquist in the opinion he wrote for the unanimous Court.

Box 2.3   Arguments in *Wisconsin v. Mitchell* (1993)

In 1993, the United States Supreme Court considered whether Wisconsin's hate crime statute violated the Constitution. Several individuals and organizations submitted amicus curiae (friend of the court) briefs either supporting or opposing the law. Most of the briefs in support of the law contained assertions that hate crimes are more harmful to victims than other crimes. Here are some examples:

The sort of bias attack that took place in this case is more than an assault on the victim's physical well-being. It is an assault on the victim's essential human worth. A person who has been singled out for victimization based on some group characteristic—such as race, religion, or national origin—has, by that very act, been deprived of the right to participate in the life of the community on an equal footing for reasons that have nothing to do with what the victim did but everything to do with who the victim is. In short, a bias attack is as much an attack on

(Continued)

Box 2.3 (Continued)

---

the victim's persona as on the victim's person. Recognition of that fact inevitably produces a sense of vulnerability, isolation and oppression that rarely disappears when the physical injuries heal.

—Brief amicus curiae of the American
Civil Liberties Union in support of petitioner

It is by now widely recognized that ordinary crimes become far more harmful and pervasive when they are undertaken as hate crimes by intentionally targeting the victim based on specified characteristics.

—Brief for members of
Congress Charles E. Schumer et al.

Bias related crimes inflict on the victim an injury wholly different in kind from the physical or financial harm caused by ordinary criminal conduct. Such crimes are an attack on the right of the victim to participate equally, if at all, in American society, and can inflict serious and long lasting injury to the dignity of the victim. Such attacks are likely to induce in the victims a sense of separation from and rejection by the larger community "that may affect their hearts and minds in a way unlikely ever to be undone." Brown v. Board of Education, 347 U.S. 483, 494 (1954)

—Brief amicus curiae in support of petitioner of the
Crown Heights Coalition, The NAACP Legal Defense and
Educational Fund, Inc., and the American Jewish Committee

How severe is the psychological trauma of being a hate crime victim? Consider these claims by Delgado (1993), who was writing only of verbal (rather than physical) abuse based on race:

The psychological responses to such stigmatization consist of feelings of humiliation, isolation, and self-hatred. . . . Racial tags deny minority individuals the possibility of neutral behavior in cross-cultural contacts, thereby impairing the victims' capacity to form close interracial relationships. Moreover, the psychological responses of self-hatred and self-doubt unquestionably affect even the victims' relationships with members of their own group. The psychological effects of racism may also result in mental illness and psychosomatic disease. The affected person may react by seeking escape through alcohol, drugs, or other kinds of antisocial behavior. (p. 91)

If verbal affronts alone can cause such dire consequences, one can only imagine that the effects of attacks upon one's property or body would be even greater. Weisburd and B. Levin (1994, p. 26) report that aside from low self-esteem and depression, victims of hate crimes suffer from "profound sadness; lack of trust in people; withdrawal; excessive fear of personal and family safety; sleep problems; headaches; physical weakness; increased use of alcohol and drugs; excessive anger; and suicidal feelings."

The problem with these assertions is that they are difficult to support empirically. Do victims of hate crimes experience psychological trauma? It seems extremely likely. But the real question is whether they experience *more* trauma than victims of "ordinary" crimes. As Jacobs and Potter (1998, p. 83) point out, "It should come as no surprise that hate crime victims report psychological and emotional effects. *All victims do*" (emphasis in the original).

There have been only a few studies that actually assess the emotional state of hate crime victims. Two studies cited fairly frequently were conducted by the National Institute Against Prejudice and Violence (NIAPV, 1986, 1989). The NIAPV concluded in both of these studies that hate crime victims suffer more than other victims. However, neither study was methodologically strong, and both included only hate crime victims, so the validity of this claim is questionable. In fact, when Barnes and Ephross (1994) examined the 1989 NIAPV data, they concluded that, for the most part, "the predominant emotional responses of hate crime victims appear similar to those of victims of other types of personal crime" (p. 250).

A report by the League for Human Rights of B'nai B'rith Canada (1993) also concluded that hate crime victims suffer increased trauma. Again, however, the methodology was weak and there were no comparison groups.

Owyoung (1999) interviewed four Asian American victims of hate crimes in the San Francisco area. All reported emotional difficulties as a result of their experiences. With a sample size of four (three of whom had experienced hate crimes only indirectly) and no comparison group, it is impossible to draw any firm conclusions from this study.

Only a very few studies have used more rigorous methodology. In a survey of gay and lesbian youths, Hershberger and D'Augelli (1995) found that victimization was moderately correlated with psychological distress. Similarly, Herek and his colleagues conducted surveys of gay, lesbian, and bisexual adults. Those respondents who had been victims of hate crimes were more likely to have higher levels of depression, anxiety, and anger than were respondents who had been victims of other types of crimes (Herek, Cogan, & Gillis, 2002; Herek, Gillis, & Cogan, 1999; Herek, Gillis, Cogan, & Glunt, 1997). McDevitt, Balboni, Garcia, and Gu (2001) sent a mail survey to victims of hate crimes in Boston, as well as a random sample of victims of nonbias assaults. In comparison with the other victims, hate crime victims reported significantly more psychological impact on some measures, including greater fear and decreased feelings of safety.

The question of whether hate crimes are more psychologically harmful is complicated by a number of factors, none of which have been studied. Are some types of victims more likely to be adversely affected than others? Are some types of hate crimes more harmful than others? Jacobs and Potter (1998) suggest that perhaps victims of violent crime are equally traumatized, regardless of whether they were bias motivated, whereas victims of minor crimes such as graffiti are more traumatized when the crime was bias motivated. This is certainly a plausible, but untested, hypothesis. The most we can conclude, it seems, is that under some circumstances, hate crimes *might* be more traumatic than other crimes. We can also conclude that statements like those in Box 2.3, although sincere, are not well supported empirically.

Victims May Suffer More Physical Trauma

A second reason why hate crime offenders deserve increased punishments, some assert, is that hate crime results in more *physical* trauma to the victims (see, e.g., J. Levin & McDevitt, 1993; Weisburd & B. Levin, 1994). The evidence here, however, is even sketchier than the evidence for increased psychological trauma. J. Levin & McDevitt

(1993, p. 11) conclude that "hate crimes tend to be *excessively brutal*" (emphasis in the original), but they cite as support only an unpublished study by McDevitt (1990). In this study, McDevitt found that, among crimes that were reported to the Boston police, the proportion of attacks against the person was higher among hate crimes than among crimes in general. More recent data concur: From 1997 to 1999, 60% of the hate crimes reported to the FBI were violent crimes (Strom, 2001), whereas in 1999, only about 26% of all crimes were violent (Rennison, 2000).

The problem in drawing conclusions based on law enforcement data is that, as I will discuss at length in Chapter 3, hate crimes are severely underreported (Gerstenfeld, 1998). Logically, nonviolent hate crimes are probably reported even less frequently than violent ones (among other reasons, because they are less likely to come to the attention of law enforcement authorities). Thus, the official data give a highly inaccurate picture. Furthermore, even if the data were accurate, they would not allow us to determine whether hate crimes are more violent than ordinary crimes of a similar type. For example, were the 1,120 bias-motivated aggravated assaults reported in 1999 (FBI, 2001b) more heinous, on average, than the 1.5 million aggravated assaults reported overall that same year (Rennison, 2000)? There is no way to tell.

Most of the other support that is given for the argument that hate crimes are more violent is anecdotal. Weisburd and B. Levin (1994, p. 23), for example, describe several particularly brutal incidents. Although some hate crimes have been undeniably vicious, so have many ordinary crimes. The murders of Matthew Shepard and James Byrd, Jr., horrified many, but it would be easy to find many killings that were equally awful that were not bias motivated.

Hate Crimes May Have a Wider Impact

Many have also argued that hate crime offenders deserve enhanced punishments because hate crimes have a wider impact than do ordinary crimes. Hate crimes, it is asserted, affect not only the victim, but also all members of the victim's group (see, e.g., Dillof, 1997; Iganski, 2001; Lawrence, 1994, 1999; B. Levin, 1999; Wang, 1997; Weisburd & B. Levin, 1994). This is a reasonable claim. I am Jewish, and when I have read about a synagogue having been spray-painted with swastikas, it has had an emotional impact on me that ordinary graffiti would not have. On the other hand, as Jacobs and Potter (1998, p. 87) point out, "Many crimes, whatever their motivation, have repercussions beyond the immediate victim and his or her family and friends." Many people become frightened when they hear about violent crimes in their community; most of us would become more vigilant if we learned of a burglary in our neighborhood.[2] The ripple effect of hate crimes is far from unique.

Noelle (2002) attempted to assess the impact of the Matthew Shepard murder on gays, lesbians, and bisexuals nationwide. Twenty-nine people were given a survey, and the nine participants who identified themselves as most affected by the crime were interviewed in detail. Noelle (2002) concluded that Shepard's murder had, in fact, traumatized these other members of the victim's group. However, given the very small sample size, as well as the fact that the participants were probably not representative of gays, lesbians, and bisexuals in general, the ramifications of this study are unclear.

Hate Crimes May Spark Retaliation and Conflict

Finally, it is argued, hate crimes deserve greater punishment because they tend to spark retaliation and further intergroup conflict (see, e.g., Lawrence, 1994; Taslitz, 1999;

Weisburd & B. Levin, 1994). Several of the amici curiae in *Wisconsin v. Mitchell* (1993) made this argument, and Justice Rehnquist appeared to agree: "According to the State and its amici, bias-motivated crimes are more likely to provoke retaliatory crimes, inflict distinct emotional harms on their victims, and incite community unrest" (*Wisconsin v. Mitchell,* 1993, p. 488).

Once again, although this proposition seems reasonable, it has extremely little empirical support. Weisburd and B. Levin (1994, p. 26) report that in the month after an infamous race-related murder of a black man in Howard Beach, New York, in 1986, police recorded twice as many hate crimes as the previous month. Of course, it is possible that the number of bias-related incidents themselves did not increase, but rather that, because of the well-publicized incident, people were more likely to report them to the police. Or perhaps the police were more sensitive to the issues involved and were therefore more likely to interpret and record a crime as bias motivated.

Although it is easy to recall events such as the riots that followed the acquittal of the Los Angeles police officers for the beating of Rodney King (a beating that was, arguably, race related), events like that are (thankfully) rare. Certainly there was no obvious retaliation and unrest after the great majority of the 9,301 hate crimes the FBI recorded in 1999 (FBI, 2001b) or the thousands more that likely occurred but went unreported.

Thus far, only one experiment has studied the potential retaliatory responses to a hate crime in a controlled manner, and it does not support the hypothesis that hate crimes spark additional retaliatory violence. Craig (1999) showed African American and white college students videotapes of assaults that either were, or were not, hate crimes. She also varied the race of the offenders and victims. After viewing one of the videos, each participant filled out a questionnaire that asked, among other things, about the likelihood that the participant would want revenge, were they in a similar situation to the victim. Craig (1999) found that there was no difference in responses to the two events—that is, participants were no more likely to seek revenge for the hate crime than for the ordinary assault.

On the other hand, the effects of hate crimes on communities may be more insidious than simple retaliatory violence. Perhaps, as Weisburd and B. Levin (1994) suggest, hate crimes attack community cohesion and the social order in general, leading to distrust, fear, and anxiety. These effects would be much harder to measure and, so far, nobody has tried.

Are hate crimes different? Do the people who commit them warrant harsher sentences than other offenders? The arguments seem plausible. Yet, despite the degree to which they appear to be accepted by scholars and courts alike, there turns out to be little data to support them. Probably the most that a cautious person could conclude is that when it comes to this issue, the jury is still out.

Deterrence

The arguments we have discussed so far are primarily retributive in nature: Hate crime laws should exist because these offenses deserve worse penalties. Another broad category of arguments in support of the laws, although related, focuses instead on the laws' supposed deterrent effects. Adherents to this point of view claim that hate crime legislation is a good idea because it will discourage people from committing these harmful offenses (see, e.g., Chang, 1994). Although it seems intuitively accurate, this argument suffers from a number of assumptions that are probably faulty.

The first assumption is that potential offenders will be aware of hate crime laws. People cannot possibly be deterred by a law if they do not know it exists! There have been no empirical studies concerning awareness of hate crime laws. Certainly, hate crimes themselves seem to get a fair amount of media attention, and an English-speaking American

---

Box 2.4   Hate Crimes in Capital Cases

Even as Matthew Shepard lay dying in a Colorado hospital, there was much public outcry over the fact that Wyoming remained one of the very few states without a hate crime law. Some people urged Wyoming to pass such legislation, whereas others pushed for more federal hate crime laws or argued for tougher hate crime laws in general (see, e.g., Crowder, 1998; Green, 1998; McCullen, 1998; Miniclier, 1998; "Tough Hate-Crime Laws," 1998). Even had such legislation existed, it is extremely implausible that it would have deterred Aaron McKinney and Russell Henderson; even without a hate crime law, both offenders faced first-degree murder charges and the possibility of a death sentence.

Ultimately, Russell Henderson pled guilty to first-degree felony murder and kidnapping, and was given two life sentences (Lane, 1999). Aaron McKinney was

---

would probably have to live in a cave to have never at least heard of hate crimes. But does the average person understand how the laws work? Do they realize that in most states bias-motivated crimes are discrete offenses or lead to enhanced sentences? Do they know what constitutes a hate crime? It seems unlikely.

Even supposing that potential offenders are relatively well-versed in the intricacies of hate crime legislation, the deterrence hypothesis further assumes that these offenders believe that there is a reasonable likelihood of being caught and prosecuted. After all, the average driver is well aware of the penalties for speeding, and yet speeds anyway—until he or she spots a patrol car. In the year 2000 in the entire state of California, 360 people were prosecuted for hate crimes, and 213 were actually convicted (California Department of Justice, 2001b). This was despite the fact that there were 2,002 reported offenses, involving 2,107 known offenders (California Department of Justice, 2001b). There were undoubtedly many more unreported offenses. Thus, a person who committed a hate crime in California that year had less than a 1 in 10 chance of being convicted. If we assume, therefore, that potential offenders logically weigh the risks and benefits of their behavior, we might also assume that they would rationally conclude that they are pretty unlikely to get caught.

But, even supposing that bigots know about hate crime laws and that they believe there is a reasonable chance that they will be punished for their behavior, the deterrence argument makes a third, even more far-fetched assumption. To conclude that hate crime laws will act as effective deterrents, you must believe that a potential offender is not already deterred by the potential of being punished for the underlying crime, but would be by the addition of a punishment for a hate crime. As we will discuss in greater detail later in this chapter, hate crime laws come into play only when people commit some other violation of the criminal code (and when the offense is motivated, at least in part, by the victim's group). Therefore, hate crimes involve only acts that are already punishable. For example, in Wisconsin, a person considering causing another severe bodily injury (aggravated battery) is faced with the potential of up to 10 years in prison and a $40,000 fine (Wisconsin Statutes §§ 940.19, 939.50). If the crime is motivated by the victim's group, the sentence can be increased by as much as 5 years, and the fine can be up to $5,000 more (Wisconsin Statutes § 939.645). Can we reasonably believe that a person would *not* be deterred by

convicted of several offenses, including felony murder. After Matthew Shepard's parents interceded to save his life, McKinney agreed to a deal in which he was given two consecutive life terms with no opportunity to appeal (Cart, 1999).

Similarly, after James Byrd, Jr., was murdered, there were calls to stiffen Texas's hate crime law. Then-Governor George W. Bush refused on the basis that all crimes are hate crimes, but the James Byrd, Jr., Hate Crimes Act was eventually signed by Bush's successor, Governor Rick Perry. Among other provisions, the new law specified which groups are protected, mandated hate crime training for police officers, and provided for civil causes of actions for hate crime victims. Of course, the law itself would have had little effect on Byrd's killers (who, obviously, were not deterred by the already-existing law against murder): Two of them were sentenced to death, and the third was given a life sentence ("Governor of Texas," 2001).

10 years and $40,000, but *would* be deterred by 15 and $45,000? That seems unlikely. The chance of hate crime laws deterring would-be offenders seems even less in murder cases (see Box 2.4).

A final assumption of deterrence theory is that, in the minds of would-be offenders, the risks of engaging in a hate crime outweigh the potential benefits. And yet the benefits may be substantial. By acting on their biases, people may receive acceptance and admiration from their peers, boost their own self-esteem, and experience excitement and a release of pent-up frustrations. All of these are powerful rewards that probably would not be outbalanced by the relatively small risks associated with being arrested.

In any case, it is unlikely that many potential hate crime offenders engage in the rational weighing of risks and benefits that deterrence theory assumes. As Uhrich (1999) points out (and as we will discuss in Chapter 4), hatred is a strong and well-ingrained part of human psychology. It is also a powerful emotion that is frequently supported by various elements within our society. Deterring people who are bent on committing hate crimes will take much more than a simple law or two.

Symbolic Effects

Even if hate crime laws are not justified by retribution or deterrence arguments, a strong third claim has been made that the laws serve a symbolic or denunciatory purpose. Many scholars have written about the symbolic value of laws in general. Melton and Saks (1986), for example, state that the law does not simply act as "a carrot or a stick" (p. 35). In addition, it "teaches the moral and social norms of the community and, in a sense, announces, reiterates, and indeed ritualizes the myths and themes of the culture" (Melton & Saks, 1986, p. 251). Similarly, the denunciatory theory holds that punishing a crime "declares, in effect, that in the society in question the offense is not tolerated" (Walker, 1980, p. 28). Furthermore, punishment may repair the tears in the social fabric that were created by the offender.

Many commentators have noted the potential symbolic effects of hate crime legislation (see, e.g., Beale, 2000; Gerstenfeld, 1992; Jolly-Ryan, 1999; Lawrence, 2000b). Hate crime laws might send a message, as follows:

Box 2.5   Selected Quotes in Support of Hate Crime Laws

Hate crime legislation sends a clear message that racism and bigotry will not be tolerated in our community.

—Friebert (1999)

Denver District Attorney Bill Ritter testified in favor of the [hate crime] measure, urging committee members to support the bill in order to "send a message" that assaults against homosexuals would be punished as severely as attacks motivated by racial hatred.

—Luzadder (1999)

There was a time in this country when the messages, from the halls of Congress to the Supreme Court, were clear. Although the Declaration of Independence proclaimed that "all men are created equal," blacks were not included in the Declaration. Blacks did not have the same rights as whites. This country has done some growing up since then. But it still sends out messages about gays and lesbians that their rights are somehow not as great as those of the rest of us. That's why hate-crime laws are necessary. They send a clear message that the rights of

First, [hate crime legislation] expresses strong social condemnation of bias crimes. Second, the condemnation of hate crimes implies a general affirmation of the societal value of the groups targeted by hate crimes and a recognition of their rightful place in society. Hate crimes legislation is seen as reinforcing the community's commitment to equality among all citizens. (Beale, 2000, p. 1254; footnotes omitted)

Of course, the crimes themselves may be symbolic as well. The messages of a swastika or a burning cross are clear: "You're not wanted here. You're not valuable human beings. We are powerful and you should fear us. Do not assert yourself or you will be harmed." The hate crime law provides a countermessage to this.

Of all the arguments made in support of hate crimes legislation, the symbolic argument is perhaps the most frequently given. It has been made by legislators, scholars, journalists, advocacy groups, and members of the general public (see Box 2.5).

Do hate crime laws have the impact that these arguments assume? Perhaps, especially if prosecutions are fairly frequent and well publicized. And it is possible that such laws will be instrumental in effecting a long-term reduction in prejudice. However, such results are difficult (if not impossible) to quantify empirically. Furthermore, as some commentators have cautioned (Beale, 2000; Gerstenfeld, 1992), there is the danger that hate crime laws will lure the public into complacency, resulting in little demand for other ways of addressing bias that might have more tangible effects.

If hate crime laws do have symbolic effects, there are corollaries that should not be overlooked. If prosecuting hate crimes sends a message that such behavior is unacceptable, then *failing* to prosecute might convey the idea that the society approves of the behavior.

everyone are guaranteed in this country, not just those who happen to
be in the majority.

—Freeman (1998)

[Hate crime laws] help to create a climate of tolerance and respect
for differences which, when combined with education, diminishes
the likelihood of violent hate crimes being committed—or accepted.

—Rosenthal (1998)

"This [hate crime bill] allows a judge to say . . . that Parliament and
Canadian society do not approve of this kind of a thing."

—Warran Allmand,
chairman of the Canadian Parliament's
justice committee; quoted in Peritz (1994)

"The jury told us afterward they wanted to send a message and they
certainly did."

—Frederick Sperling, attorney for the plaintiff in a
federal hate crimes lawsuit in which a black woman was
awarded $1.75 million for an attack by three white
supremacists; quoted in Rossi (1992)

As we will discuss in Chapter 3, there might also be symbolic effects if the laws protect
some groups but not others.

Of the pillars that support hate crime legislation—retribution, deterrence, and symbolic
effect—the latter is probably the strongest. By the 1980s, advocacy groups and some legislators frequently mentioned all three as support for their arguments that existing federal and
state laws were not sufficient, and that special hate crime legislation should be enacted.

THE BIRTH OF HATE CRIMES

In 1977, a neo-Nazi group called the National Socialist Party of America (NSPA) wished
to hold a demonstration in front of the village hall in Skokie, Illinois. Skokie had a large
Jewish population, many of whom were Holocaust survivors. The village first obtained an
injunction against the event and then passed a series of ordinances that would have prohibited the NSPA from obtaining the necessary permits. The NSPA sued (with the assistance of the American Civil Liberties Union), and eventually won the right to demonstrate
(Downs, 1985). Although the NSPA ultimately never did march in Skokie, the controversy
received national attention, even to the extent of being satirized in the movie *The Blues
Brothers* (Landis, 1980).[3]

One organization that paid special attention was the Anti-Defamation League of B'nai
B'rith (ADL), a group that combats anti-Semitism and other forms of bigotry. Beginning
in 1978, the ADL started tracking anti-Semitic incidents across the United States. Between

1978 and 1981, the number of reported incidents increased from 49 to 974 (Gellman, 1991; Greenberg, 1993). Alarmed by what it interpreted as a disturbing trend, and frustrated with existing federal and state laws, the ADL drafted a model ethnic intimidation statute in 1981. Together with allies such as the National Gay and Lesbian Task Force, the National Institute for Prejudice and Violence, and the Southern Poverty Law Center, the ADL began lobbying states to pass the statute (see Box 2.6; for more on the hate crime law movement, see Jenness & Grattet, 1996).

Box 2.6   The Anti-Defamation League's Model Ethnic Intimidation Statute

2. Intimidation

A. A person commits the crime of intimidation if, by reason of the actual or perceived race, color, religion, national origin, or sexual orientation of another individual or group of individuals, he violates Section _____ of the Penal code (insert code provisions for criminal trespass, criminal mischief, harassment, menacing, intimidation, assault, battery and/or other appropriate statutorily proscribed criminal conduct).

B. Intimidation under this code provision is a _____ misdemeanor/felony (the degree of criminal liability should be at least one degree more serious than that imposed for commission of the underlying offense).

Note: The ADL model statute also contains provisions concerning institutional vandalism, civil actions, and police reporting and training. This is the original language of the statute, which has since been revised; the current language is available at www.adl.org/ 99hatecrime/text_legis.html.

SOURCE: Anti-Defamation League [ADL] (2001).

When it was passed, the model statute contained four provisions. The first of these (which is not reproduced in Box 2.6), Institutional Vandalism, was aimed primarily at people who targeted cemeteries, community centers, and places of worship. It was not a new idea, as some states already had similar laws on the books, but the ADL and its allies clearly wanted more states to follow suit.

The second provision was more revolutionary. Although it was likely inspired by existing civil rights laws, it was quite different in scope. Under this provision, a person would be found guilty of an "intimidation" if he or she violated some existing criminal law, and if he or she committed the crime because of the victim's group (or perceived group). Although only certain types of groups were protected—race, color, religion, national origin, and sexual orientation—this was a significant expansion upon civil rights laws. The model statute also specified that it would act as a penalty enhancer, essentially bumping up the seriousness of the underlying crime by one degree. Thus, someone who committed what ordinarily would have been a Class C Felony, for example, could now be sentenced for a Class B Felony.

The ADL later modified this section of the model statute slightly. It renamed it "Bias-Motivated Crimes," and it also added gender to the list of protected groups (ADL, 2001). The substance, however, remained the same.

The third provision of the model statute creates a civil cause of action so that victims of institutional vandalism and bias crimes may sue their attackers. Antihate groups have found civil actions to be a particularly useful strategy. See Box 2.7 for some examples of successful lawsuits against hate groups and their leaders.

Box 2.7   Suing Hatemongers

- 1981, Mobile, Alabama: Two members of the United Klans of America abducted and killed Michael Donald as he walked to the store. Donald was African American. Donald's mother sued the United Klans and, in 1987, won a judgment of $7 million. The group was forced to turn over its headquarters to Donald's mother.
- 1987, Forsyth County, Georgia: An interracial group attempted to march in honor of Dr. Martin Luther King, Jr. They were prevented by rock- and bottle-throwing members of the Klan. The Southern Poverty Law Center sued two Klan groups and 11 Klan members, receiving a $1 million verdict.
- 1988, Portland, Oregon: Members of a Skinhead gang beat to death Mulugeta Seraw, a student from Ethiopia. The Skinheads had been organized and trained by a recruiter from the White Aryan Resistance (WAR). Seraw's family sued Tom Metzger, the founder and head of WAR; Metzger's son; and two of the Skinheads who killed Seraw. In 1990, the family won a final judgment of $12.5 million.
- 1995, South Carolina: Members of the Christian Knights of the Ku Klux Klan burned the Macedonia Baptist Church, which had a mostly African American membership. In 1998, the church sued the Klan group and won $37.5 million.
- 2000, Hayden Lake, Idaho: A Native American woman and her son were assaulted by security guards on the premises of property belonging to the Aryan Nations. They sued the Aryan Nations and won $6.3 million in damages. The Aryan Nations were forced to auction the property to pay the judgment. The woman bought the property at the auction and sold it to a man who planned to raze the buildings and build a retreat center.

Finally, the model statute provides for collection of law enforcement data on bias crimes and for specialized training of police officers.

HATE CRIME LAWS TODAY

State

The ADL's efforts to persuade states to pass hate crime legislation were successful. Very soon after the ADL drafted the model statute in 1981, a few states, such as Oregon and Washington, passed similar laws. Other states quickly followed suit. By 1994, 34 states and the District of Columbia had some kind of penalty-enhancement–type law (ADL, 1994) and, by 2000, only seven states did not (ADL, 2000).[4]

Although many of the states used the ADL model as a guide, they frequently made substantial changes when they first passed their laws, as well as in subsequent amendments. Other states created their statutes from scratch rather than using the model. As a result, existing hate crime legislation today is quite diverse. Some laws act as pure penalty enhancers, increasing the sentence for crimes motivated by bias. The increase can be substantial: In some situations, the maximum sentence may be doubled, tripled, or increased by even more. These laws can also result in misdemeanors being reclassified as felonies, which can have serious legal implications for the offender. Other states created a separate substantive offense, for which the offender may be convicted in addition to receiving a conviction for the underlying crime. In any of these cases, the practical effect is that the defendant faces more severe penalties than he or she would for an ordinary crime.

There are other differences between the states as well. Some permit any underlying crime to qualify as a hate crime, whereas some limit their definition to certain specific offenses, such as harassment, assault, and vandalism (see Jacobs & Potter, 1998, p. 31). Some include language that the victim must have been chosen "because of" or "by reason" of his or her group, whereas others simply require that the crime "evidence" or "demonstrate" prejudice or that prejudice be "manifest." The states also differ substantially as to which groups are enumerated in the statute. All statutes include at least race, religion, and ethnicity (or national origin), but only 23 include sexual orientation, 21 include gender, 23 include mental or physical disability, 4 include political affiliation, and 4 include age. See Box 2.8 for a sampling of state statutes.

Box 2.8   A Sampling of State Hate Crime Laws

CODE OF ALABAMA § 13A-5-13

Hate crimes

. . .

(c) A person who has been found guilty of a crime, the commission of which has been shown beyond a reasonable doubt to have been motivated by the victim's actual or perceived race, color, religion, national origin, ethnicity, or physical or mental disability shall be punished as follows:

(1) Felonies:

a. On conviction of a Class A felony that was found to have been motivated by the victim's actual or perceived race, color, religion, national origin, ethnicity, or physical or mental disability, the sentence shall not be less than 15 years.

b. On conviction of a Class B felony that was found to have been motivated by the victim's actual or perceived race, color, religion, national origin, ethnicity, or physical or mental disability, the sentence shall not be less than 10 years.

. . .

(2) Misdemeanors:

On conviction of misdemeanor that was found beyond a reasonable doubt to have been motivated by the victim's actual or perceived race, color, religion, national origin, ethnicity, or physical or mental disability, the defendant shall be sentenced for a Class A misdemeanor, except that the defendant shall be sentenced to a minimum of three months.

CALIFORNIA PENAL CODE § 422.6

(a) No person, whether or not acting under color of law, shall by force or threat of force, willfully injure, intimidate, interfere with, oppress, or threaten any other person in the free exercise or enjoyment of any right or privilege secured to him or her by the Constitution or laws of this state or by the Constitution or laws of the

United States because of the other person's race, color, religion, ancestry, national origin, disability, gender, or sexual orientation, or because he or she perceives that the other person has one or more of those characteristics.

(b) No person, whether or not acting under color of law, shall knowingly deface, damage, or destroy the real or personal property of another person for the purpose of intimidating or interfering with the free exercise or enjoyment of any right or privilege secured to him or her by the Constitution or laws of this state or by the Constitution or laws of the United States because of the other person's race, color, religion, ancestry, national origin, disability, gender, or sexual orientation, or because he or she perceives that the other person has one or more of those characteristics.

(c) Any person convicted of violating subdivision (a) or (b) shall be punished by imprisonment in a county jail not to exceed one year, or by a fine not to exceed $5,000, or by both that imprisonment and that fine, and the court shall order the defendant to perform a minimum of community service, not to exceed 400 hours, to be performed over a period not to exceed 350 days, during a time other than his or her hours of employment or school attendance. However, no person shall be convicted of violating subdivision (a) based on speech alone, except upon a showing that the speech itself threatened violence against a specific person or group of persons and that the defendant had the apparent ability to carry out the threat.

FLORIDA STATUTES § 775.085

Evidencing prejudice while committing offense; reclassification

(1)(a) The penalty for any felony or misdemeanor shall be reclassified as provided in this subsection if the commission of such felony or misdemeanor evidences prejudice based on the race, color, ancestry, ethnicity, religion, sexual orientation, national origin, mental or physical disability, or advanced age of the victim.

. . .

MICHIGAN COMPILED LAWS § 750.147B

Ethnic intimidation, guilt; examples

(1) A person is guilty of ethnic intimidation if that person maliciously, and with specific intent to intimidate or harass another person because of that person's race, color, religion, gender, or national origin, does any of the following:

(a) Causes physical contact with another person.

(b) Damages, destroys, or defaces any real or personal property of another person.

(c) Threatens, by word or act, to do an act described in subdivision (a) or (b), if there is reasonable cause to believe that an act described in subdivision (a) or (b) will occur.

Classification of crime, punishment. (2) Ethnic intimidation is a felony punishable by imprisonment for not more than 2 years, or by a fine of not more than $5,000.00, or both.

Civil cause of action, recovery. (3) Regardless of the existence or outcome of any criminal prosecution, a person who suffers injury to his or her person or damage to his or her property as a result of ethnic intimidation may bring a civil cause of action

(Continued)

Box 2.8 (Continued)

---

against the person who commits the offense to secure an injunction, actual damages, including damages for emotional distress, or other appropriate relief. A plaintiff who prevails in a civil action bought pursuant to this section may recover both of the following:

(a) Damages in the amount of 3 times the actual damages described in this section or $2,000.00, whichever is greater.

(b) Reasonable attorney fees and costs.

TEXAS CODE OF CRIMINAL PROCEDURE ARTICLE 42.014

Finding that an offense was committed because of bias or prejudice

In the punishment phase of the trial of an offense under the Penal Code, if the court determines that the defendant intentionally selected the victim primarily because of the defendant's bias or prejudice against a group, the court shall make an affirmative finding of that fact and enter the affirmative finding in the judgment of that case.

Finally, in addition to the penalty-enhancement–type laws, some states also have additional, related laws. Most (40 states) have laws prohibiting institutional vandalism. Almost half (24 states) have statutes regarding collection of hate crime data. A few (10 states) have laws relating to specialized training for law enforcement personnel. And many (28 states) also authorize civil actions. A handful of states—California, Illinois, Louisiana, Massachusetts, Minnesota, and Washington—have all five types of laws. Only one state—Wyoming—has no hate crime provisions at all.

Federal

Having experienced success in their lobbying efforts with the states, the ADL and its allies turned to the federal government. The gains there, although still substantial, were slower and more qualified.

The first federal law relating specifically to hate crimes was enacted in 1990. Called the Hate Crimes Statistics Act (HCSA), it required the United States Department of Justice to collect data on hate crimes from local law enforcement agencies and to publish the results. The law itself was very strongly supported in Congress. Only four senators voted against it, all on the basis that it included sexual orientation within its enumerated categories. Although the HCSA was originally set to expire in 1995, it was later extended through fiscal year 2002. As we will see in Chapter 3, there have been problems relating to the collection of hate crime data, as well as the interpretation of the data that are produced.

In 1994, Congress passed the Hate Crimes Sentencing Enhancement Act. This law essentially acted as a federal penalty enhancement statute. It ordered the United States Sentencing Commission to revise sentencing requirements for situations where a person was being tried for a federal crime, and where the defendant intentionally selected the victim because of the victim's group. The penalty was to be increased at least three levels. It is important to note that this law is of limited potential scope because, as we have already discussed, federal criminal prosecutions are relatively rare, at least in comparison with state prosecutions. In 1999, this law was used in only 58 cases (Chorba, 2001).

Also in 1994, Congress passed the Violence Against Women Act (VAWA). Among other things, this law provided that victims of gender-based crimes could sue their attackers and receive compensatory and punitive damages. This law could be seen as an extension of existing hate crimes legislation, none of which included gender. However, as already mentioned, the Supreme Court declared a portion of VAWA unconstitutional in 2000 because it exceeded Congress's authority under the Commerce Clause.

Another specialized federal hate crime law is the Church Arson Prevention Act, passed unanimously in 1997. This law was a reaction to a reported spike in the numbers of church burnings in the mid-1990s; many of the affected churches had predominantly African American congregations. The law contains a number of provisions, including facilitating federal prosecutions and increasing penalties for damaging places of worship.

One additional piece of federal legislation, the Hate Crimes Prevention Act, has been bouncing around Congress for several years now. It was originally proposed in 1998 but was then put aside as Congress became busy with the impeachment hearings of then-President Clinton. In 1999, two competing bills were introduced by Senators Kennedy and Hatch. Both would have extended federal jurisdiction in hate crime cases. The primary difference between them was that Kennedy's bill included crimes based on sexual orientation, whereas Hatch's did not (Peek, 2001). Both bills died in committee. Kennedy's bill was revised and reintroduced in 2001 as the Local Law Enforcement Enhancement Act; in early 2002, the bill was once again killed in committee.

HATE SPEECH

All of the laws that we have discussed in this chapter assume that the offenders commit some underlying criminal act. But what if they do not? What if, for example, I want to display a swastika on my own property, or call someone else a racially disparaging name? Can I be punished for that?

A general definition of hate speech is words or symbols that are derogatory or offensive on the basis of race, religion, sexual orientation, and so on. The critical distinction between hate *crime* and hate *speech* is that the former requires some underlying criminal act; the latter does not. In the case of hate speech, it is the speech itself that is punished.

Many commentators have argued that hate speech is dangerous and harmful. According to these people, not only does hate speech have many of the same effects as hate crimes (psychological trauma, adverse impact on the community, etc.), but it also may foster an atmosphere in which bias-motivated violence is encouraged, subtly or explicitly (see, e.g., Delgado, 1993; Greenawalt, 1995; Lawrence, 1992; Matsuda, 1993; Smolla, 1992; Tsesis, 2000).

As a result of these arguments, many college campuses created codes prohibiting certain kinds of speech. These codes varied widely in scope and quality; some were so broadly written as to encompass a great deal of legitimate academic discourse. Critics claimed that not only were these codes a violation of the First Amendment freedom of speech, but also that they interfered with academic freedom and hobbled scholarship and the exchange of ideas (see, e.g., Haiman, 1993; Heumann & Church, 1997; Holzer, 1994; Strossen, 1990).

Box 2.9   *R. A. V. v. City of St. Paul*

R. A. V. v. City of St. Paul, Minnesota
Supreme Court of the United States
505 U.S. 377 (1992)
Excerpts from Justice Scalia's opinion:
In the predawn hours of June 21, 1990, [R. A. V.] and several other teenagers allegedly assembled a crudely made cross by taping together broken chair legs. They then allegedly burned the cross inside the fenced yard of a black family that lived across the street from the house where [R. A. V.] was staying. Although this conduct could have been punished under any number of laws,[5] one of the two provisions under which respondent St. Paul chose to charge petitioner (then a juvenile) was the St. Paul Bias-Motivated Crime Ordinance, . . . which provides:

> Whoever places on public or private property a symbol, object, appellation, characterization or graffiti, including, but not limited to, a burning cross or Nazi swastika, which one knows or has reasonable grounds to know arouses anger, alarm or resentment in others on the basis of race, color, creed, religion or gender commits disorderly conduct and shall be guilty of a misdemeanor.

We conclude that, even as narrowly construed by the Minnesota Supreme Court, the [St. Paul] ordinance is facially unconstitutional. Although the phrase in the ordinance, "arouses anger, alarm or resentment in others" has been limited by the Minnesota Supreme Court to reach only those symbols or displays that amount to "fighting words," the remaining, unmodified terms make clear that the ordinance applies only to "fighting words" that insult, or provoke violence, "on the basis of race, color, creed, religion or gender." Displays containing abusive invective, no matter how vicious or severe, are permissible unless they are addressed to one of the specified topics. Those who use "fighting words" in connection with other ideas—to express hostility, for example, on the basis of political affiliation, union membership, or homosexuality—are not covered. The First Amendment does not permit St. Paul to impose special prohibitions on those speakers who express views on disfavored subjects.

Hate speech restrictions were created in a few other situations as well. For example, St. Paul, Minnesota, enacted a city ordinance that prohibited offensive symbols or displays in public places.

The debates about hate speech codes were, to say the least, lively. Eventually, of course, these debates entered the legal arena. In cases in Michigan and Wisconsin, federal courts struck down campus speech codes as impermissible restrictions on freedom of speech. In 1992, the United States Supreme Court entered the fray when it decided the case of *R. A. V. v. City of St. Paul.* In that case, the Court held that St. Paul's ordinance was unconstitutional (see Box 2.9; for an extensive discussion of the *R. A. V.* case, see Cleary, 1994). A year later, as we will discuss in Chapter 3, the Court held that hate *crime* laws are permissible. Thus, the distinction between hate speech and hate crime is extremely important: One cannot be punished, and the other can.

As you might imagine, the line between hate crime and hate speech is often quite thin. Laypeople quite frequently use the terms interchangeably. Perhaps the simplest way to distinguish them is to ask whether the perpetrator has violated any other criminal law. If

In its practical operation, moreover, the ordinance goes even beyond mere content discrimination, to actual viewpoint discrimination. Displays containing some words—odious racial epithets, for example—would be prohibited to proponents of all views. But "fighting words" that do not themselves invoke race, color, creed, religion, or gender—aspersions upon a person's mother, for example—would seemingly be usable ad libitum in the placards of those arguing in favor of racial, color, etc., tolerance and equality, but could not be used by those speakers' opponents. One could hold up a sign saying, for example, that all "anti-Catholic bigots" are misbegotten; but not that all "papists" are, for that would insult and provoke violence "on the basis of religion." St. Paul has no such authority to license one side of a debate to fight freestyle, while requiring the other to follow Marquis of Queensberry rules.

Finally, St. Paul and its amici defend the conclusion of the Minnesota Supreme Court that, even if the ordinance regulates expression based on hostility towards its protected ideological content, this discrimination is nonetheless justified because it is narrowly tailored to serve compelling state interests. Specifically, they assert that the ordinance helps to ensure the basic human rights of members of groups that have historically been subjected to discrimination, including the right of such group members to live in peace where they wish. We do not doubt that these interests are compelling, and that the ordinance can be said to promote them. But the "danger of censorship" presented by a facially content-based statute requires that that weapon be employed only where it is "necessary to serve the asserted [compelling] interest." The existence of adequate content-neutral alternatives thus "undercuts significantly" any defense of such a statute, casting considerable doubt on the government's protestations that "the asserted justification is in fact an accurate description of the purpose and effect of the law." The dispositive question in this case, therefore, is whether content discrimination is reasonably necessary to achieve St. Paul's compelling interests; it plainly is not. An ordinance not limited to the favored topics, for example, would have precisely the same beneficial effect.

Let there be no mistake about our belief that burning a cross in someone's front yard is reprehensible. But St. Paul has sufficient means at its disposal to prevent such behavior without adding the First Amendment to the fire.

I trespass onto your property and burn a cross, that is a hate crime. If I burn the cross on my own property (assuming I have not violated any burning ordinances or the like), that is hate speech.

Often, the distinction between hate crime and hate speech is as simple as the distinction between conduct and expression. Hate crime and hate speech become especially hard to tell apart in two situations: verbal acts and symbolic expression. A verbal act occurs when words themselves constitute some element of a crime. For example, a person can be convicted of the crimes of harassment or making terroristic threats based solely on the words he or she has spoken or written. This is not a violation of freedom of speech. The problem comes when we try to distinguish general statements of rancor, which may be distasteful but which are protected, from actual threats, which are not protected. If I say, "I hate you and I wish you would die," that is probably okay; if I say, "I hate you and I'm going to kill you," I can go to prison.

The second problematic situation occurs when a person does not say a word, and yet still expresses a thought or idea. The Supreme Court calls this symbolic expression or

NARRATIVE PORTRAIT

THE ADL PERSPECTIVE ON HATE CRIMES

The following pieces were written by two attorneys for the Anti-Defamation League. Michael Lieberman has been the Washington Counsel for the Anti-Defamation League since January 1989. He has written widely about the impact of hate crimes and has been actively involved in efforts to secure passage of a number of federal and state hate crime statutes. Mr. Lieberman has participated in seminars and workshops on response to violent bigotry and has served on advisory boards for hate crime projects funded by the Department of Justice, the Department of Education, and the Department of Housing and Urban Development. Karen Zatz is the Assistant Western States Counsel for the ADL. She works in the San Francisco ADL office responding to acts of hate and bias and training law enforcement personnel and attorneys on effective responses to hate.

By Michael Lieberman

All Americans have a stake in an effective response to violent bigotry. Hate crimes demand a priority response because of their special emotional and psychological impact on the victim and the victim's community. The damage done by hate crimes cannot be measured solely in terms of physical injury or dollars and cents. Hate crimes may effectively intimidate other members of the victim's community, leaving them feeling isolated, vulnerable, and unprotected by the law. By making members of minority communities fearful, angry, and suspicious of other groups—and of the power structure that is supposed to protect them—these incidents can damage the fabric of our society and fragment communities.

ADL has long been in the forefront of national and state efforts to deter and counteract hate-motivated criminal activity. Hate crime statutes are necessary because the failure to recognize and effectively address this unique type of crime could cause an isolated incident to explode into widespread community tension. Since 1913, the mission of ADL has been to "stop the defamation of the Jewish people and to secure justice and fair treatment to all citizens alike." Dedicated to combating anti-Semitism, prejudice, and bigotry of all kinds, ADL has played a national leadership role in the development of innovative materials, programs, and services that build bridges of communication, understanding, and respect among diverse racial, religious, and ethnic groups. ADL developed the model hate crimes statute as a way to counteract hate. The League's Web site, www.adl.org, contains excellent background on legal and legislative response to hate violence, as well as educational resources on prevention of bias that can lead to violence. The Partners Against Hate Web site, www. partnersagainsthate.org, serves as a comprehensive clearinghouse of hate crime-related information. In addition, the Web site includes access to the finest database of hate crime laws that form the basis of criminal enforcement in the states, and counteraction tools.

The urgent national need for both a tough law enforcement response as well as education and programming to confront violent bigotry has only increased. In the aftermath of the September 11 terrorism, the nation has witnessed a disturbing increase in attacks against American citizens and others who appear to be of Muslim,

Middle Eastern, and South Asian descent. Perhaps acting out of anger at the terrorists involved in the September 11 attacks, the perpetrators of these crimes are irrationally lashing out at innocent people because of their personal characteristics—their race, religion, or ethnicity. Law enforcement officials are now investigating hundreds of incidents reported from coast to coast—at places of worship, neighborhood centers, grocery stores, gas stations, restaurants, and homes—including vandalism, intimidation, assaults, and several murders. As the nation witnessed a series of disturbing attacks against individuals perceived to be Middle Eastern, Arab, or Muslim, in the aftermath of the September 11 terrorist incidents, we were reminded of the need to directly confront the prejudice and intolerance that can lead to hate crimes—in our communities, in our houses of worship, in our schools, and, especially, in our homes. Our civic leaders set the tone for national discourse and have an essential role in shaping attitudes. On September 26, 2002, at a meeting with Sikh leaders at the White House, President Bush pledged "our government will do everything we can not only to bring those people to justice, but also to treat every human life as dear, and to respect the values that made our country so different and so unique. We're all Americans, bound together by common ideals and common values." ADL will continue to advocate for hate crimes laws as an effective means to prevent and combat acts motivated by hate.

By Karen Zatz

Clearly hate crime numbers do not speak for themselves. Behind each and every one of these statistics is an individual or a community targeted for violence for no other reason than race, religion, sexual orientation, disability, or ethnicity. Every day, ADL staffers respond to victims of hate crimes or incidents. When we talk to them, we hear not just of the pain inflicted by the underlying crime, but also of the devastating effect of being targeted because of something you cannot change about yourself. If you are the victim of an assault while walking to the store, you can blame the route, time of day or what you were wearing. When someone calls you a "dirty spic," "a damn faggot," or a "stupid kike" and assaults you on the same route, the impact on you as a victim is quite different. You may be afraid that in the future, wherever you go, and whatever you are doing, you will be targeted for violence based on a fundamental characteristic of yourself that you cannot change (an immutable characteristic). This fear may make you try and hide your identity; it may make you not want to associate with other members of your race or religion. When members of your community hear about this incident, they may also be frightened that they will be attacked. Other minority communities may fear that an attack on one group means that their group is in danger. One act of hate in a community may impact hundreds or thousands of individuals.

A few years ago a synagogue in Reno, Nevada, was targeted with a firebombing, damaging the building. Within the following year, it was targeted a second time. When I met with members of the congregation, they were alarmed by these repeat attacks. They wondered if their children were safe at Sunday School and whether they were safe attending services. Some congregants resigned their memberships out of fear, and individuals told me that they would no longer gather publicly with other Jews to pray. Some community members told me how angry these attacks made them; older community members were reminded of their experiences during the holocaust. Jews who were members of other congregations in Reno also felt that they were in danger. They told me that they were afraid for their businesses and for their families. Even when I spoke with members of other minority groups, including members of Christian and Muslim community groups, they spoke of their concern that they would be the next targets. But what I was heartened to see were that hundreds of people from the community—from church groups, civic organizations, law enforcement, and from the government, gathered together at community rallies to speak out against this act of hate. Over and over again I heard from the Jewish community, that it was this support from the community in general that made them feel safe to send their kids to religious school, to keep attending services, and to decorate their houses for Chanukah. It is hard to overstate the importance of outspoken leadership in opposition to all forms of bigotry, whether the leadership comes from elected officials, newspaper editorial boards, school administrators, or your next-door neighbor.

I think that hate crime laws speak for a fundamental value of American society, that we should be able to live our lives free from violence, whether as people of faith, or people from a particular country, or race, or sexual orientation or disability. Philosophical promises of America's constitution are our freedoms of religion and association, and these can only occur meaningfully when individuals cannot be legally targeted for violence based on who they are. I can comfort victims of hate because I know that my entire community stands against hate, as evidenced by the public statement made when hate crime laws are enacted. I can advocate for hate crime laws because I know that they are fundamental to the kind of society Americans have always envisioned for themselves—one where we are free to live our lives without hate based violence.

expressive conduct. In a series of cases over the last four decades, the Court has held that actions such as burning a flag or wearing a black armband express ideas, and therefore are protected by the first amendment. In *R. A. V.,* the Court held that cross-burning fell in this category. On the other hand, the Court has also stated that actions such as burning a draft card and dancing nude were merely conduct, and not protected. Clearly, it can sometimes be difficult to classify speech and conduct.

Entire books could be (and have been: see, e.g., Coliver, 1992; Heumann & Church, 1997; Holzer, 1994; Matsuda, Lawrence, Delgado, & Crenshaw, 1993; Walker, 1994) written on the subject of hate speech alone. Because the issues involved are significantly different from those involved with hate crimes, we will leave a more involved discussion of hate speech for other venues.

CONCLUSION

Although laws addressing some bias-related acts have existed since the Civil War, hate crime legislation as such is a modern phenomenon. Over the past two decades, it has sprung up rather quickly in most states, as well as within the federal codes. Furthermore, there is good reason to believe that hate crime laws will continue to evolve in the near future.

The need for special hate crime laws has been asserted by many people and has been supported by a number of cogent and rational arguments. What remains missing, for the most part, is empirical evidence to substantiate those arguments. Of course, some might argue, such empirical evidence is not really necessary; even if the laws are not particularly required, they can do little harm. However, there are a number of serious legal and policy challenges associated with the implementation of hate crime laws. We will examine those challenges in the next chapter.

DISCUSSION QUESTIONS

1. The Federal Civil Rights Acts were aimed at specific behaviors and situations. What kinds of hate crime don't they address?

2. How would you design an experiment to study whether hate crimes have more severe effects than other crimes?

3. Summarize the arguments on why hate is different. Do you find them convincing? Is there sufficient evidence to support them?

4. It has been argued that hate crimes affect communities more adversely than ordinary crimes. Why might this be so?

5. Under what circumstances do you believe a hate crime law might serve as an effective deterrent?

6. This chapter argues that hate crime laws may serve a symbolic function. Can you think of other examples of laws whose effects are largely symbolic? Is it worth passing legislation for this purpose?

7. Organizations such as the ADL played a major part in the passage of hate crime legislation. In general, what role should special interest groups play in the creation of laws?

8. Draft a hate crime statute of your own. Which parts of the ADL's model law would you keep, and which parts would you change?

9. Why do so many people argue that the federal government should enact additional hate crime legislation? Do you agree? What practical effect would such laws likely have?

10. If you lived in a jurisdiction without hate crime laws, would you support their passage?

11. Would you support a hate speech code on your campus? What do you see as the pros and cons of such a code?

INTERNET EXERCISES

1. Go to the Anti-Defamation League's Web page on hate crimes: www.adl.org/main_hate_crimes.asp. Scroll down to the second section, "Hate Crimes Laws," and click on the link labeled "Map: State Hate Crimes Statutory Provisions." Now find your state on the map and click on it. What hate crime provisions does your state have? How do they compare with those of neighboring states?

2. The American Psychological Association has an online position paper on hate crimes at www.apa.org/pubinfo/hate/. Review the APA's recommendations to Congress. If you were an APA member, are there any that you would add, subtract, or change?

NOTES

1. There have also been a number of arguments *against* hate crime legislation. These are discussed in Chapter 3.

2. Recently, someone broke into my garage and stole some tools and other small items. Since then, not only has my family been especially careful about keeping doors and cars locked, but so has our entire neighborhood. The neighbor across the street even installed additional locks around his house.

3. Here are the relevant lines of dialogue:

Ellwood Blues (Dan Akroyd) (having come upon a neo-Nazi rally): Illinois Nazis.

Jake Blues (Jim Belushi): I hate Illinois Nazis.

4. Those seven states were Arkansas, Hawaii, Indiana, Kansas, New Mexico, South Carolina, and Wyoming.

5. R. A. V. was also prosecuted separately under Minnesota's hate crime statute. He didn't challenge that charge.

Chapter 3

THE HATE DEBATE

Constitutional and Policy Problems

Every year, I issue a challenge to the students in my Hate Crimes class. Any student who can convince me that hate crime laws either are, or are not, constitutional, receives an automatic A in the class. So far, no student has successfully met the challenge, and, despite having studied the laws for more than a decade, I remain firmly on the fence. I am not the only one who has had difficulty with this issue; organizations such as the American Civil Liberties Union have suffered from internal divisions concerning their stand on hate crime laws, and scholars and even judges have expressed their own ambivalence (see Box 3.1). This ambivalence is due not to lack of thought on the matter, but rather to the strength of the arguments both for and against the constitutionality of hate crime legislation. In this chapter we will explore those arguments, and perhaps you will be able to reach your own conclusions.

The constitutionality of hate crime laws, though it has sparked much vigorous debate, is not the only dilemma regarding these laws. There are also a number of serious policy concerns about the scope and use of the laws, and in this chapter we will explore those issues as well. Among the major issues I discuss are which groups to protect with the laws, problems with identifying and prosecuting hate crimes, and the potential paradoxical effects of hate crime legislation.

HATE CRIME LAWS AND THE CONSTITUTION

Motive

Some of the strongest arguments against hate crime laws have focused on the First Amendment's protections of speech and association. The problems are related to a unique aspect of hate crime laws: They are the only kind of laws for which motive is an element of the crime.

Many laypeople, and even some legal scholars, confuse motive and intent. Most crimes contain some intent requirement, also known as mens rea. For example, in California (and in most other states), a conviction for murder requires proof that the offender acted with "malice aforethought," whereas vehicular manslaughter requires that the offender drove

Box 3.1   Ambivalence About Hate Crime Laws

Many people have had a difficult time deciding whether hate crime laws are constitutional. In the first quote presented here, a justice of the Wisconsin Supreme Court, dissenting from that court's decision to overturn Wisconsin's hate crime law, expressed her own ambivalence. The Wisconsin Supreme Court's decision was later overruled by the United States Supreme Court, which agreed with Justice Abrahamson that the law met constitutional standards:

Had I been in the legislature, I do not believe I would have supported this [hate crime] statute. I do not think this statute will accomplish its goal; I would direct the state's efforts to protect people from invidious discrimination and intimidation in other directions. As a judge, however, after much vacillation, I conclude that this law should be construed narrowly and should be held constitutional. (Justice Shirley Abrahamson [dissenting], *Wisconsin v. Mitchell,* 485 N.W.2d 807, 818 [Wis. 1992]; *rev'd* 508 U.S. 476, 1993)

As civil rights attorney Susan Gellman asserts here, Justice Abrahamson is not alone in her indecision:

The debate over [hate crime] laws is occurring not merely between traditional allies, but between one side and itself. Moreover, whenever either viewpoint prevails, whether in the legislature, the courts, or even in a purely academic argument, its proponents do not seem very happy about it. They can see very well their opponents' point of view, and in fact largely agree with it. (Gellman, 1991, p. 334)

Even organizations have experienced difficulty in reaching a consensus, as described as follows by law professor James B. Jacobs and attorney Kimberly Potter:

Civil libertarians, torn between commitments to equality and free speech, are divided on this question. On the one hand, for example, the national American Civil Liberties Union (ACLU), in its amicus brief to the U.S. Supreme Court in *Wisconsin v. Mitchell,* defended the extra punishment imposed by the Wisconsin hate crime law. . . . On the other hand, the Ohio chapter of the ACLU and the Center for Individual Rights submitted amicus briefs opposing the statute. (Jacobs & Potter, 1998, pp. 111-112)

with "gross negligence." Essentially, intent refers to the degree to which a person meant to commit a particular action or cause a particular result.

On the other hand, motive refers to the reason *why* a person commits a particular act. Motive is virtually never made an element of crimes (Candeub, 1994; Gardner, 1993; Gaumer, 1994). If someone robs a bank, it is bank robbery, regardless of whether the robber wanted to use the money to buy medicine for a dying mother or to finance a pornography habit. His or her motive, be it compassion or greed and prurience, is immaterial as far as proving the crime itself is concerned (although in some cases, motive might be taken into consideration during sentencing).[1]

Hate crimes are the exception to this rule. If I punch a person because I dislike his choice of football teams, it is ordinary battery. In California, I could receive a fine of up to $2,000, or six months in jail. However, if I punch a person because I dislike his religion, it is a hate crime, and I face an additional $5,000 fine and another year in jail. In both cases my intent, or mens rea, is the same: I willfully used force against my victim. But the reason why I used force—in other words, my motive—is different.

Punishing motive leads to some particularly thorny problems. From the constitutional perspective, it raises the arguments that hate crimes amount to thought crimes and that hate crime laws impermissibly penalize speech and group affiliation. As I will discuss in the "Identifying and Prosecuting Hate Crimes" section later in this chapter, it also causes policy dilemmas, because motive is difficult to determine and prove.

Are Hate Crimes Thought Crimes?

Discussion of Hate Crimes as Unconstitutional

A basic principle of American jurisprudence is that thoughts alone, no matter how abhorrent, cannot be punished. The Constitution does not explicitly protect thought, but at least as early as 1929, the Supreme Court recognized that the First Amendment's provisions imply freedom of thought (see Box 3.2). Expression of those thoughts—again, even when they are repugnant—is also protected, as was explicitly held by the Supreme Court in *R. A. V. v. St. Paul.*

Box 3.2   The Supreme Court on Thought Crimes

If there is any principle of the Constitution that more imperatively calls for attachment than any other it is the principle of free thought—not free thought for those who agree with us but freedom for the thought that we hate. (Justice Holmes, dissenting, in *United States v. Schwimmer,* 1929)

At the heart of the First Amendment is the notion that an individual should be free to believe as he will. (Justice Stewart, in *Abood v. Detroit Board of Education,* 1977)

If there is a bedrock principle underlying the First Amendment, it is that the government may not prohibit the expression of any idea simply because society finds the idea itself offensive or disagreeable. (Justice Brennan, in *Texas v. Johnson,* 1989)

Many commentators, as well as some judges, have argued that hate crime laws constitute punishment of thought crimes and therefore violate the First Amendment. As in my previous example, imagine that I have punched Able because of his football preference and Baker because of his religious preference. The acts are identical; the only difference between the scenarios is the thoughts that are going through my head as I attack. But I am liable to receive harsher punishment for hitting Baker than for hitting Able. Does this not amount to punishing my thoughts?

Numerous legal scholars have made this argument (see, e.g., Brooks, 1994; Degan, 1993; Fleisher, 1994; Gaumer, 1994; Gellman, 1991, 1992; Gey, 1997; Jacobs & Potter, 1997, 1998; Redish, 1992). At least two courts have agreed. In 1992, the Wisconsin Supreme Court struck down its state's hate crime law. The court held, "The hate crime statute violates the First Amendment directly by punishing what the legislature has deemed to be offensive thought. . . . Without a doubt, the hate crime statute punishes bigoted thought" (*Wisconsin v. Mitchell,* 1992, pp. 811-812). This decision, as we will see, was later overturned by the U.S. Supreme Court.

The Ohio Supreme Court also declared its state's law unconstitutional: "Enhancing a penalty because of motive . . . punishes the person's thought, rather than the person's act or criminal intent." (*Ohio v. Wyant,* 1992, p. 454). This court also discussed why it is dangerous to punish bigoted thought: What, then, would stop the legislature from enhancing penalties for crimes committed because of opposition to abortion or war, or any other viewpoint?

Discussion of Hate Crimes as Constitutional

Other commentators and courts, on the other hand, have strongly disagreed with these assertions (see, e.g., Grannis, 1993; Lawrence, 1999; Micacci, 1995; Selbin, 1993). They argue that although hate crime laws may appear on their face to punish thought, in fact they do comply with the Constitution. Several points are made in support of this view.

First, although laws that consider motive as an element of a crime are rare, law does consider motive in some situations. One such scenario is a sentencing decision, in which the sentencing authority (usually the judge) may take into account the defendant's motive. In fact, in 1983 in *Barclay v. California,* the U.S. Supreme Court held that a trial judge properly considered the defendant's racial animus toward his victim when sentencing the offender to death.[2] If motive can properly be considered by a judge, why can the legislature not essentially mandate that certain motives will increase punishment? Some courts have been persuaded by this argument; for example, both a Florida appeals court (*Dobbins v. Florida,* 1992) and a California appeals court (*California v. Joshua H.,* 1993) upheld hate crime laws partially on this basis.

The other situation in which motive matters is in civil antidiscrimination cases, such as under the federal Title VII. If an employer fires someone because she does not like his taste in music, it violates no law. If she fires him because of his race, it does.

Of course, the laws' critics have countered these arguments. Determining the appropriate sentence for someone who has already been convicted of a crime is significantly different from convicting someone of a crime itself. Furthermore, the impact of aggravating circumstances such as biased motives in sentencing decisions is tempered by the fact that the sentencer may also consider mitigating circumstances. For example, suppose the perpetrator had been the victim of repeated racial attacks himself, and thus had developed fear of and animosity toward the victim's race. Hate crime laws would leave no opportunity for him to present this evidence.

As for the comparison to antidiscrimination laws, critics point out that there are important differences between civil and criminal law. Nobody will go to prison or lose citizenship rights for discriminating in housing or employment or education; they might for committing a hate crime. Because of the unique gravity of criminal sanctions, our legal system traditionally places greater restrictions on what criminal sanctions the government may impose, and how. Furthermore, as one scholar pointed out (Fleisher, 1994), it is often exceedingly difficult to determine whether a person acted because of the plaintiff's or victim's race—in other words, it is often difficult to determine motive. (I will discuss this point in greater detail later in this chapter.) In civil discrimination cases, a plaintiff is often aided by the fact that the defendant has exhibited a pattern of discriminatory acts against other people. In hate crime cases, it would be rare that a prosecutor would be able to find such a pattern on the part of a defendant.[3] Therefore, to delve too deeply into a defendant's motive in a criminal case is to raise a matter that criminal law, with its great need for certainty of guilt, should not consider.

The second argument that the supporters of hate crime legislation make is that these laws do not punish thought at all, but rather conduct. Although thought and expression are protected by the First Amendment, conduct is not. Florida's Fifth Circuit Court of Appeal, for example, explained that it was not the defendant's opinions that were punished, but rather his act of choosing his victim because of the victim's group. The Oregon Supreme Court agreed (*Oregon v. Plowman,* 1992). In other words, the defendant is free to think all the bigoted thoughts he wishes; it is only when he acts on those thoughts that he will be penalized. From the critics' perspective, however, the fact remains that the actual observable behavior committed by the offender is identical to behaviors that were not motivated by bias, and yet only the offender with bigoted thoughts will receive an enhanced sentence.

The laws' supporters counter this with a third argument: Hate-motivated crimes, they assert, are not identical to other crimes. They differ in other dimensions, most notably their effects on the victims and the community. This is the same issue discussed at length in Chapter 2. As we have seen, there are strong and credible claims that hate crimes are qualitatively different from other crimes, but those claims are largely unsupported by empirical evidence.

The Law's Answer to the Puzzle: Wisconsin v. Mitchell

Perhaps now the source of many people's ambivalence on the constitutionality of hate crime laws is clear. The arguments on both sides seem very persuasive, and clearly there is much room for reasonable people to differ in their views. I believe this problem is a bit like one of those figure-ground perceptual puzzles commonly found in Psychology textbooks. Is the drawing a vase or two faces? A rabbit or a duck? There is no "right" solution to these puzzles: The answer lies in an individual viewer's own perceptions. A hate crime law can be logically interpreted as either punishing thought or punishing behavior.

Unlike perceptual puzzles, the law gives us (as it must) definitive answers to these dilemmas. In the case of hate crime laws, the answer on this issue was given by the U.S. Supreme Court when it ruled on *Wisconsin v. Mitchell* in 1993. The events of the case unfolded as follows:

One evening in Kenosha, Wisconsin, a group of African American teenagers was discussing a scene from the movie *Mississippi Burning* (Zollo, Colesberry, & Parker, 1988) in which a white man beats a black child. Todd Mitchell, 19, asked his friends, "Do you all feel hyped up to move on some white people?" Shortly afterward, 14-year-old Gregory Riddick, who was white, happened to walk by on the other side of the street. Mitchell said, "You want to fuck somebody up? There goes a white boy; go get him," and he pointed at Riddick. The group ran at Riddick and beat him severely enough to leave him in a coma for 4 days. It was possible that he suffered permanent brain damage. They also stole his tennis shoes.

Mitchell was convicted of aggravated battery, which in Wisconsin brings a maximum sentence of 2 years. However, because the jury found that Mitchell was motivated by

Riddick's race, he was subject to the hate crime penalty enhancement statute. He was sentenced to 4 years in prison.

When Mitchell appealed his conviction, the Wisconsin Supreme Court declared the hate crime law unconstitutional. However, a year later the U.S. Supreme Court heard the case. That Court unanimously held that Wisconsin's law (which was similar to the ADL Model Hate Crime Law) met constitutional muster. Essentially, the Court was persuaded by the arguments discussed previously in this chapter: It found that motive is an element of antidiscrimination laws and sentencing considerations, that the law punished conduct rather than thought, and that bias-motivated offenses were worse than other crimes (see Box 3.3).

Box 3.3   Excerpts From *Wisconsin v. Mitchell*

Justice Rehnquist, writing for a unanimous Court:

In *Barclay* we held that it was permissible for the sentencing court to consider the defendant's racial animus in determining whether he should be sentenced to death, surely the most severe "enhancement" of all. And the fact that the Wisconsin Legislature has decided, as a general matter, that bias-motivated offenses warrant greater maximum penalties across the board does not alter the result here. For the primary responsibility for fixing criminal penalties lies with the legislature.

. . . Whereas the ordinance struck down in *R. A. V.* was explicitly directed at expression (*i.e.,* "speech" or "messages") . . . the statute in this case is aimed at conduct unprotected by the First Amendment.

Moreover, the Wisconsin statute singles out for enhancement bias-inspired conduct because this conduct is thought to inflict greater individual and societal harm. For example, according to the State and its *amici,* bias-motivated crimes are more likely to provoke retaliatory crimes, inflict distinct emotional harms on their victims, and incite community unrest. . . . The State's desire to redress these perceived harms provides an adequate explanation for its penalty-enhancement provision over and above mere disagreement with offenders' beliefs or biases. (pp. 486-487)

Do Hate Crime Laws Have a "Chilling Effect"?

Aside from the thought crimes issue, the consideration of motive in hate crime cases creates another First Amendment problem. Naturally, jurors cannot read defendants' minds to determine whether their acts were precipitated by bigotry. Therefore, motive must frequently be determined through circumstantial evidence. This evidence most often takes one or more of these forms: first, that there was no other apparent motive; second, that the defendant uttered slurs around the time of the crime; and third, that the defendant was affiliated with a hate group. It is the latter two kinds of evidence that raise First Amendment issues.

As we have already discussed, even distasteful expression such as burning a cross is protected by the Constitution. Under the First Amendment's freedom of association clause, people also have the right to join groups, even hate groups such as the skinheads or the Ku Klux Klan. Of course, speech and group membership are frequently used as evidence in criminal trials. For example, a prosecutor might prove premeditation in a murder case by showing that prior to the killing, the offender said to a witness, "I'm going to kill him." The problem with hate crimes, however, is that their motives are proven almost exclusively by the defendants' speech and groups. Consider, for example, the cases in Box 3.4; in all of these cases, the hate crime conviction rested almost entirely on the defendant's biased words.

Box 3.4   A Sampling of Hate Crime Cases

OHIO V. WYANT, 1992

The defendant, who was white, and the victim, who was black, had rented adjoining campsites. The victim complained to park officials about the volume of Wyant's radio, and the officials made Wyant turn it down. Later, the victim heard Wyant say to a companion, "We didn't have this problem until those niggers moved in next to us. . . . I ought to shoot that black motherfucker. . . . I ought to kick his ass."

OREGON V. PLOWMAN, 1992

The defendant, along with three companions, approached two men outside a convenience store. One of Plowman's friends asked one of the men if he had any cocaine. When the man said no and started to walk away, Plowman and his friends began beating the other men. During the attack, they shouted such things as, "Talk in English, motherfucker," "White power," and "They're just fucking Mexicans."

CALIFORNIA V. JOSHUA H., 1993

Kiley, the victim, had been involved in an extended feud with his neighbors, the H. family. The H. family had sued Kiley after his tenant's dog bit Mrs. H. and the H. family was upset because when Kiley mowed his lawn, grass clippings blew onto their driveway. Kiley felt that the family was harassing him because he was gay. One afternoon (with video camera rolling) Kiley mowed his lawn. A member of the H. family later left a pile of dirt and grass on his front porch, which Kiley threw onto their driveway. That evening, 17-year-old Joshua H. accosted Kiley, demanding he clean up the grass. He then began punching and kicking Kiley. During the attack, Joshua and his family shouted out such things as, "Come on, let's get it on you faggot queer," and "Where are you going, faggot, you going to suck some faggot dick?"

ILLINOIS V. NITZ, 1996

The Nitz family, which was white, lived across the street from the Gaines family, which was black. The families were involved in frequent altercations with each other. Among other things, the children of the two families got into shouting matches with each other, Mr. Nitz accused Mrs. Gaines of calling the police to get his car towed, and Mr. Nitz and his son used racial slurs when talking to and referring to the Gaines family. In a 1-year span, the police responded to 65 incidents involving the two households. Mr. Nitz was charged with several crimes, including disorderly conduct and contributing to the delinquency of a minor.

Critics argue that to rest a hate crime conviction nearly exclusively on proof of constitutionally protected activities comes perilously close to punishing those activities themselves. Furthermore, some claim that hate crime laws will have a chilling effect, in that people will be afraid to utter (constitutionally protected) unpopular words or join (constitutionally protected) unpopular groups for fear they may later be subjected to hate crime prosecution. One commentator, for example, wrote, "When ordinary speech can be used to show racial or ethnic prejudice allegedly betraying an impermissible motive, as it can under [the hate crime] statute, there will be a chilling effect on first amendment rights," (Mazur-Hart, 1982, p. 212). The Wisconsin Supreme Court, too, found that "opprobrious though the speech may be, an individual must be allowed to utter it without fear of punishment by the state" (*Wisconsin v. Mitchell,* 1992, p. 816). Finally, these critics point out that heavy reliance on such things as racial slurs and group membership to determine motive is poor policy; many people may utter slurs during the heat of an argument, regardless of their actual motivation, and even avowed racists are not always motivated by racism.

Others, including the U.S. Supreme Court, have not been convinced by these arguments. Evidence of a defendant's prior speech is common in criminal trials of all kinds, and there is a significant difference between making speech an element of a crime and using speech to prove a crime. Furthermore, in *Wisconsin v. Mitchell,* the Supreme Court found it "unlikely" and "speculative" that hate crime laws would have any chilling effect on protected speech and association with certain groups.

Other Constitutional Issues

Although the First Amendment arguments against hate crime laws are probably the strongest, additional claims have been made as well, with varying degrees of success. These have primarily been based on two provisions within the Fourteenth Amendment: the equal protection clause and the due process clause.

Fourteenth Amendment: Equal Protection Clause

The equal protection clause was intended to protect people from discrimination by the government. Of course, in reality, the government discriminates all the time. For example, only citizens over the age of 18 may vote, and only men are required to register for the draft. In practice, unless a law discriminates on the basis of certain "suspect categories," such as race or religion, the Supreme Court has interpreted the clause to allow differentiation, as long as the law is supported by a legitimate state interest. This is called the "rational relationship" test.

In a few cases, defendants have claimed that hate crime laws violate the equal protection clause because they give greater protection to some victims (those chosen on the basis of their group) than others. Only one defendant has prevailed on this claim (in *Oregon v. Beebe,* 1984), and even then only in the trial court; the appellate court later overturned the trial court's decision, finding that the law was based on a legitimate interest in preventing bias-based crime because that type of crime might have greater impact upon the community (Gerstenfeld, 1992). For the purposes of this court's analysis, it was not necessary to prove that hate crimes actually are more harmful, but rather that they reasonably might be.

Fourteenth Amendment: Due Process Clause

The Fourteenth Amendment's due process clause is, from a legal standpoint, quite complicated. In a nutshell, it is meant to ensure that laws are fair, both in substance and in implementation. Two types of claims have been made about hate crime laws under the due process clause. The first of these is that the laws are so vague as to lead an ordinary person to be uncertain of their meaning (Gellman, 1991; Gerstenfeld, 1992). Sometimes the claims concern the terminology that is used within the laws, such as "color," "intentionally selects," and "harass"; at other times, the issue centers on the interpretation of the law as a whole: Must the defendant be entirely motivated by the victim's group, or are mixed motives included? What if the defendant believes that the victim belongs to a particular group but is, in fact, mistaken?[4] What if the offender commits a crime not because of the victim's group, but rather because of whom the victim has been associating with?

Like the equal protection challenges, the vagueness challenges have, by and large, not been successful. For the most part, the terms and interpretations seem open to reasonable understanding. One exception to this, perhaps, is Florida's law, which enhances a penalty when, during the commission of a crime, the offender "evidences prejudice." Despite at least one jury's confusion as to what this phrase meant (*Richards v. Florida,* 1992) and the judge's refusal to further explain or define it, the Florida Supreme Court found that the law was not impermissibly vague (*Florida v. Stalder,* 1994). It did so by interpreting the law narrowly to include only those types of offenses in which the perpetrator was actually motivated by the victim's group, as opposed to those committed for some other reason, and in which the perpetrator utters slurs or otherwise shows bias.

The second type of due process claim—that facts that result in sentence enhancements must be proven to a jury beyond a reasonable doubt—arose in a case from New Jersey (*Apprendi v. New Jersey,* 2000). Charles Apprendi fired his rifle at the home of a black family that had recently moved into an all-white neighborhood. Apprendi pled guilty to several counts of unlawful possession of weapons. Under New Jersey's law, the judge, rather than the jury, was to determine whether he was motivated by bias, and the standard of proof was the more lenient preponderance of the evidence, rather than reasonable doubt. The only evidence of bias was a statement Apprendi made to a police officer, and later retracted, that he had fired the shots because he did not want the family to live there. The judge found that it had been a hate crime, and Apprendi was sentenced to 12 years in prison. Without the hate crime enhancement, he would have received 10 years.

After the New Jersey courts rejected his appeal, Apprendi's case went to the U.S. Supreme Court. In a 5-4 decision, the Court held that New Jersey's law was unconstitutional. Due process, the Court held, requires that any fact, other than a prior conviction, that increases the penalty for a crime beyond the statutory maximum must be submitted to a jury and proved beyond a reasonable doubt.

The *Apprendi* decision is interesting from several standpoints. You will remember that one rationale the Court used in *Mitchell* to uphold hate crime laws was that motive is traditionally considered by sentencers when determining a punishment. In *Apprendi,* however, the Court significantly reduced this power of sentencers. From a practical standpoint, the *Apprendi* decision is apt to have little effect on most hate crimes, because in most states it was already the jury's job, rather than the judge's, to determine motive. However, the decision is likely to have significant impact on most states' sentencing procedures in general (Huigens, 2002; Smith, 2001).

Clearly, there are many constitutional issues concerning hate crime laws. For a summary of relevant Supreme Court decisions, see Table 3.1. The arguments are, at times, complex. Reasonable minds can, and do, differ concerning these matters. But constitutional issues are by no means the only potential problems with the laws. In the next sections, we will discuss several policy problems, many of which are at least as difficult to resolve as the constitutional ones.

Table 3.1   U.S. Supreme Court Cases on Hate Crimes

Barclay v. Florida (1983)	Defendant's racial animosity toward the victim may be considered when determining whether to sentence the defendant to death.
R. A. V. v. St. Paul (1992)	Hate speech laws violate the First Amendment.
Dawson v. Delaware (1992)	Defendant's racist beliefs cannot be considered at sentencing if they were unrelated to the crime.
Mitchell v. Wisconsin (1993)	Hate crime laws are constitutional.
Apprendi v. New Jersey (2000)	Motive must be determined by jury, rather than judge, and standard of proof is beyond a reasonable doubt.

WHICH GROUPS SHOULD BE PROTECTED?

As discussed in Chapter 2, hate crime laws come into effect when a person commits a crime and when that crime is motivated by the victim's group. Each state lists the types of groups that are protected under these laws and, of course, not all kinds of groups are included. For example, suppose I decide to rob a person because I believe she is wealthy or because her job (e.g., she is a convenience store clerk) makes her vulnerable. This robbery would generally not be considered a hate crime because neither socioeconomic status nor type of employment are usually protected categories. And, most likely, these types of crimes would not fall into most people's subjective definitions of "hate" crimes. So what kinds of groups should be included within this legislation? This issue has turned out to be a contentious one.

All states with hate crime laws include at least crimes based on race, ethnicity (or national origin), and religion (Gerstenfeld, 1992). Some states include only those categories. Others, on the other hand, include many additional groups as well (Grattet et al., 1998). Iowa, for example, lists 10 protected categories, and Oregon lists 12 (see Table 3.2).

The decision about which groups to include is important and often contentious. Recall that Chapter 2 concluded that one of the primary values of hate crime laws is symbolic: They send a message that certain types of behavior are intolerable. What message does it send, then, when a particular group is excluded from the list?

Sexual Orientation

When it comes to deciding which groups should be protected by hate crime laws, probably the issue that has been the subject of the most debate is whether to include sexual orientation. Currently, according to the National Gay and Lesbian Task Force (NGLTF, 2001), 16 states have laws that do not include sexual orientation. Several other states (e.g., Florida and Missouri) did not originally include this category but later added it to their statutes.

Inclusion of sexual orientation as a protected group has also been hotly debated in Congress. When the Hate Crime Statistics Act was being considered in 1989, a group of Senators led by Jesse Helms vehemently opposed the inclusion of sexual orientation (Jacobs & Potter, 1998; Peek, 2001). Remember, this law did nothing but require the Department of Justice to collect data from local law enforcement agencies; it did not actually create a federal crime. But some members of Congress did not want hate crimes

Table 3.2 Groups Protected by Hate Crime Laws in Selected States

State	Groups Protected
California	race, color, religion, ancestry, national origin, disability, gender, or sexual orientation
Idaho	race, color, religion, ancestry, or national origin
Iowa	race, color, religion, ancestry, national origin, political affiliation, sex, sexual orientation, age, or disability
Louisiana	race, age, gender, religion, color, creed, disability, sexual orientation, national origin, or ancestry
Missouri	race, color, religion, national origin, sex, sexual orientation, or disability
New York	color, national origin, ancestry, gender, religion, religious practice, age, disability, or sexual orientation
North Dakota	sex, race, color, religion, or national origin
Ohio	race, color, religion, or national origin
Oregon	race, color, religion, national origin, sexual orientation, marital status, political affiliation or beliefs, membership or activity in or on behalf of a labor organization or against a labor organization, physical or mental handicap, age, economic or social status, or citizenship of the victim

against gays and lesbians to even be monitored. Not surprisingly, then, one of the contributing factors to the repeated failure of passage of the federal Hate Crime Prevention Act has been this issue. In 1999, as mentioned in Chapter 2, two competing hate crime laws were introduced in Congress, one sponsored by Edward Kennedy and one by Orrin Hatch. The Kennedy bill explicitly included crimes based on sexual orientation, whereas the Hatch bill did not (Peek, 2001).

Why is there such controversy over this issue? It is not because offenders are rarely motivated by sexual orientation; in fact, as Chapter 6 discusses, gays and lesbians are among the most frequent victims of hate crimes. Nor is it due to lack of advocacy on behalf of gays and lesbians. Gay and lesbian organizations were among the earliest and strongest supporters of hate crime legislation (Jenness & Broad, 1997). Instead, the problem has to do with the symbolic nature of hate crime laws. Those who oppose homosexuality are afraid that including sexual orientation within the hate crime laws will send a message that the government approves of homosexuality, and that doing so will even open the door to inclusion of gays and lesbians within federal civil rights acts and other laws. "Today hate crimes, tomorrow same-sex marriage" seems to be the fear of some.

Some members of Congress have been quite unambiguous in voicing these fears. During the debate on the Hate Crime Statistics Act, Senator Helms wrote an amendment that would have "condemned homosexuality, rejected it as a lifestyle, condemned government support for extending civil rights to homosexuals, and called for strict enforcement of state sodomy laws" (Jacobs & Potter, 1998, p. 71). In response, and as a compromise, several other senators authored a second section of the bill that emphasized the importance of family life, and stated that the Act should not be construed so as to promote or encourage homosexuality. The bill eventually passed both the Senate and House with the compromise amendment included.

Aside from the moral issues surrounding homosexuality, another reason that has been given for excluding this category from hate crime legislation is that sexual orientation is a choice. Only immutable characteristics such as race and ethnicity, this argument goes, should be protected. There are serious flaws in this line of reasoning. As Lawrence (1999, pp. 18-19) points out, there exists credible scientific evidence that sexual orientation is not, in fact, a choice, but rather a function of genes and environment. Second, if sexual orientation is a choice, then so is religion. Yet nobody seriously contends that religion should not be protected by hate crime laws. Some laws include other categories as well that are certainly at least as mutable as sexual orientation: political affiliation, labor organization activity, and economic status. The general weakness of this argument suggests that it is merely a (rather transparent) front for the morals-based claim against homosexuality.

Regardless of the strength of these arguments, however, the fact remains that they have frequently been successful. As a result, in many jurisdictions, gay-bashing is not a hate crime. The repercussions of this could be serious. As Lawrence (1999, p. 20) asserts, "Failure to include sexual orientation implies that gays and lesbians are not as deserving of protection as racial, religious, or ethnic minorities, and that sexual orientation is not as serious a social fissure line as race, religion, and ethnicity." Even worse, it is possible that the calculated exclusion of this category puts an implicit governmental seal of approval on violence against gays and lesbians (see, e.g., Herek & Berrill, 1992b; Perry, 2001, pp. 179-223).

Gender

The second category that has inspired much debate is gender. The arguments here are quite different, however. Although it is debatable whether sexual orientation is a choice, people clearly do not select their own gender (unless one counts transsexuals and, even then, gender identity is, arguably, not determined by the individual). Furthermore, although conservative politicians feel comfortable in stating their opposition to homosexuality and governmental support of homosexuality, very few would openly speak against women, who are the most common victims of gender-based hate crimes.

The Anti-Defamation League (ADL) changed its model hate crime law in 1996 to include crimes based on gender. Some states followed suit but, by 2001, only 19 states included gender or sex in their laws (ADL, 2001). Gender-based hate crimes are not included within the federal Hate Crime Statistics Act, but Congress did pass the Violence Against Women Act (a portion of which, as discussed in Chapter 2, was subsequently declared unconstitutional by the U.S. Supreme Court).

Several commentators have argued that gender-based crimes fall into the same pattern as crimes based on race and religion (see, e.g., Chen, 1997; Goldscheid & Kaufman, 2001; Lawrence, 1999; McPhail, 2002; Pendo, 1993; Weisburd & B. Levin, 1994) and so ought to be included within hate crime legislation. Weisburd and B. Levin (1994), for example, point out that gender, like race and religion, is included in antidiscrimination laws, and that gender-based crimes such as spousal abuse and rape may, like race-based crimes, be intended to maintain a particular group's subordinate status. Perry (2001) argues,

By leaving gender out of the hate crime equation, legislators are recreating the myth that gendered violence is an individual and privatized form of violence, unequal to the public and political harm suffered by racial or religious minorities, for example. (p. 210)

On the other hand, several arguments have been made against including gender, as is clear from the fact that relatively few statutes contain this category. One claim is that it is unnecessary to include gender-based crimes because these offenses are already punished by laws against rape and domestic violence. One flaw in this argument is that for *all* hate crime cases, the underlying offense is already punishable by other laws. It is true that a relatively small percentage of, say, assaults or vandalisms are motivated by the victim's group. The question is whether all rapes and cases of domestic violence are motivated by the victim's gender. This is a difficult question. In most rape cases, the man likely would not have attacked the victim had she not been female, but does this mean he attacked her *because* she is female? Similarly, a heterosexual man would likely not even be in an intimate relationship with a person unless she was female. But does that mean he strikes his wife *because* she is female? Certainly, these are questions for philosophers and psychologists to ponder!

A second flaw in this argument is that not all gender-based crimes are rapes or domestic abuse. In 1989, Marc Lepine murdered 14 women in an engineering class at the University of Montreal, all the while shouting, "I hate feminists!" (Chen, 1997, p. 277). Obviously, Lepine's crime was neither rape nor domestic abuse. In fact, it is difficult to see how this incident was substantially different from that of Buford Furrow, who in 1999 shot five people at a Jewish Community Center near Los Angeles and then murdered a Filipino American postal carrier.[5]

A second argument that is made against including gender is the "floodgates" argument: Because of the frequency of rapes and domestic assaults (the FBI reported more than 90,000 forcible rapes in 2000, compared with about 9,500 hate crimes), the criminal justice system will be overwhelmed with these types of hate crimes. Hate crime recording and tracking will suffer, as will prosecutions of other types of hate crimes. The ADL (2001), however, concluded that in those jurisdictions where gender is included, the reporting system has not been overwhelmed, nor have prosecutors been distracted from prosecuting other kinds of hate crimes, because prosecutors have used their discretion in determining the types of gender-related crimes for which hate crime charges are appropriate.

Interestingly, some feminists have also argued against including gender in hate crime laws. Instead of fearing that rapes and domestic violence will engulf other types of hate crimes and diminish their attention, they are worried about just the opposite: that rape and domestic violence will be subsumed under the larger rubric of bias crime, and thus will be largely forgotten. Again, however, there is no evidence that this has been the case.

Some critics have claimed that gender-based crimes should not be included as hate crimes because those who commit such offenses do not really "hate" women, at least in the same sense that white supremacists hate blacks and Jews. However, some gender-based offenders clearly *do* hate women in exactly that sense. Marc Lepine is a good example of this. Moreover, despite the name "hate crime," most hate crimes do not technically require hate, either: They require only that the offender choose the victim because of the victim's group, not that the offender actually *hate* the victim. For instance, several states include disability or handicap in their law, and some include age. If a person specifically chose to rob disabled or elderly people because they were disabled or elderly, and thus seemed to be easy targets, those acts would still be covered by hate crime legislation, regardless of whether the person had any actual animosity toward his or her victims.[6]

Finally, opponents of gender-based hate crime laws maintain that gender-based crimes are unlike other kinds of bias crimes because the gender-based victim is individualized. She is not the victim of happenstance, interchangeable with all others of her sex. Frequently, she and the offender know each other well. This is all undoubtedly true for many victims of domestic violence and some victims of rape. It was not true, however, for the 14 women whom Marc Lepine killed. Nor for the 14 women killed by George Henard in 1991, when he drove his pickup through the window of Luby's Cafeteria in Killeen, Texas, and then

Table 3.3   Does Your State Protect Sexual Orientation and Gender?

State	Sexual Orientation	Gender
Alabama	N	N
Alaska	N	Y
Arizona	Y	Y
Arkansas	N	N
California	Y	Y
Colorado	N	N
Connecticut	Y	Y
Delaware	Y	N
District of Columbia	Y	Y
Florida	Y	N
Georgia	N	N
Hawaii	Y	Y
Idaho	N	N
Illinois	Y	Y
Indiana	N	N
Iowa	Y	Y
Kansas	Y	N
Kentucky	Y	N
Louisiana	Y	Y
Maine	Y	Y
Maryland	N	N
Massachusetts	Y	N
Michigan	N	Y
Minnesota	Y	Y
Mississippi	N	Y
Missouri	Y	Y
Montana	N	N
Nebraska	Y	Y
Nevada	Y	N
New Hampshire	Y	Y
New Jersey	Y	Y
New Mexico	N	N
New York	Y	Y
North Carolina	N	Y
North Dakota	N	Y
Ohio	N	N
Oklahoma	N	N
Oregon	Y	N
Pennsylvania	N	N
Rhode Island	Y	Y
South Carolina	N	N
South Dakota	Y	Y
Tennessee	Y	N
Texas	Y	Y
Utah	N	N

State	Sexual Orientation	Gender
Vermont	Y	Y
Virginia	N	N
Washington	Y	Y
West Virginia	N	Y
Wisconsin	Y	N
Wyoming	N	N

opened fire with a semiautomatic rifle (J. Levin & McDevitt, 1993, p. 93).[7] Nor for the two dozen or more women Ted Bundy raped and murdered.

Moreover, undoubtedly some victims of "traditional" hate crimes also knew the offender or were not selected at random. For example, as Weisburd and B. Levin (1994, p. 37) point out, Emmett Till was not randomly chosen to be lynched; he was intentionally selected because he had transgressed social norms (he was African American and had whistled at a white woman). Jacobs and Potter (1998, p. 73) cite the murders of Martin Luther King, Medgar Evers, and Meyer Kahane as further examples of hate crimes in which the victim was particularly selected.[8]

Levin and McDevitt (1993) have proposed a compromise on the issue of gender-based crimes. They suggest that such crimes be considered hate crimes when, and only when, the victim is "interchangeable"—that is, whenever the perpetrator was looking to harm any woman, not a particular woman, and the victim was simply a stranger picked at random. No state has explicitly adopted this idea. However, it may very well be that this is essentially the route that many prosecutors do take in practice when deciding which charges to bring against a defendant. It would be interesting if empirical research were to be conducted on this subject.

Although Congress has not yet passed a federal hate crime law, it did pass a law aimed specifically at gender-based violence, the Violence Against Women Act (VAWA). As discussed in Chapter 2, the Supreme Court held that this Act exceeded Congress's authority. Conversely, although most states have some sort of general hate crime law, few, if any, have the equivalent of the VAWA (although, of course, all have statutes concerning rape and domestic violence). Table 3.3 summarizes state hate crime legislation on sexual orientation and gender.

Factors That Affect States' Decisions About Which Groups to Protect

Why is there so much differentiation between states as to which groups are protected by hate crime laws? Several researchers have studied the factors that influence whether and when a state creates hate crime legislation. These factors include economic, political, and sociodemographic conditions within the state, as well as the activities of neighboring states (Grattet et al., 1998; Soule & Earle, 2001). For example, states that are wealthier, that have Democratic majorities in the legislature, or that have had much media attention relating to hate crimes are faster to pass hate crime laws (Jenness & Grattet, 1996; Soule & Earle, 2001).

Grattet et al. (1998) also found that these factors influence the content of hate crime laws. So does timing: States that waited longer to enact hate crime legislation tend to protect more categories. This is probably because the earliest laws that were passed protected relatively few groups, whereas later laws successively added more categories. Therefore, states that entered relatively late in the game had more complex laws on which to model their own.

There are clearly important regional characteristics of hate crime laws. States along the West Coast and in the Northeast are much more likely to protect sexual orientation than are states in the Southeast, Midwest, and Mountain regions. This may be attributable, in part, to the relatively liberal politics prevalent in the West and Northeast. Soule and Earle (2001) found that, in general, states that have repealed their sodomy laws were more likely to adopt hate crime laws and did so earlier than other states.

Activist groups and social movements also play a part in determining who is protected by hate crime laws (Grattet & Jenness, 2001a; Jenness & Broad, 1997; Jenness & Grattet, 1996).[9] There is value in being recognized as a victimized group, just as there might be risk in being excluded (e.g., exclusion may send a message that a particular group is not worth protecting). Jacobs and Potter (1998) argue that the process of deciding who will be protected, which they call "identity politics," is divisive and counterproductive:

> By redefining crime as a facet of intergroup conflict, hate crime laws encourage citizens to think of themselves as members of identity groups and encourage identity groups to think of themselves as victimized and besieged, thereby hardening each group's sense of resentment. That in turn contributes to the balkanization of American society, not its unification. (p. 131)

As I will discuss in the "Paradoxical Effects of Hate Crime Laws" section later in this chapter, there are other reasons, as well, to suspect that hate crime laws may actually be more harmful than helpful.

IDENTIFYING AND PROSECUTING HATE CRIMES

After a state has decided which groups to include in its hate crime law and the constitutional issues have been settled by the courts, the way is clear to successfully find and prosecute people who victimize others out of bias, right? Wrong! Consider Table 3.4, which shows hate crime data from California for the past several years. California data are included here because California is one of the few states that issues a comprehensive annual hate crime report.

Each year in California, complaints are filed (that is, criminal charges are brought) for only about 10% to 15% of all known hate crime offenders, and 85% to 90% of these offenders are never charged with a hate crime. Furthermore, even when a complaint is filed, only about half of the cases result in a hate crime conviction. The result? Only about 5% of the people whom police report as committing a hate crime are ever convicted of one.[10] The situation is even worse when we consider that probably far fewer than half of all hate crimes are even reported to the police.

It is clear that if a person chooses to commit a crime because of the victim's group, that person has a very small likelihood of being punished for a hate crime. This is not because most hate crime offenders are criminal masterminds capable of pulling off the perfect

Table 3.4   Reported Hate Crimes and Hate Crime Convictions in California

Year	Offenses Reported to Police	Suspects	Hate Crime Complaints Filed by Prosecutors	Hate Crime Convictions
2000	2,002	2,107	360	213
1999	2,001	2,021	372	174
1998	1,801	1,985	244	131
1997	2,023	2,206	313	223
1996	2,321	2,441	182	87
1995	1,965	2,225	187	107

SOURCE: California Attorney General (2001).

crime. Nor is it due to lack of appropriate legislation—as we have already seen, most states now have hate crime laws. Instead, the problems lie in the nature of hate crimes themselves and with the process by which the laws are enforced. We will begin with the first people who are usually aware that a hate crime has occurred: the victims.

Victims' Reporting of Hate Crimes

Several years ago, the house where one of my students lived was vandalized. Someone spray-painted slurs related to her sexual orientation on the garage and front door. She called the police, and two officers soon arrived and made a report. They were polite, but before they left, one of them turned to her and said, "You know, if you don't want this to happen, you shouldn't be so obvious about being a lesbian."

Speaking to me about this incident some months later, my student was angry at the police. Not only did she believe that they had insulted her, but she also thought that they were blaming her for being a victim. She felt victimized twice: once by the vandal and then again by the police. "If that ever happens to me again," she told me, "I won't call the police."

I do not know how typical her case was of police responses in her city, and I do not know if she was ever placed in that position again. However, hers is a good example of a well-documented phenomenon: Many victims of hate crimes do not report the crime to the police. And, almost always, if the crime is not reported, the police remain unaware that it happened, it never gets included in the official hate crime data, and the perpetrator is never punished.

Of course, it is impossible to know exactly what percentage of hate crimes get reported to police. In general, people frequently do not report crimes. In 2000, for example, the Department of Justice estimated that only 47.9% of violent crimes, and only 35.7% of property crimes (Department of Justice, 2002a), were reported. Some have estimated that the rates are even lower for hate crimes. Perry (2001, p. 216), for example, states that less than 20% of hate crimes against gays and lesbians are reported, and B. Levin (1999) estimates that fewer than one in three hate crimes in general is reported.

There are a variety of reasons why victims of hate crimes may not report the offenses to the police. In 2001, the California Attorney General's Civil Rights Commission on Hate Crimes issued a report on issues involved in reporting hate crimes (California Attorney General, 2001); the report was based on a series of forums that were held throughout the state. The Commission identified a number of reasons why hate crimes often go unreported:

1. Lack of knowledge about what hate crimes are and how the laws are applied;

2. Denial by the victim(s) that a hate crime was perpetrated;

3. Fear of retaliation by the perpetrator for reporting;

4. Fear of being revictimized by law enforcement or a belief that law enforcement does not want to address hate crimes;

5. Shame for being a victim of hate crime;

6. Cultural or personal belief that one should not complain against misfortunes;

7. Fear of being exposed as gay, lesbian, bisexual, or transgendered to one's family, employer, friends, or the general public;

8. Lack of English language proficiency and knowledge of how to report hate crimes;

9. Fear of being identified as an undocumented immigrant and being deported;

10. Fear on the part of people with disabilities who use caregivers that the caregivers who have committed hate crimes against them will retaliate and leave them without life-supporting assistance; and

11. Inability of some people with disabilities to articulate when they have been a victim of hate crime (California Attorney General, 2001, p. 11).

Based on these findings, the Commission made several recommendations (see Box 3.5).

Box 3.5   Recommendations of the California Attorney General's Civil Rights Commission on Hate Crimes

---

1. The state Department of Justice should launch a multilingual public information campaign about hate crimes, and about community, civil, and criminal resources.

2. The state Department of Justice should establish a toll-free hotline for reporting hate crimes and an online incident reporting form. Victims who use these resources should be referred to the appropriate local law enforcement agency.

3. Legislation and support should be created for local human rights commissions to sponsor hate crime prevention and response networks.

4. Training and resources should be provided for educational institutions.

5. Training on hate crimes should be required for all law enforcement officers as well as dispatchers and other staff.

6. Funding should be provided to local law enforcement agencies so they can partner with local community agencies to prevent and respond to hate crimes.

7. Training should be required for corrections personnel.

8. Training should be provided for prosecutors.

SOURCE: Adapted from California Attorney General (2001).

Other commentators have largely agreed with the Commission's assessments on reporting (see, e.g., Perry, 2001). People who are victims of hate crimes are disproportionately likely to be relatively powerless. In fact, arguably, a primary reason why offenders commit these crimes is to maintain their victims' powerless position (see, e.g., Wang, 2002). These victims are also often members of groups that have traditionally had poor relations with the police. And these victims frequently have language or cultural barriers that keep them from communicating with law enforcement. For a large majority of the hate crimes that occur in the United States, the existence of hate crime laws is simply a nonissue because the government will never become aware that they occurred.

Police Responses to Hate Crimes

Assuming a victim of a hate crime does actually report the incident to the police, many potential barriers exist between the reporting of the crime and the offender's eventual conviction. One barrier is that the responding police officer may not interpret or report the crime as hate motivated. This is an important problem because police officers are vested with more discretion than any other actors within the criminal justice system and because in most cases they act as gatekeepers to that system (Bell, 1997, 2002a; Franklin, 2002; Walker & Katz, 1995).

Again, a personal experience might help illustrate the issue. Several years ago, I was engaging in a research project that involved contacting law enforcement agencies throughout the country and asking for their hate crime data. The results were quite illuminating. Although some agencies were able to quickly provide very complete information, others were not. For example, an officer in Idaho (which was home to several hard-core white extremist groups) informed me that they did not have any hate crimes because "we don't really have any minorities in Idaho." Several officers with whom I spoke had no idea what a hate crime was.

The data that I did collect were often subject to odd inconsistencies. In 1993, San Francisco reported 319 hate crimes, whereas neighboring San Jose reported only 27 (Gerstenfeld, 1998). San Jose actually has slightly more people than San Francisco, and both cities have diverse populations. Is San Francisco a hotbed of intolerance, whereas San Jose is a paragon of brotherhood and peace? A more likely way to explain these data, however, is to note that in 1993, San Francisco had a hate crime unit, but San Jose did not. Milwaukie, Wisconsin, with a population roughly two thirds of San Francisco's, reported only two hate crimes in the third quarter of 1993. Like San Jose, Milwaukie did not have a hate crime unit. San Francisco's numbers were probably high because that city had had officers trained to recognize hate crimes and because dealing with these crimes was an explicit priority of the police department.

Several potential explanations exist for why a police officer may not record an incident as a hate crime, even when the victim believes that it was. The first of these is that the officer himself or herself may be biased against the victim. Unfortunately, prejudice exists as much within law enforcement agencies as within the rest of society. Well-publicized incidents of police-officer–perpetrated bigotry, such as the sexual assault of Haitian immigrant Abner Louima by several New York City officers, abound in towns and cities across the country (Perry, 2001; Walker & Katz, 1995). Furthermore, there have been incidents in which individual police officers have been identified as members of white supremacist groups (Novick, 1995). Assuming, however, that the officers are relatively unbiased, there are several other reasons why their recording of an incident may differ from the victims' interpretations. As we have already discussed, the determination of motive is problematical and subjective; what I consider to have been a theft provoked by my race, you might consider just an ordinary robbery (Bell, 1997; Garofalo & Martin, 1993). Officers may also have received little or no training in identifying a hate crime. And, even when police believe that a hate crime has in fact occurred, they may have personal or departmental reasons for wanting to avoid recording it as such: Perhaps they wish to avoid the additional bureaucratic requirements attendant to hate crimes in many jurisdictions, perhaps they believe that hate crimes are a legal category not worth pursuing, or perhaps their department wishes to underplay the prevalence of bias in their city (Bureau of Justice Assistance, 1997; Maroney, 1998).

Several researchers have examined the factors that influence police hate crime reporting. Martin (1995) examined the behavior of the Baltimore County, Maryland, Police Department. She found that reporting behavior varied a great deal from officer to officer, and she identified a number of variables that were associated with whether a suspected hate crime case was later verified, including the race of the victim and the victim's own perception of the event. She also discovered that this department's interpretation of hate crimes may have been overzealous. One person she interviewed said that the department "has trained the officers to the point that if there's graffiti, they assume it's [a hate crime]" (Martin, 1995, p. 308). As a result, "a sizable proportion of the cases contain bias that is a secondary motivation, . . . an additional motivation, . . . or even an afterthought" (Martin, 1995, p. 317). Martin concluded that "what is defined as 'bias motivated' is arbitrary and results in statistical reports that are uninterpretable and may be misleading" (p. 323).

Boyd, Berk, and Hamner (1996) also questioned the reliability and validity of hate crime data. They looked at hate crime reporting in two divisions of a large metropolitan police department. Consistent with Martin's (1995) results, they found significant differences between the way the two divisions classified hate crimes, with one focusing more on the offender's motive and the other on any facts of the case that were potentially related to bias. For example, an incident occurred in the first division in which the victim, who was driving his truck, was followed by a group of five teenagers. When he pulled over, they taunted him with racial slurs and hit him with a steering wheel lock. Because the victim admitted to police that he may have accidentally cut off the suspects' car while driving, the division concluded that this was not a hate crime (Boyd et al., 1996, p. 836). On the other hand, someone in the second division drove golf carts recklessly across a golf course, damaging the carts and some landscaping. The complainant said that this incident happened on the eve of Rosh Hashanah and that two years earlier, a similar incident had also occurred on that day. This was recorded as a hate crime (Boyd et al., 1996, p. 843).

The Boyd et al. (1996) study also highlighted a number of other potential problems with law enforcement and hate crimes. Many of the officers disliked the additional paperwork required by the department's hate crime policy, and they questioned whether hate crimes should be a special crime category at all. Some officers were reluctant to call anything a hate crime unless it was as obvious as a cross-burning on the property of an African American family, whereas others would classify domestic violence as a hate crime because the husband and wife hated each other. In general, acts by juveniles were likely to be interpreted as simple irresponsibility rather than hate crimes. And some officers felt that hate crimes were not real crimes at all but simply human nature. One officer said, "A couple of fruits get bashed—that's not a crime. That's normal. There are just two types of crime—dope and cars. The rest is just stupidity. I say dope and stolen cars are the only calls worth my time. Not fruit bashing and not these domestic calls" (Boyd et al., 1996, p. 827).

Using focus interviews and surveys, Nolan and Akiyama (1999) looked at the individual and departmental variables that influenced hate crime reporting (see Box 3.6). Like previous researchers, they found important differences from officer to officer and agency to agency.

Box 3.6   Factors Influencing Police Hate Crime Reporting

INDIVIDUAL ENCOURAGERS:

Departmental policy mandates reporting

Belief that early identification of problem is key to effective solution

Belief that it's an important part of the job

Belief that it will help prevent problems

Belief that reporting will prevent officer liability

Belief that hate crimes are morally wrong

Encouraged to report by department officials

Encouraged and supported by supervisors and colleagues

Clear, understood, and accepted department policy

Belief that reporting hate crimes benefits victims and communities

Internal checks ensure officers don't misidentify hate crime

Being recognized by other officers as good at investigating and recording hate crime

Desire to be considered a good police officer

Reporting hate crimes is encouraged and rewarded by the department

Personal desire to comply with departmental policy

INDIVIDUAL DISCOURAGERS:

Belief that reporting hate crimes is not viewed as important by department officials

Too much additional work

Sometimes runs counter to officer's personal beliefs

Belief that hate crimes are not serious

Belief that hate crimes should not be treated as special

Little concern for some minority groups

Not the job of police (more like social work)

Not recognized or rewarded for reporting hate crimes

Informally encouraged to adjust complaints because of the large number of calls for service

Lack of common definition of hate crime

Fear that incident will be blown out of proportion

(Continued)

Box 3.6 (Continued)

Officers already too busy

Personally opposed to supporting gay and minority political agendas

Lack of training

Victims do not want to assist in prosecution

SOURCE: Adapted from Nolan & Akiyama (1999).

The most recent study on police reporting of hate crimes was conducted by Bell (2002a) in a large metropolitan police department with a hate crimes task force. Bell concluded that officers did not rely on the perpetrators' language alone to determine whether a crime was hate motivated, contrary to the fears of some First Amendment scholars. Instead, the two most important factors were whether the victim and offender were of different groups and whether bias, rather than some other emotion, appeared to have motivated the crime. The use of these criteria is likely to result in a rather conservative approach because they would weed out cases in which the offender had multiple motives or was acting out bias toward his own group. Based on these criteria, the police might not report cases in which the offender and the victim were the same on one identity dimension (e.g., race) but different on another (e.g., sexual orientation).

In some jurisdictions (generally larger cities), police reporting of hate crimes is further complicated because it requires a two-step process. Officers who initially respond to the call must first identify a crime as being hate motivated. The case is then referred to a special bias crime unit, which verifies it as a hate crime. In these jurisdictions, the opportunities for an offense to be disqualified from the hate crime reporting system are increased, and there is also the possibility of friction between or disparate goals among the ordinary patrol officers and the bias crime unit.

A variety of programs, initiatives, and policies have been created to try to improve police reporting of hate crimes. One common route is for a police department to create a specialized hate crimes unit or task force. This might consist of a single officer in smaller jurisdictions or several officers in larger ones. Typically, officers in this unit have specialized training and are called in to investigate whenever a hate crime is suspected. The exact number of bias crime units nationwide is unknown. In 1990, the Department of Justice estimated that there were approximately 450 of these units in the country, but a study by Walker and Katz (1995) concluded that this number was likely overinflated. Moreover, they also found that only half of the departments they surveyed that had bias crime units or special procedures provided the officers with any formal training on hate crimes. Thus, in some cases at least, there may be "far less to [bias crime units] than meets the eye" (Walker & Katz, 1995, p. 42).

A second, and related, strategy is for a department to have special procedures for handling suspected hate crime cases (Bell, 1997). The ADL (1988) published a collection of procedures, including model guidelines and reporting forms. Although official procedures and policies may help standardize reporting, there is also the danger that they will backfire. If police believe that hate crime reporting procedures are too unwieldy or time-consuming, they may be hesitant to invoke them (Boyd et al., 1996; Bureau of Justice Assistance, 1997).

Finally, many jurisdictions require training in hate crimes for police officers. California, for example, has required that all future police officers receive some training on hate crimes while they are at the law enforcement academy (California Attorney General, 2001). In addition, several national agencies assist in police officer training. The Office for Victims of Crime has made available a detailed bias crime training manual (McLaughlin, Brilliant, & Lang, 1995). The Southern Poverty Law Center offers an online course on hate crimes for law enforcement officers and publishes a free quarterly magazine on hate crimes and hate groups. The Simon Wiesenthal Center conducts training sessions for police officers.

Training requirements are hardly universal, however. Although California requires some hate crime training in the academy, it does not mandate how much. And it does not require any ongoing training or education for officers who went through the academy prior to 1993 or for other support personnel such as dispatchers. Most states have no hate crime training requirement at all.

Furthermore, training alone is probably not the answer. A study by Sloan, King, and Sheppard (1998) concludes that simply providing more training to police will not necessarily improve reporting. The available training materials themselves do not always accurately reflect the true legal and practical definitions of hate crimes (Walker & Katz, 1995). Moreover, improved training will not necessarily overcome all the other obstacles to reporting, such as those found in the Nolan and Akiyama (1999) study.

Even when police do report hate crimes, it is important to note that extreme caution must be used in interpreting the data. As the research shows, reporting tends to be extremely inconsistent from officer to officer and place to place. In addition, the data can be influenced by a variety of external factors. These can include police-related variables such as the creation of or activities of a bias crime unit, and also local and national events (Gerstenfeld, 1998). After the Rodney King incident, many cities experienced a surge in race-related violence and, after the September 11 attacks, there was a sudden increase in attacks on Arab Americans (and even those who were mistaken for Arab Americans; Gerstenfeld, 2002). Using police reports to determine patterns or trends in hate crimes is therefore problematic to say the least.

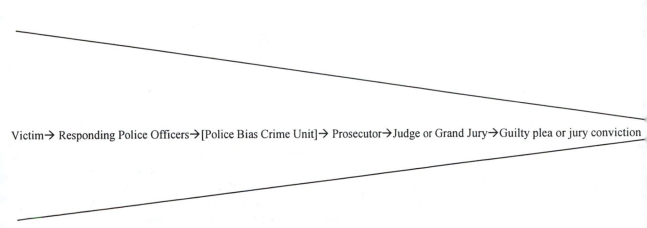

Victim→ Responding Police Officers→[Police Bias Crime Unit]→ Prosecutor→Judge or Grand Jury→Guilty plea or jury conviction

Figure 3.1   Barriers to Hate Crime Convictions

Box 3.7   The Case of Randal Lee Krager

Randal Lee Krager lived in a racially diverse neighborhood in Portland, Oregon. On the night of April 17, 1992, the girlfriends of Krager and his roommate walked to a nearby convenience store. When the women did not return when expected, Krager (who was white and 18 years old) and his roommate left their apartment to look for them. They found them involved in a confrontation with two black men, one of whom was 34-year-old Jacob Johnson. The testimony at trial was conflicting; Krager's roommate said he saw the women being struck, whereas Johnson's friend said the women were slapping them.

The four men soon began arguing with each other. Johnson (who had a blood alcohol level of .11) waved around a beer bottle. Krager's friend pushed him. Krager claimed that Johnson continued to threaten him with the bottle, although there was

---

Prosecuting Hate Crimes

Of all the hate crimes that occur, very few make it past the first two hurdles to conviction (the victim and the police). But those that do are faced with additional barriers (see Figure 3.1): The prosecutor has to decide to file hate crime charges; a judge has to determine that probable cause exists to try the defendant;[11] and the defendant has to either plead guilty to those charges or be found guilty by a jury. All of these barriers are difficult to overcome.

Prosecutors, like police, have a large amount of discretion. There are few controls on what charges they choose to bring, whether they choose to bring charges at all, and what plea-bargaining processes they engage in (Maroney, 1998). In most jurisdictions, District Attorneys are elected officials. Their offices, therefore, may be sensitive to political and social pressures as well as the need to maintain high conviction rates. In some places, they may also be faced with overburdened or limited resources.

Furthermore, prosecutors may have even less training on hate crimes than police officers do. It is not a topic that receives much (if any) coverage in most law school curricula, nor is it a common subject in continuing legal education seminars. Prosecutors in some large cities may have had experience with previous hate crime cases and may even be part of a hate crime task force. San Francisco is an example of one such city. In smaller jurisdictions, however, because of the rarity of hate crime cases and the absence of special units or task forces, prosecutors may have little or no familiarity with these kinds of cases.

It is also important to point out that prosecutors, like police officers, may be influenced by their own biases (Hernandez, 1990). Consciously or unconsciously, their decisions about which charges to bring may be affected by their personal feelings about the victim and the perpetrator.

In a sense, hate crime cases are no different from other kinds of crime, all of which have to survive the same series of decision makers. And there are other types of infrequent offenses, such as capital murder, in which the attorneys involved may have had little or no relevant experience. However, hate crimes have one important distinction: They require that each of the decision makers determine the offenders' motive (Franklin, 2002).

Determination of another person's motive is inherently subjective. None of us can read the offenders' minds and, except in rare cases, perpetrators are savvy enough not to admit that they committed the crime because of the victim's group. Unfortunately, most of the

also testimony that Johnson dropped the bottle after being pushed. Krager then punched Johnson in the face. Johnson fell and struck his head on the ground, fracturing his skull. He was in a coma for several days and may have suffered permanent brain damage.

When police arrested Krager, they learned he was a skinhead. One of his hands had a swastika tattoo. A grand jury indicted Krager for second-degree assault but refused an indictment on hate crime charges. Krager was subsequently convicted of second-degree assault. The judge, however, doubled his sentence to three years, largely because the judge was "satisfied the incident was motivated by bigotry and anti-black racism."

SOURCE: "Jury to Get Complicated Assault Case" (1992) and "Judge Doubles Sentence" (1992).

time, most of us are quite poor at interpreting the reasons behind other people's behavior. Our interpretations are subject to our own biases and expectations.

What is worse, we frequently tend to be inaccurate in interpreting our *own* motives (Gerstenfeld, 1992). Psychological literature is full of examples of people failing to understand their own behavior, and some therapists spend careers trying to help their patients overcome this difficulty. Even the causes of simple acts can be unclear. Why did I eat that candy bar this afternoon? Was I hungry? Bored? Addicted to chocolate? Influenced by hormones? Compensating for some emotional distress? If I cannot identify the motive behind such an uncomplicated act as a midday snack, how can we expect people to determine the motives of other people in such complex behaviors as criminal offenses—and to do so, as the law demands, beyond a reasonable doubt?

Of course, in some cases, it is not nearly so difficult to determine the offender's motive. If someone paints a swastika on a synagogue or lights a cross on the lawn of the only black family in the neighborhood, it is hard to think of any reasonable explanation other than bias. As we have already mentioned, however, many potential hate crime cases are not so clear. Consider, for example, the case of Randal Lee Krager (see Box 3.7).

It is impossible to know whether Krager would have hit Johnson had Johnson been white. Perhaps he would have stepped into any altercation involving his girlfriend. On the other hand, perhaps his anger was fueled, at least in part, by Johnson's race. In fact, Krager himself likely could not know whether he would have reacted differently if Johnson had been white.

If you were the victim of this crime, do you think you would believe that this was a hate crime? How would your view change if you were the responding police officer? The prosecutor? A member of the grand jury? The trial judge? A member of the community in which Krager and the victim lived? Would anybody even have suspected that this was a hate crime had Krager not belonged to a white supremacist group?

In cases as ambiguous as Krager's, reliable determination of the offender's motive becomes impossible. Jurors and others who are required to undertake this task will inevitably have their perceptions colored by their own biases and by extraneous factors that, legally, should not be considered. Although it is not feasible to study the influences on jurors' decisions in real hate crime cases, several studies have attempted to address this issue.

Marcus-Newhall, Blake, and Baumann (2002) conducted a series of experiments in which they asked participants to read a brief scenario describing a possible hate crime. The race of the defendant and of the victim was varied, as was the degree of peer support the defendant had received for his behavior. The researchers also collected data on the participants' race, as well as their political affiliation. Participants were asked to state how certain they were of the defendant's guilt and were asked to choose a sentence for him. The researchers found that the participants' decisions were significantly affected by the race of the offender and victim, by the race of the participants, and also by the participants' political views. All of these are extralegal factors that should not influence the outcome of a case.

I conducted a similar mock-juror study (Gerstenfeld, in press). In this study, participants read a somewhat lengthier scenario of an ambiguous case (adapted, in fact, from Krager's case). I varied the offenders' and victims' races, and also measured the participants' level of racism. Contrary to my own expectations, as well as the results of the Marcus-Newhall et al. (2002) study, neither race nor racism generally affected participants' decisions. I did find, however, that people were significantly less likely to determine the act was a hate crime if the victim and offender were of the same group than if they were of different groups. Although this might make sense—people might reasonably be expected to be more biased against other groups than their own—it does ignore the fact that hate crimes can, and do, occur between people of the same race.

Other experiments have also shown an influence of extralegal factors in perceptions of what constitutes a hate crime. Craig (1999) and Craig and Waldo (1996), for example, found that reactions to and interpretations of hate crimes were affected by participants' race and the race of the victim.

It should not be surprising that race may play a part in decision making in hate crime cases. Ample research has demonstrated that race affects decisions in criminal cases in general, not just among juries, but among other components of the criminal justice system as well, such as prosecutors and judges (see, e.g., Baldus & Woodworth, 1998; Nickerson, Mayo, & Smith, 1986; Pfeifer & Ogloff, 1991; Sommers & Ellsworth, 2000; Ugwuegbu, 1979). Not only is there no good reason to suspect that hate crimes are any different than other crimes in this regard, but, in light of the particular subjectivity involved in determining motive, it is possible that hate crime cases may be even *more* influenced by factors such as race (Maroney, 1998).

Much of the discussion about motive and hate crimes has centered on the constitutional issues discussed at the beginning of this chapter. First Amendment problems, however, are not the only problems created when a determination of motive is required. Among other things, it makes it much more difficult to obtain accurate convictions of hate crime offenders. When a potentially high rate of inaccurate convictions combines with serious reporting difficulties among both victims and police officers, it calls into question the wisdom and utility of hate crime laws.

PARADOXICAL EFFECTS OF HATE CRIME LAWS

Most advocates of hate crime legislation are genuinely concerned with the insidious effects of bigotry and see these laws as a viable and even necessary means of combating those effects. Many critics of the laws share the same concerns, but some of them worry that hate crime laws might actually result in the paradoxical effect of harming members of minority groups.

Hate Crime Laws Might Inspire Complacency

One way that hate crime laws might be counterproductive is that they might inspire complacency among policymakers. By the relatively simple act of enacting hate crime legislation, politicians may feel that they have done their part to combat prejudice. It is politically expedient for them to pass this legislation: They can satisfy civil rights activists while simultaneously appearing tough on crime (Maroney, 1998). And advocacy groups might also find themselves so busy addressing the relatively rare problem of hate-motivated crime that they largely overlook the much more common (albeit subtle) problems of bias in employment, education, housing, and other facets of everyday life. This dilemma is made worse by the fact that, as discussed in Chapter 2, hate crime laws will probably have very little direct effect on hate. Gellman (1991) is one author who has made this point:

> If enacting a largely ineffective ethnic intimidation statute allows us to feel that we have taken steps to eliminate bigotry and bias-related crime and thus reduces somewhat or even entirely our feeling of the urgency of doing more, the enactment of the law ultimately *slows* the process of combating bigotry. (Gellman, 1991, p. 389)

Some empirical evidence supports this argument. Soule and Earle (2001) found that states that had initially enacted hate crime data collection or civil legislation were significantly slower to adopt hate crime laws. These authors refer to this as a "buffer" effect and conclude that "data collection and private civil redress statutes, when unaccompanied by criminalization, serve to deflect pressure for hate crime laws while not actually providing important protections for potential hate crime victims" (Soule & Earle, 2001, p. 294). The federal government may be a good example of this; although it created the Hate Crime Statistics Act in 1990, 13 years later Congress has still not passed a meaningful federal hate crime law (see Chapter 2).

Hate Crime Laws Might Cause Resentment of Minorities

A second risk of hate crime laws is that they will inspire resentment of minorities (Gerstenfeld, 1992). This phenomenon is similar to the way in which children often dislike the "teacher's pet" (Gellman, 1991). Members of the general public, who are usually uninformed about the realities of how the laws work, may feel that certain groups are getting special treatment.

White supremacist groups appear to be taking advantage of this angle. Several of their Web sites, for example, state (incorrectly) that the laws protect only minorities, not white people. Some of them (e.g., David Duke's site at www.davidduke.com) claim that the laws result in the persecution of white Christians. Some also decry what they call the "real" hate crimes—crimes committed by blacks against whites—and claim that the government and the media are ignoring those happenings. It is unclear how many people are actually convinced by this rhetoric, but the extremist groups seem to think it is worth their while to post these messages.

Hate Crime Laws Might Disempower Minorities

A third, more troubling possibility is that hate crime laws could be used to disempower minorities. The government is hardly a neutral bystander when it comes to bias; in fact, it has a long history of perpetuating and encouraging bias (Maroney, 1998). To do so is self- serving because it protects the status of those in power. As Perry (2001) argues, "The state is embedded in the processes of legitimizing and defining difference, and of constructing a racialized and genderized hegemonic formation" (p. 182). Is it not possible that hate crime laws could be used to disenfranchise the people they were meant to protect?

Maroney (1998) argues that hate crime laws arose very quickly and rapidly became assimilated into criminal justice institutions. As a result, "anti-hate-crime measures now reflect the culture and priorities of those institutions and therefore inadequately alter those institutions' treatment of hate crime and its victims" (Maroney, 1998, p. 568).

Lest this sound like radicalism or some sort of conspiracy theory, consider historical events. For example, in 1956, the Alabama Attorney General used state incorporation requirements to attempt to oust the NAACP from Alabama. The Attorney General enjoyed 8 years of success in the courts in this endeavor until the U.S. Supreme Court finally ruled the effort unconstitutional (*NAACP v. Alabama,* 1964). Alabama also tried to require the NAACP to turn over the names and addresses of all its members (*NAACP v. Alabama,* 1958). Also in 1956, Louisiana tried to expel the NAACP via the state laws on registering organizations (ultimately, the state failed, *Gremillion v. NAACP,* 1961). In 1957, Virginia sought to use its statutes prohibiting solicitation by lawyers to restrain the NAACP from providing legal assistance in civil rights cases (this was also held unconstitutional in *NAACP v. Button,* 1963).

Clearly, states have not hesitated to use what appear to be neutral statutes to try to weaken individuals or groups who question the status quo. Had hate crime statutes existed during the 1960s, is it not possible that civil rights activists could have been charged with hate crimes? After all, is not a sit-in at a segregated lunch counter arguably a trespass committed because of race (see Box 3.8)?

Box 3.8   The Case of Michael Hamm

I first became interested in hate crimes when I read an article on the front page of the August 25, 1991, issue of the *New York Times.* Police in Punta Gorda, Florida, had been called to a domestic disturbance. When white officer Stephen Keyes arrived at the scene, Michael Hamm, who is black, said to him, "I'll shoot you, white cracker." Hamm was subsequently charged with violating Florida's hate crime law, which could have increased his sentence from 1 to 3 years.

This case raised a number of interesting questions. Was this an example of the state using hate crimes legislation to disempower a person of color? Was this the type of incident for which these laws were created? How is prejudice or bias to be determined in a case like this? When Hamm called Keyes a "cracker," was that a manifestation of prejudice? (The *Times* article pointed out that Senator Bob Graham of Florida once proudly campaigned as a "Graham cracker," and Atlanta once had a minor league baseball team called the Crackers.)

Eventually, amid much public outcry, the prosecutor dropped all charges against Hamm for lack of evidence.

SOURCE: Rohter (1991).

Again, there is some empirical support for the argument that hate crime laws could be used to disempower members of minority groups. Official hate crime data have consistently shown that, although African Americans are disproportionately likely to be the victims of hate crimes, they are also disproportionately likely to be identified as the perpetrators (Franklin, 2002; Gerstenfeld, 1998). For example, in Minnesota, approximately 2.2% of the population is African American. In 1993, however, 36.9% of the reported hate crime victims were black, as were 34% of the reported offenders (Gerstenfeld, 1998, pp. 40-41). In 2000, according to the FBI, 18.7% of the offenders were black; about 12% of the United States population is African American (Federal Bureau of Investigation, 2001b).[12]

The reasons for this pattern are unclear. It is possible that, due to factors such as economic deprivation and anger over racism, blacks actually do commit more hate crimes. On the other hand, it is possible that victims, witnesses, and police officers are more likely to interpret a crime as hate motivated when the offender is white than when he is black (Gerstenfeld, in press). It could also be a statistical fluke: Because whites are the majority in most places, a person of any race who chooses a crime victim at random is probably going to choose a white victim. Incidents between a white offender and white victim are unlikely to be interpreted as hate crimes, whereas those between a black offender and a white victim might. Whatever the explanation (and, clearly, this issue merits more research), this is a troubling trend.

One author (Fleischauer, 1990), acknowledging that minorities may be disproportionately subject to hate crime penalty enhancement, suggests that this problem be prevented by exempting minorities from prosecution under hate crime laws. This proposal, however, is untenable. Not only would it open the laws to legitimate (and probably successful) challenges based on the equal protection clause of the Fourteenth Amendment, but it also ignores the fact that many hate crimes are committed by members of one minority group against members of another (Perry, 2001).

Hate Crime Laws Might Increase Prejudice

A final risk of hate crime laws is that, rather than reducing prejudice through their symbolic message, they will actually increase it. On the level of the individual offender, this seems quite likely; a defendant will probably not reform his or her bigoted ways after being convicted of a hate crime. To the contrary, the conviction may actually increase his or her standing among peers by making him or her into a martyr and a hero. The perpetrator may also blame the group to which the victim belongs—after all, if it were not for that group, he or she would not have been found guilty of a crime. Moreover, incarcerating a racist will hardly reform him or her. Prisons and jails are among the most prejudice-ridden institutions in our society. Prison gangs are usually organized along racial and ethnic lines, and they sometimes have ties to external extremist organizations.

On a larger scale, it is questionable whether hate crime laws will reduce prejudice in the community in general. Two psychological theories might predict the opposite. First, the theory of attitudinal inoculation predicts that if people have never been exposed to weak counterarguments to beliefs they hold, they are especially vulnerable to strong arguments later. It is akin to being vaccinated: Initial exposure to a weak form of the virus provides resistance to the virus in its full form (Cohen, 1964). Hate crime laws might discourage people from openly expressing biased beliefs. People who never hear these types of beliefs might later fall under the influence of a particularly persuasive, bigoted speaker.

Cognitive dissonance theory might also predict a paradoxical effect of hate crime laws. According to this theory, when people are given a small reward for acting in a way contrary to their beliefs, they may change those beliefs. The explanation for this phenomenon is that they conclude that the reward was not large enough to justify their actions, so therefore they assume that their actions must really reflect their beliefs. Similarly, if people are

(text continues on p. 68)

NARRATIVE PORTRAIT

PROSECUTING HATE CRIMES

Wendy L. Patrick is a San Diego Deputy District Attorney. As a member of the Special Operation/Hate Crime Unit, she is responsible for prosecuting hate crimes. She also belongs to the San Diego Hate Crime Coalition and the Hate Crime Registry Management Team, and she has made numerous presentations on the topic. In this piece, she discusses the satisfactions and frustrations of her job.

Introduction

The most satisfying part of being a hate crime prosecutor is the same as the most frustrating part: striving to achieve justice for victims. It is satisfying when we can identify and prosecute a hate crime suspect; it is frustrating when, despite our best efforts, we cannot.

Hate crimes are message crimes. They affect more than the particular individual who was targeted—they affect the entire community. Hate crime victims cannot take precautionary measures to defend themselves because they are targeted solely because of who they are.

Hate Crime Laws

Most hate crime laws take the form of enhancements that increase punishment for an underlying crime. Hate crime enhancements require us to prove that the crime was committed *because of* the victim's *actual or perceived* race, color, religion, nationality, country of origin, ancestry, disability, gender, or sexual orientation. Legally, we need to prove that the suspect's bias was a *substantial factor* behind the commission of the crime. Hate crime enhancements make lesser crimes (misdemeanors) punishable as more serious crimes (felonies). Serious crimes that are already felonies may be punishable by an additional term in custody if we can prove the hate crime allegation.

Hate Crimes and Hate Incidents

Hate crimes are *criminal acts* committed against someone because of their membership in one of the protected classes. Hate incidents are noncriminal acts of discrimination, such as speaking racial slurs or distributing racist leaflets. Although hate incidents are not punishable as crimes, we encourage law enforcement to document them for tracking purposes and future use.

A Voice for Victims

Many hate crime victims are reluctant to report their victimization. Some are unaware of hate crime laws; others don't trust the police and don't believe they will receive any help. Some fear retaliation; others fear deportation. In all cases, they suffer from a range of negative emotions that are unique to hate crime victims.

Hate crime victims are targets. When people are victims of hate crimes, the fear of revictimization may lead to feelings of helplessness and isolation. They feel degraded, frustrated, and worst of all, they are afraid. These emotions ripple through the particular community targeted by hate crime offenders, leading to outrage, blame, retaliation, and collective fear.

As a hate crime prosecutor, I can help. If we are able to charge a suspect with the hate crime, we become the voice the victim never had. We become not only the victims' advocate, but also a spokesperson for their community, sending the message that bigotry will not be tolerated. Herein lies the most rewarding part of my job. The frustrating part, as explained next, is that we cannot always find the perpetrator.

Unsolved Hate Crimes

Statistically, most hate crimes go unsolved. The reason is that most hate crimes are attacks on strangers that result in little or no information about the suspect. The victims of these unsolved cases thereby suffer the additional adverse emotion of frustration that their offender will not be brought to justice. This frustration, like the outrage of the crime itself, also ripples through the victim's community. We in law enforcement often find ourselves in the unenviable position of having to explain to an upset hate crime victim or community member that, despite our best efforts, we do not have a suspect.

The Power of Partnerships

The partnerships between prosecutors, law enforcement, community groups, and victim advocacy groups bring hope to hate crime victims even if we cannot solve their crime. We track all reported hate crimes and hate incidents, and share information and intelligence with law enforcement agencies. This cooperation, combined with increased public participation, increases our chances of locating and prosecuting hate crime offenders. Our partnership with community groups increases the visibility and public awareness of hate crimes, and our partnerships with victim advocacy groups allow us to offer hate crime victims support and sometimes even financial assistance, even if we cannot find the perpetrator. Some hate incidents that cannot be prosecuted criminally are referred to private attorneys, who may pursue a civil rights cause of action on behalf of the victim.

The Future

Although we cannot realistically expect to eradicate hate crimes, we do hope to make a difference. We will work tirelessly to educate our community that they have a right to be free from hate crimes, and we seek to send a message to hate crime offenders that we will prosecute them to the fullest extent of the law. In the balance, the opportunity to help hate crime victims regain their dignity outweighs the frustration of unsolved cases. I am looking forward to the opportunity to continue making a difference—the most rewarding part of my job.

threatened with only small punishments for certain behaviors, they may conclude that they avoided those behaviors because they actually did not want to perform them. On the other hand, when people are given large rewards (or threatened with large punishments), their beliefs do not change. They assume that their behavior is due to the reward (or punishment) rather than intrinsic motivations (see, e.g., Carlsmith, Collins, & Helmreich, 1966; Deci, 1975; Festinger & Carlsmith, 1959; Lepper & Greene, 1975). This is referred to as the overjustification effect (Tang & Hall, 1995).

Because hate crime laws increase the penalties for bias-motivated behaviors, they may trigger the overjustification effect. That is, individuals who have biased beliefs may conclude that their restraint from committing hate crimes is due not to lack of desire to commit them but rather to the possibility of harsh punishment. They will not internalize unbigoted ideologies and may therefore continue to act out their prejudice in ways that are not illegal. Furthermore, if the laws are ever repealed, their rationale for not committing hate-motivated acts will disappear, and they may be even more likely to commit those acts.

At this point, it is merely conjecture whether hate crime laws have the kinds of effects that attitudinal inoculation and cognitive dissonance theories would predict. Nobody has conducted an empirical examination of these particular questions. However, both theories are well supported by several decades of research, so their potential effects are certainly worth considering.

CONCLUSION

The constitutional dilemmas concerning hate crime laws are surely interesting and can lead to some lively debates. Among legal scholars, these issues remain quite contentious. Although the philosophical questions remain, the practical questions have largely been answered by the Supreme Court. Hate crime laws, as long as they comport with certain substantive and procedural requirements, are constitutional. However, the discussion about the wisdom, utility, and implementation of these laws is far from over because many important policy concerns still remain.

DISCUSSION QUESTIONS

1. If you were a judge who had to rule on the constitutionality of hate crime laws, how would you rule and why? Which arguments do you find most convincing? If you were a state legislator, would you support hate crime legislation?

2. Whose job do you think it should be to determine a defendant's motive in hate crime cases: the jury's or the judge's? What leads you to this conclusion? What do you think the standard of proof should be?

3. If you were writing a hate crime law, which groups would you include? What are the dangers of over- or under-inclusion?

4. Assess the assertion that by excluding sexual orientation from hate crime laws, the government implicitly condones violence against gays and lesbians.

5. Discuss the relationship between hate crimes and the crimes of rape and domestic assault. Are all rapes and domestic assaults hate crimes? Should gender-based crimes be included in hate crime legislation? Would doing so diminish the attention paid to rapes and domestic assaults? Conversely, would inclusion of gender-based crimes lessen the focus on hate crimes based on categories such as race and religion?

6. Jacobs and Potter (1998) argue that hate crime laws actually increase intergroup conflict. What is the basis for this argument? Do you agree? How would you design a study to examine this question?

7. Why are so few hate crimes reported to the police? What strategies or programs might help increase reporting? Do you think the recommendations of the California Attorney General's Civil Rights Commission on Hate Crimes will be followed and, if so, do you think they will be effective?

8. Based on the research presented in this chapter, what policies or procedures would you recommend to improve police reporting of hate crimes?

9. Describe the many barriers to obtaining a hate crime conviction. In general, do you believe that these barriers are good, in that they may act to screen out unmeritorious cases? Given the very small percentage of cases that actually make it through, do you believe hate crime laws still have merit?

10. What factors might affect a jury's decision in a hate crime case? How would you design a study to examine these factors? If you found that juries' decisions were heavily influenced by extralegal factors such as race, what would be the policy implications?

11. Some critics have argued that hate crime laws might actually hurt the people they were intended to protect. What is the basis of this argument? How convincing do you find it? What kinds of empirical evidence would you look for to support or refute it?

12. If you were a policymaker, what questions about hate crimes would you want to see pursued? Who should pursue them? Scientists? The government? Advocacy groups? What are the potential advantages and disadvantages of each of these?

INTERNET EXERCISES

1. Visit a few Web sites of police agencies to see what information, if any, they have about hate crimes. Here are a few places to start: the Los Angeles Police Department (www.lapdonline.org), the New York City Police Department (www.ci.nyc.ny.us/html/nypd/home.html), and the Boston Police Department (www.ci.boston.ma.us/police/default.asp). How complete, accurate, and helpful is the information available on these sites? You might also try visiting some prosecutors' sites, such as the Los Angeles County District Attorney's office (http://da.co.la.ca.us).

2. The FBI publishes annual hate crime reports (www.fbi.gov/ucr/ucr.htm), as does the state of California (caag.state.ca.us/cjsc/pubs.htm). In addition to looking at data from a single year, you can examine trends over several years. After you read these reports, and considering what you now know about hate crime reporting, what conclusions do you draw?

3. The Bureau of Justice Statistics funded a large study on hate crime statistics. The study, conducted by the Center for Criminal Justice Policy Research and the Justice Research and Statistics Association, is called "Improving the Quality and Accuracy of Bias Crime Statistics Nationally: An Assessment of the First Ten Years of Bias Crime Data Collection." You can view the entire report at www.dac.neu.edu/cj/crimereport.pdf or a summary at www.dac.neu.edu/cj/executives.pdf. The authors of the report make a variety of recommendations. With which do you agree or disagree, and are there any that you would add?

NOTES

1. The distinction between motive and intent is, legally, quite complex. For an extensive discussion, see Gardner (1993).

2. However, in a later case, the Court held that the defendant's racist beliefs could not be considered during sentencing if they did not relate specifically to the crime (*Dawson v. Delaware*, 1992).

3. Absent previous hate crime convictions (which would probably be inadmissible), such a pattern could generally only be shown through evidence of the defendant's hate group membership and/or bigoted speech. As I will discuss later in this chapter, the use of such evidence raises important First Amendment questions in and of itself.

4. Some statutes deal directly with this issue by stating that an act is a hate crime if it is motivated by the offender's perception of the victim's group, as well as the victim's actual group.

5. Actually, there was one important difference between these two cases. After shooting the women, Lepine turned his gun on himself. Furrow, however, did not. He ultimately pled guilty to 16 federal charges and was sentenced to life without parole (Anderson, 2001).

6. As an example, in the three weeks preceding August 16, 2002, the Associated Press reported that five men over 70 years of age had been robbed in the neighborhood near the California Capitol (Associated Press, 2002).

7. Henard blamed women for the ills of society and had once tried to file a civil rights complaint against the "white women of the world."

8. King and Evers were both African American civil rights leaders who were murdered by white men. Kahane was the leader of a militant Jewish group; he was killed by an Egyptian man in 1990.

9. Interestingly, however, Soule and Earle (2001) and Grattet, Jenness, and Curry (1998) found no relationship between the presence of an ADL office in a particular state and the likelihood or speed of that state passing a hate crime law.

10. It should be noted that these data do slightly underestimate the effect of hate crime laws in that they do not include offenders who were initially charged with a hate crime but who, pursuant to a plea bargain, pled guilty to the underlying offense. The existence of the hate crime charge, and the threat of penalty enhancement, may encourage some defendants to plead guilty when they otherwise would not have or to plead guilty to a more serious charge than they would have otherwise.

11. In a few states, as well as in federal cases, this decision is made by a grand jury instead of a judge.

12. The FBI report does not specify how many of the victims were black. However, 36.2% of the offenses were classified as antiblack.

Chapter 4

COMMITTING HATE
Who and Why

What images come to your mind when you read the words *hate crimes*? Red-necked, white-robed men burning crosses? Gangs of skinheads attacking people of color? Swastika-bedecked neo-Nazis toting copies of *Mein Kampf* and shouting "Seig Heil!"? All of these are part of the reality of bias in America. The truth, however, is that they make up only a small part of the hate crimes that occur. Contrary to popular notions, most hate crimes are not committed by members of organized white supremacist groups (J. Levin, 2002; J. Levin & McDevitt, 1993). Most, in fact, are committed by people very similar to those you might see sitting in almost any high-school or college classroom: teenagers and young adults who are unaffiliated with any racist organization. In fact, hate crimes on campus are a serious problem (Downey & Stage, 1999; J. Levin & McDevitt, 1993; see Box 4.1).

What leads people to perpetrate acts of violence and intimidation against others? In any individual case, there is no one single cause of bigotry, and the origins of prejudice vary from person to person. In this chapter, I will address this issue by looking at a variety of explanations for prejudice and hatred.

Scholars in many different fields—psychology, sociology, and economics, among others—have spend several decades studying the causes, correlates, and effects of prejudice. Many books have been written on the subject and, by necessity, this chapter will consist of only an introduction to that literature. In recent years, a small body of work has accumulated concerning the specific roots of hate crimes, and we will explore that work in more detail. This chapter begins with an examination of research on the profile of the typical hate crime offender. The discussion then turns to the psychology of prejudice and to situational factors that influence bigotry in individuals. Finally, we will consider why some people join hate groups.

As you read this chapter, it is very important to note that none of us is free of prejudices. Different people have different biases, and some people are more strongly biased than others. Most of us probably cannot imagine being so bigoted as to actually commit a crime against someone. However, these are differences in kind and degree only, and there is no magic boundary that separates you or me from the people who commit hate crimes.

Box 4.1   Hate on Campus

In a report issued in 2001, the Department of Justice concludes that hate crimes on college campuses are a serious problem. Very few hate crimes on campus get reported, so their true extent is unknown, but they occur at colleges and universities of all sizes and in all locations. The report gives the following examples of recent events on campuses:

- A man in Utah detonated a pipe bomb in the dorm room of two African American students. He later left a threatening note on the door of another African American student.
- A former student in California e-mailed threats to 59 other students, most of whom were Asian Americans.
- A student in Maine yelled antigay slurs at another student, threatened him, and choked him.
- A college student in Massachusetts made several anti-Semitic comments, threatened several fellow students, and delivered photos of Holocaust victims to another.

The report also concludes that noncriminal acts of prejudice and harassment are common on campuses.

SOURCE: Bureau of Justice Assistance (2001).

THE OFFENDER PROFILE

The "Typical" Offender

In recent years, a consistent, if incomplete, picture of the "typical" hate crime offender has emerged. He is young, white, and male; he does not come from an especially impoverished background; he has little or no previous contact with the criminal justice system; and he does not belong to an organized hate group (Craig, 2002; Dunbar, in press). In 1999, for example, more than 68% of reported hate crime offenders were white (Nolan, Akiyama, & Berhanu, 2002), and in 2000, of the 6,091 offenders of known race, the FBI reported that 67.5% were white (Federal Bureau of Investigation, 2001b). J. Levin (2002) estimated that no more than 5% of hate crimes are committed by members of hate groups; in a study of 58 convicted hate crime offenders in Los Angeles, fewer than 14% had belonged to a hate group (Dunbar, in press). However, this is not to say that hate crime offenders act alone; in fact, they are more likely than other kinds of offenders to act in small groups (Craig, 2002; Dunbar, in press). B. Levin (1993) reported that more than half of all hate crimes involve multiple offenders, whereas only about one quarter of ordinary violent crimes do.

Although the data give us some useful information about hate crime perpetrators, some commentators have warned that it is not wise to draw too firm a profile of the typical offender. Craig (2002), for example, argues,

Whether it is useful to attempt construction of a profile of the typical hate crime offender beyond predicting that he is male, is debatable. . . . Hate crime perpetrators though overwhelmingly male, hail from a diversity of ethnic and racial backgrounds, and with the exception of their hate-motivated activity, have little in common and are otherwise unremarkable. (p. 97)

Interestingly, however, this same author and a colleague found that college students, especially those of color, tended to associate hate crimes with white, male perpetrators (Craig & Waldo, 1996).

Offender Motivations

Attempts have been made to develop a typology of hate crime offenders. J. Levin and McDevitt (1993) examined the case files of the Boston Police Department from 1991 and 1992. They found that offender motivations could be divided into three classifications: thrill-seeking crimes, reactive crimes, and mission crimes.

Thrill-Seeking Crimes

Thrill-seeking crimes were the most common type, constituting two thirds of the total (McDevitt, J. Levin, & Bennett, 2002). These were cases in which the offenders, almost always young and in small groups, were "just bored and looking for some fun," (McDevitt et al. 2002, p. 307). In other words, like many young people, they were looking for a little excitement, only they decided to have it at someone else's expense. J. Levin & McDevitt (1993) state that in many of these cases, most of the perpetrators may not even have been especially biased toward the victim, but they were following a leader who was. They were unwilling to refuse to go along, perhaps because they feared rejection from their friends or because they wished to be able to brag about their prowess later. In almost all of these cases (91%), the offenders left their own neighborhood and purposely sought out a victim somewhere else, such as at a gay bar, at a synagogue, or in a minority neighborhood (McDevitt et al., 2002).

Other research has confirmed that thrill-seeking seems to be a common motive among hate crime offenders. Byers and Crider (2002) conducted interviews with eight young men who had committed acts of "claping" (assaults, property damage, and harassment) against Amish victims. The researchers found that generating excitement and alleviating boredom were common explanations given by the offenders for their behaviors. For example, one participant gave the following explanation for why he claped: "It is fun, an adrenaline rush, tradition I suppose you could say. Just something to do. They [the Amish] were there. . . . [We were] fifteen or sixteen year old kids that had nothing [else] to do" (Byers & Crider, 2002, p. 124). Another participant explained, "Because I was probably with a lot of my friends and we were just looking for something to do and they were doing it and I was doing it so we were all doing it" (Byers & Crider, 2002, p. 124). There were other motives as well—for example, the Amish were perceived as easy targets or as deserving of ill treatment—but excitement appeared to be a major theme.

In another study, Franklin (2000) surveyed 489 community college students about their attitudes and behaviors toward gays and lesbians. Ten percent of the participants admitted to having threatened or physically assaulted people they believed were homosexual, and an additional 23.5% reported having engaged in verbal harassment. Most of the incidents occurred at school or work, and more than 70% of the offenders were with friends at the time of the incident, whereas most of the victims were alone. Franklin (2000) found that four factors accounted for most of the variance in offenders' motivation. In order of importance, these were peer dynamics (i.e., the desire to fit in with and impress friends), antigay

Box 4.2   Howard Beach and Bensonhurst

In the late 1980s, there was much media attention on two deaths that occurred in New York, both precipitated when whites encountered African Americans in what the whites considered "their" neighborhoods.

The first incident occurred in Howard Beach, a largely Italian American, working-class community. On December 19, 1986, a car belonging to three young black men broke down in Howard Beach. As these men searched for a way to return home to Brooklyn, they were verbally accosted by some young white men. The black men later went to a local pizzeria. One of the white men, Jon Lester, went to a birthday party and rounded up nearly two dozen friends. Armed with baseball bats and other impromptu weapons, the mob accosted the "intruders" outside the restaurant. They then gave chase. One of the black men escaped; a second was badly beaten. The third, Michael Griffith, was struck by a car and killed as he tried to escape.

Lester and two of his colleagues were eventually found guilty of manslaughter and assault, and received sentences ranging from 10 to 20 years. In 1999, the

ideology, thrill seeking, and self-defense. Both Franklin's (2000) peer-dynamics and thrill-seeking factors would fall under J. Levin and McDevitt's (1993) thrill-seeking category.

One important finding of all these studies is that, in the case of thrill-seeking hate crimes, the offenders generally do not have a particularly strong animosity toward their victims. To call these "hate" crimes, then, is somewhat of a misnomer. This is not to say that the offenders are completely without bias, nor does it alleviate the psychological and physical damage done to the victims. However, it does suggest that, given the prevalence of these kinds of crimes, prevention programs aimed at simply reducing bigotry may not be as effective as hoped.

Reactive (Defensive) Crimes

The second type of hate crimes identified by J. Levin and McDevitt (1993) was originally called "reactive" crimes, but these researchers later renamed them "defensive" crimes (McDevitt et al., 2002). These are crimes in which the perpetrator is reacting to what he considers to be an intrusion—that is, some incident triggers an expression of anger. The trigger incident need not be objectively significant. For example, a black person might simply enter an all-white neighborhood, as happened in 1986 in Howard Beach, New York, and in 1989 in the Bensonhurst section of Brooklyn (see Box 4.2). Or perhaps a would-be offender spots an interracial or same-sex couple, or a person of a different group than the offender receives a promotion at work.

Defensive crimes accounted for one quarter of the total in the McDevitt et al. (2002) study. They differed from thrill-seeking crimes in that the offenders usually did not leave their own neighborhoods to seek out the victims; instead, the victims happened upon them. In addition, although the offenders' primary feelings tended to be those of having their territory invaded, there was frequently an economic theme as well. As an

street in Brooklyn where Michael Griffith had lived was renamed after him.

A strikingly similar crime occurred in the Bensonhurst section of Brooklyn, another mostly white neighborhood, on August 23, 1989. Four black teenagers walked to Bensonhurst to look at a car they had seen advertised for sale. They were soon surrounded by a group of bat-wielding white youths, who began chasing them and yelling racial slurs. One of the whites, Joey Fama, pulled out a gun and fired several times. One of the blacks, 16-year-old Yusuf Hawkins, was shot twice and died soon afterward.

Eight people were tried for Hawkins's death. Three were acquitted, two were convicted but given probation, and three were sent to prison. By 1999, only one of the defendants, Fama, remained in prison. During several marches that were held in Bensonhurst after the killing to protest the violence, spectators yelled racial slurs, waved watermelons, threw bananas at the marchers, and even spat in the face of Hawkins's father.

SOURCES: Donahue (1999); Hajela (1999); J. Levin & McDevitt (1993); Santiago (1999).

example, J. Levin and McDevitt (1993) cite events at Galveston, Texas, in 1981. There was a growing number of shrimp fishermen who were Vietnamese immigrants, and the white fishermen felt that the newcomers were competing unfairly. The white fishermen invited the Ku Klux Klan to help them drive out the Vietnamese, and the Klan engaged in a several-month campaign of intimidation, property destruction, and violence (Stanton, 1992).[1] In most cases of defensive hate crimes, the offenders have no previous criminal history, and were not previously involved in hate-related activities.

Mission Crimes

The third, and rarest, type of hate crime identified by J. Levin and McDevitt (1993) is the mission hate crime. Only one mission crime occurred in Boston during the two years of the study. In a mission crime, the offender, usually acting alone, seeks to rid the world of a particular kind of people whom he views as evil. Here are some examples of offenders and the mission hate crimes they committed:

- Marc Lepine, who killed 14 women in a Montreal classroom in 1989 (see Chapter 3)
- Patrick Purdy, who took an AK-47 to an elementary school in Stockton, California, in 1989 and sprayed 60 rounds of bullets, killing five Southeast Asian children and wounding 30 other children and adults (J. Levin & McDevitt, 1993)
- Timothy McVeigh and Terry Nichols, who bombed the federal building in Oklahoma City in 1995, killing 168 people
- Buford Furrow, a white supremacist who shot five people in a Los Angeles Jewish Community Center in 1999 and then shot and killed a Filipino American postal worker
- Benjamin Smith, a former member of the racist World Church of the Creator, who went on a shooting spree in Indiana and Illinois in 1999, aiming at blacks, Jews, and Asian Americans, killing two and injuring nine

Three of the killers (Lepine, Purdy, and Smith) in these examples subsequently committed suicide. Furrow pled guilty to murder and was given two life sentences, McVeigh was executed for his crime in 2001, and Terry Nichols, who was convicted of involuntary manslaughter and conspiring with McVeigh, is serving a life sentence.

People who commit mission hate crimes are usually deeply troubled and sometimes even psychotic. They see others as having perpetrated some sort of conspiracy against them and desire revenge (J. Levin & McDevitt, 1993). In addition, they may have had some previous affiliation with organized hate groups. Although mission hate crimes are very uncommon, they do deserve attention due to the extreme amounts of violence involved; mission offenders, unlike those in the first two categories, will not be satisfied with simple acts of vandalism or intimidation. It is likely, however, that the psychological factors underlying mission criminals' behaviors are significantly different from those of other hate crime offenders.

Retaliatory Crimes

In 2002, McDevitt, J. Levin, and Bennett reexamined the original Boston data and determined that it would be appropriate to add a fourth category, retaliatory crimes. These crimes accounted for 8% of the total. Retaliatory crimes are those in which a person hears a report or rumor of a hate incident against his or her own group and takes revenge by committing a crime against a member of the initial supposed offending group. An example of this may have occurred in Brooklyn's Crown Heights neighborhood in August 1991. A car containing Hasidic Jews jumped the curb and struck two young African American children, killing one of them, 7-year-old Gavin Cato. Rumors immediately began that a Jewish ambulance service had declined to help the child. A series of confrontations and attacks soon erupted between blacks and Jews in the neighborhood. About 3 hours after the children were struck, a group of young black men came across Yankel Rosenbaum, an Australian Jew who had come to New York to study. One of the young men stabbed Rosenbaum, who died several hours later. The rioting and violence continued for several more days, with each side blaming the other and claiming that city officials were not properly serving justice.

The Validity of This Typology

McDevitt, J. Levin, and Bennett's typology of hate crime offenders is the most complex that has been offered, and it has been used as a guide by law enforcement agencies. However, although it is supported in part by Franklin's (2000) and by Byers and Crider's (2002) research, the robustness of this typology is unclear. It is based on data from only a single city, and those data are now more than a decade old. No large-scale efforts have yet been made to replicate the analysis in other places and at other times.

Moreover, there are clearly some incidents that do not fit neatly into any of these four categories. Consider, for example, Benjamin and James Williams. In 1999, these brothers set fire to three synagogues in the Sacramento, California, area. A week later, they burned an abortion clinic. And several days after that, they shot and killed a gay couple in Shasta County as the couple lay in their own bed. These crimes have some of the attributes of mission crimes, including the elder brother's later confession that he had committed the crimes to become a "Christian martyr" and to spark more crimes against Jews, gays, and minorities (Bailey, 2001).[2] On the other hand, they seem unlike other mission crimes in that there was more than one offender (although they were close relatives), and the scale of the violence seems somewhat milder, albeit still serious. The crime spree also lasted longer than in other mission crimes.

Consider also the case of Todd Mitchell, which we discussed in Chapter 3. His crime could be considered retaliatory, in that he and his friends were apparently reacting to a scene in a movie they had recently watched. However, this crime also had some of the earmarks of a thrill-seeking crime in that a group of young people was standing around with nothing much to do, perhaps looking for some excitement.

Finally, assuming that this typology proves to be valid in contexts other than that in which it was created, its utility and meaning remain to be found. Is it useful in solving hate crimes or in assigning appropriate levels of culpability to offenders, as its authors claim? Are there regional variations in the proportion of offenders of each type? Exactly what situational and personal factors differentiate the types of offenders? Do prevention and enforcement strategies differ in effectiveness for each type? What, exactly, is the role of organized hate groups in motivating or supporting each type of offender? Clearly, these are questions that merit additional research.

THE PSYCHOLOGY OF PREJUDICE

The Development of Prejudice

No baby is born with prejudices against other people. Considering that the average 2-year-old finds talking purple dinosaurs unremarkable, it is not surprising that very young children are pretty accepting of human beings of all shapes, colors, abilities, and beliefs. Yet we have already seen that most hate crime offenders are young people in their teens or early twenties. In the space of less than two decades, how does a person change from an accommodating (at least in the realm of intergroup relations) toddler into a violent bigot?

The first stage in the development of prejudice is the ability to engage in social categorization: the division of human beings into groups on the basis of gender, race, age, and so on (Allport, 1954). Half a century of social psychological research suggests that social categorization is an innate and inescapable human trait, necessary for us to make sense out of a complex world (Bruner, 1957; Fiske, 1993; Tajfel, 1969). Social categorization begins early in life. Ninety percent of 3-year-olds, for example, can correctly classify people in photographs as male and female (Thompson, 1975), and other research shows that children of that age can classify people according to race as well (Brigham, 1971; Katz, 1983; Milner, 1975). Children are particularly likely to sort people by certain categories—gender, race, and age (Brewer, 1988; Fiske & Neuberg, 1990)—although there are endless other ways to sort people as well.

As young children engage in social categorization, two additional processes take place. The first of these is self-identification: Children learn which groups they belong to (the ingroups) and which they do not (the outgroups) (Jones, 1982; Tajfel, 1969). In other words, they learn to distinguish between "us" and "them." Most children between ages 3 and 7 correctly identify their own ethnic group (Brown, 1995).

At about the same time, children begin to learn the stereotypes associated with different groups (Goodman, 1964; Katz, 1976). Allport (1954, p. 190) defined a *stereotype* as "an exaggerated belief associated with a category." It is a mental picture that is attached to a particular group. Stereotypes are often negative—e.g., "women are bad drivers"—but they can be positive as well—e.g., "blacks are good at sports." Whether positive or negative, they are often erroneous and are potentially dangerous because they lead us to make judgments about an individual based only on that individual's group affiliation rather than on information we have actually learned about that individual.

As they learn stereotypes, children also learn racial slurs. They may not fully realize what the words mean, but they recognize their potency and ability to bring about strong reactions from adults (Allport, 1954). Children may use racial slurs (albeit often incorrectly) because of the lure of the forbidden at that age.

Although preschoolers are aware of stereotypes, research indicates that children do not actually internalize the attitudes associated with them until about age 7 (Goodman, 1964; Katz, 1976). This is when children begin to show a preference for their own group. Observe, for example, third graders on the playground, and you will see the girls mostly playing with other girls, and the boys with other boys. The internalization of stereotypes leads to *prejudice,* which Allport (1954) defines as "an avertive or hostile attitude toward a person who belongs to a group, simply because he belongs to that group, and is therefore presumed to have the objectionable qualities ascribed to that group" (p. 7).

Allport (1954) also pointed out an interesting facet of children's behavior. Although grade-school–age children often engage in total verbal rejection of the outgroup, that rejection is often not reflected in their behavior: They will continue to associate with members of the outgroup. However, by the time children reach middle school, they have learned that verbal rejection is usually not socially acceptable, so they engage in politically correct speech. However, by then they have adopted society's prejudices, and they will usually reject members of the outgroup behaviorally. You can verify this by visiting any ethnically diverse middle or high school at lunch time, where you will see the students mostly sitting with members of their own race or ethnicity.

If babies are not born knowing stereotypes, and yet they learn them in only a few short years, where do these ideas come from? Are they some sort of germ they pick up in preschool, like chicken pox? Yes, metaphorically speaking. Children are exposed to stereotypes from all aspects of their society and culture, including family, teachers, friends, and the media (Allport, 1954; Brown, 1995; Nelson, 2002). These stereotypes are pervasive and can be quite insidious. Recently, for example, I ordered a fast-food meal for my daughter. The restaurant employee asked whether I wanted a "girl's" meal or a "boy's" meal. It turned out that, in addition to hamburger and fries, the girl's meal contained a small doll (wearing a frilly dress and flowered hat), whereas the boy's meal came with a toy fire truck.

A particularly interesting aspect of stereotypes is their consistency. In 1933, Katz and Braly asked white college students to indicate which attributes they believed were typical for each of 10 different ethnic groups (Jews, Blacks, Italians, Turks, and so on). There was a great deal of consensus on some attributes. For example, 75% of the students believed that blacks were lazy (Brown, 1995, p. 83). The experiment was repeated a generation later (Gilbert, 1951) and again a generation later (Karlins, Coffman, & Walters, 1969). In each of these cohorts there was a great deal of similarity in the content of the lists. There were also some differences, most notably a "fading" of many of the more negative characteristics. This fading has been seen in more recent studies as well (Dovidio & Gaertner, 1991). However, some research has indicated that this is due not so much to a reduction in prejudice, as to a change in the social desirability of expressing it. People today may still have bigoted attitudes (consciously or not) while at the same time valuing equality. The theories to explain this phenomenon have referred to it as modern racism (McConahay, 1986), symbolic racism (Kinder & Sears, 1981), or aversive racism (Gaertner & Dovidio, 1986) (see Box 4.3).

Box 4.3   Modern Racism

When researchers first began studying prejudice, stereotypes, and racism in the 1920s, it seemed to be a fairly simple task. Researchers like Katz and Braly (1933), for example, could simply ask people what traits they associated with various groups, and people would willingly list very negative traits. As the studies continued over the next several decades, however, subjects showed less and less overt prejudice against others. Did this mean that people really were becoming less biased? Perhaps.

Some scholars, however, proposed that what was really happening was that the form of prejudice was changing and that the new type, which they called *modern, symbolic,* or *aversive racism,* was more subtle. Technically, modern, symbolic, and aversive racism are all slightly different. Under all three, however, a person still has negative feelings about others, but now those feelings are expressed indirectly. The problem is, how can this form of prejudice be measured accurately? McConahay (1986) devised the Modern Racism Scale to accomplish this task. Instead of asking people to list traits of various groups, people are presented with statements and asked to indicate the degree to which they agree or disagree with them; then, the results are analyzed. Here are a few items from that scale:

- I don't think any black people in my state miss out on jobs or promotions because of racial discrimination.
- Minorities who receive money from welfare programs could get along without it if they tried.
- Blacks should not push themselves where they are not wanted.
- Minorities are getting too demanding in their push for equal rights.
- Discrimination against blacks is no longer a problem in the United States.

The Modern Racism Scale is not without its critics. Some, for example, argue that it measures political conservatism more than racism. Others claim that scores on this scale, too, are affected by factors such as social desirability, and that reaction-time studies (i.e., studies that examine the time it takes people to respond to certain stimuli) are more accurate at assessing true levels of prejudice (Fazio, Jackson, Dunton, & Williams, 1995). All modern researchers, however, would agree that measuring prejudice today is considerably more complicated than it was 80 years ago.

Despite the changes in the way people express prejudice today, everyone is still widely exposed to stereotypes. Every year on the first day of my hate crimes course, I ask my students to complete (anonymously) a series of questions that begin, "What do people say about . . ." I include a variety of different groups: African Americans, Anglos, Latinos, Jews, women, police officers, and Asian Americans. I collect the sentences and put them onto posters. On the second day of class, I hang these posters on the walls, together with those from previous classes. The consistency of these posters from year to year is remarkable, and it is rare for a stereotype to appear with which most of the class was not already familiar. The associations are so strong and well recognized that I do not even have to put the names of the groups on the posters; by the time the class reads the second or third item on each list, it is always obvious to everyone to which group the list of stereotypes refers. All of this holds true not only for the groups with which my students have had a great deal of contact and familiarity, but also for those whose membership is very small in our area. Furthermore, the content of these posters is not very different from the lists generated by Katz and Braly's (1933) subjects 70 years ago.

How Stereotypes Affect Us

What is the impact of this early and frequent exposure to stereotypes upon the way that we view other people? Research suggests that it is quite significant.

A *schema* is a mental model of a thing, containing all of our knowledge about and attitudes toward that thing (Nelson, 2002). For example, your schema about California may include the information that it is a state located on the West Coast of the United States, that it has the largest population of any state, and that its capital is Sacramento; your schema may also contain your impressions that it is full of palm trees, that it is too liberal or too trendy, that it has beautiful weather, or that everyone there is a movie star. Schemas are essential to enable us to deal with an extremely complex world in an efficient manner. We have schemas not only about things but also about people, and one component of those schemas is our stereotypes about those people. Thus, your mental model of attorneys may be that they are well educated (knowledge), that they can help you if you are having legal troubles (expectation), and that they are dishonest (stereotype).

When we see a person, the schema about that person's group is automatically activated (Devine, 1989; Macrae, Stangor, & Milne, 1994). This is true regardless of our own degree of prejudice or the extent to which we actually believe those stereotypes. It takes conscious effort to nullify the stereotypes (Devine, 1989). We may be unable to nullify them because we are unaware that the schema has been activated or because our cognitive processes are occupied with other tasks. Or we may be unwilling to nullify them because we are biased.

Once a schema has been activated, it acts as a filter, affecting the way in which we process information about the person. One effect of this process is called *outgroup homogeneity*. This refers to the fact that the schema for our ingroup is more complex than those for outgroups (Linville, Salovey, & Fischer, 1986). In other words, "we" are very different as individuals, but "they" are all the same.

One possible result of outgroup homogeneity is polarized appraisals of members of the outgroup. We tend to see them as either really wonderful or really awful, whereas we see members of our own group in many more shades of gray (Linville & Jones, 1980; McConahay, 1986). Thus, a white American might conclude that all blacks are either heroes, like Martin Luther King, or dangerous, like the African American gang members depicted so often in movies and on television, and fail to see that most blacks, like most members of other groups, are somewhere in between.

Stereotypes affect how we attribute others' characteristics and actions. In general, and not surprisingly, we tend to think more positive things about members of our ingroup than about members of outgroups. Causal attributions are also affected: For members of the ingroup, we view positive things as having an internal cause, and negative things as having an external cause. For members of outgroups, this pattern is reversed (Hewstone & Jaspars, 1984). Therefore, I might conclude that a member of my own race is successful because she is smart and hard-working, whereas a member of another race is successful because she has been lucky or has been given unfair breaks. Conversely, I might believe

NARRATIVE PORTRAIT

GROWING UP RACIST, I

Larry L. King grew up in a small town in West Texas during the Depression. After serving in the Army just after World Word II, he attended college at Texas Tech, became a newspaper reporter, an aide to a U.S. Congressman, and, eventually, a writer.

In this excerpt from *Confessions of a White Racist,* King recounts his experiences with racism as a young child:

On our porch in the slow summer evenings where the undistinguished dogs and small boys wondered at their inability to catch even one of a million darting fireflies, I often sat on my father's knee while he sang "The Nigger Preacher and the Bear" in a high, comic falsetto that never failed to please. The black preacher, out hunting quail and hare on the Sabbath against the natural laws of Heaven, is treed by an avenging bear. He is shot through with cowardice and malapropisms, and is so deserted by the presumably white God to whom he frantically prays that the bear is eventually permitted to squeeze the life from him—although, indeed, the black preacher gives a fair account of himself thanks to a congenital talent for expertly flourishing the razor. One of our family's private jokes, never told in the presence of company, concerned the horrifying moment when my older brother was discovered in the act of taking alternate bites off an apple with a Negro boy whose family paused briefly in Putnam. I remember the scandalized whispers when an eight-year-old cousin was caught playing bridegroom to a little black bride in a backyard mock wedding. . . .

Quite without knowing how I came by the gift, and in a complete absence of even the slightest contact with black people, I assimilated certain absolutes: the Negro would steal anything lying around loose and a high percentage of what was bolted down; you couldn't hurt him if you hit him on the head with a tire tool; he revered watermelon above all other fruits of the vine; he had a mule's determination not to work unless driven or led to it; he would screw a snake if somebody held its head.

Even our speech patterns were instructional: as we youngsters surreptitiously smoked cedar bark or dried grapevine in Cousin Kenneth's backyard storm cellar, we displayed generous contempt for the amateur who "nigger-lipped" the noxious offerings. To give participants in games of "Hide and Go Seek" time enough to conceal themselves, the hunter thrice or more intoned "Eenie meenie miney moe/ Catch a nigger by the toe/ If he holler make him pay/ Fifteen dollars every day." One's menial labors could leave one "dirty as a nigger" or possibly "sweating like a nigger at election." Get a shade sunburned at the ole swimmin' hole, and your mother was sure to pronounce you "black as a nigger" even in the presence of your lobster-red qualities. Two objects instead of being as identical as "peas in a pod" were likely "as much alike as nigger soldiers." "I wouldn't feed it to a dog" was easily interchangeable with "It would choke a nigger." If you had an exceptional pal, you might boast that the two of you were "as close as runaway niggers." David may have slain Goliath with a slingshot, but in Putnam we warred on frogs, birds, and alley cats with "nigger-shooters." I don't remember that we employed our demeaning expressions in any remarkable spirit of vitriol: we were simply reciting certain of our cultural catechisms, and they came to us as naturally as breathing.

SOURCE: King (1969, pp. 4-7).

someone of my own race is poor because of bad luck or a conspiracy against him, whereas I might believe a member of another race is poor because he is lazy and stupid.

Stereotypes also influence the ways in which we remember information about other people. We tend to recall things better when they are consistent with our stereotypes (Fiske, 1993; Rothbart, Evans, & Fulero, 1979). A consequence of this is that stereotypes tend to be self-perpetuating. We also recall negative information more easily about out-group members than about ingroup members (Howard & Rothbart, 1980). And we recall subcategory information better about ingroup members than about outgroup members, which helps produce the outgroup homogeneity effect. For example, in one experiment (Mackie & Worth, 1990), college students read short scenarios about other students. Later, the male subjects were better at remembering the major of the male student, and the female subjects better remembered the female student's major.

What all of this research leads us to conclude is that we are never neutral observers, but instead our knowledge, expectations, and beliefs are lenses through which we perceive others. And those lenses alter our perceptions in many ways.

WHAT MAKES A BIGOT?

In the previous section, we established that knowledge of stereotypes begins very early, that those stereotypes are pervasive throughout our culture, and that the content of the stereo-types is quite stable from place to place and time to time. Moreover, the cognitive effects of those stereotypes are universal: Human brains all operate more or less the same way.

All of this begs the following question: If we are all exposed to the same stereotypes, and if our minds are all identically affected by those stereotypes, why are we all not equally prej-udiced? Why is it that a small minority of people are biased enough to actually commit crimes against others because of their groups, another small minority of people devote their lives to fighting bigotry, and the rest of us lie somewhere in the large gap separating the other two?

Another way we could ask the same question is to ask, what makes a bigot? Not sur-prisingly, there is no simple answer. Research in this area has tended to focus on two rather distinct forces, both of which I discuss in this section. Some of the research, especially the early work, has concentrated on the role of the personality. These studies have attempted to determine whether there is a constellation of individual characteristics that make up a "prej-udiced personality." They examine the factors that might combine to shape a personality that is intolerant of others, and the traits of a person who is wont to act on that intolerance. In contrast, other research has looked at the role of external factors, such as the behaviors and attitudes of others, economic and social forces, and the larger cultural milieu.

The Role of the Family

The previous section discussed the fact that stereotypes and prejudice can be found in almost every segment of our society. Of these segments, perhaps none has more influence on an individual than that individual's family. Allport (1954) argued that prejudice can be learned from parents in two ways: It can be taught, or it can be caught.

Clearly, some children have parents or other family members who are bigoted and who consciously teach those bigoted values. For example, Tom Metzger heads a white supremacist organization called White Aryan Resistance (WAR). It is probably no sur-prise that his son, John Metzger, is also very active in the group. On WAR's Web site, in response to a question about when parents should teach their children about racism,

Tom Metzger replies, "Immediately! A baby records all from the very beginning. Its mind is like a blank tape and ready to go. As soon as you are training a child not to touch a hot stove is the right time to teach RACISM" (White Aryan Resistance, n.d.).

A fictional, but probably pretty accurate, portrayal of this process was given in the movie *American History X* (Kaye, 1998). In this film, a white father occasionally voices bigoted views, but he is apparently not a member of a hate group. After the father, who is a fireman, is shot and killed by a black person while putting out a fire, his older son becomes an avowed racist. He joins a skinhead gang and is carefully taught the dogma of racism and groomed to be a leader by an older man. When he is sent to prison for the brutal killing of a black man, his younger brother, who clearly idolizes him, soon follows in his footsteps.

Not all bigots, however, were intentionally tutored in bigotry by their parents, and few have family members who belong to hate groups. These children, instead, catch their parents' attitudes as they listen to their jokes, watch the way they interact with other people, and so on (Katz, 1983; Rohan & Zanna, 1996). In addition, early researchers argued that parenting style also affects the likelihood that a child will grow up prejudiced (Adorno et al., 1950; Allport, 1954). As Allport (1954) put it,

> The parents may or may not express their own prejudices (usually they do). What is crucial, however, is that their mode of handling the child (disciplining, loving, threatening) is such that the child cannot help but acquire suspicions, fears, hatreds that sooner or later may fix on minority groups. (p. 297)

Authoritarianism

What type of parents are likely to raise prejudiced children? Adorno et al. (1950) and Allport (1954) referred to them as having an *authoritarian* personality. These parents tended to treat their children in a way that was neglectful, rejective (suppressive or overly critical), or inconsistent. Moreover, when asked to state their opinions, these parents were more likely to agree with the statements contained in Box 4.4. Allport argued that this parenting style resulted in a person who learns that the world is not equal, but hierarchical, and that power and authority are the important factors in human relationships. Moreover, children of authoritarian parents learn to fight impulses within themselves that they have been taught to view as evil, and they project those impulses onto others, which means they cannot trust those others. Finally, because these children fear showing aggression toward their parents, they displace that aggression onto others whom they view as vulnerable.

Box 4.4   Opinions of the Authoritarian Parent

Obedience is the most important thing a child can learn.

A child should never be permitted to set his will against that of his parents.

A child should never keep a secret from his parents.

I prefer a quiet child to one who is noisy.

(In the case of temper tantrums) Teach the child that two can play at that game, by getting angry yourself.

SOURCE: Allport (1954, p. 298).

Box 4.5   The Special Case of Homophobia

One topic related to prejudice that has received special attention is prejudice against gays and lesbians, or homophobia. This kind of hatred is particularly common, as we will discuss in Chapter 6, and often also particularly violent.

A study by Herek and Capitanio (1999) suggests some antigay animosity is related to AIDS. The study showed that most heterosexuals continue to associate AIDS with gay men. The stigma of AIDS negatively affects attitudes in general toward gay men. On the other hand, antigay attitudes certainly predated the AIDS epidemic and may have largely contributed to the stigmatization of AIDS itself.

Other scholars have long hypothesized that homophobia may be linked to individuals' own unconscious sexual preferences. The argument here is that some men are attracted to other men but are unable to accept these feelings. They so violently reject their own desires that they also lash out at others whom they perceive to have those same desires. This theory is supported by several cases in which men who have

---

Allport's (1954) research was subject to quite a bit of criticism (Brown, 1995). Although more recent research did support the link between authoritarian parenting and prejudiced children, later researchers pointed out that in reality, the type of parenting described in the early work would be more appropriately called "Right-Wing Authoritarianism" (RWA), and, although RWA might lead to prejudice, so might its opposite, "Left-Wing Dogmatism" (LWD) (Altemeyer, 1981; Brown, 1995; Nelson, 2002; Rokeach, 1956). Altemeyer (1998) devised a scale to measure RWA and found that scores on this scale were strongly correlated with levels of political conservatism, punitiveness, and prejudice. Examples of items on this scale include "What our country really needs is a strong, determined leader who will crush evil and take us back to our true path" and "Our country will be destroyed someday if we do not smash the perversions eating away at our moral fiber and traditional beliefs" (Altemeyer, 1998). Most white supremacist Web sites and literature closely echo these sentiments.

Although there is limited research available on hate crime offenders, some of it does support the view that hate crime offending may be related to parenting style and personality factors. Ezekiel (1995, 2002), for example, conducted extensive interviews with members of Klan and neo-Nazi groups. The young men who joined a Detroit neo-Nazi group were from poor neighborhoods, had fathers who left or died during their sons' childhoods, had experienced a series of cold or abusive stepfathers or mothers' boyfriends, and had family histories of alcoholism or violence. Dunbar (in press) found a high rate of indicators of psychopathy among convicted hate crime offenders. Some research also suggests that homophobia, in particular, may be related to particular personality factors in the offenders (see Box 4.5).

Situational Factors

Despite the quite extensive work that has been done on parenting styles and authoritarianism, most psychologists today avoid attempting to define the prejudiced personality. Instead, the focus has shifted to the role of situational forces (Reynolds, Turner, Haslam, & Ryan, 2001). As Brown (1995) states,

murdered gay men were shown to have a history of same-sex encounters themselves as well as either denial of those encounters or extreme emotional turmoil associated with them. Furthermore, several men who have been tried for murdering gay men have used the *homosexual panic* or *homosexual threat* defenses, claiming that their violence was a response to the victim's sexual propositions or demands (Dunbar, 1999).

This hypothesis was tested by Adams, Wright, and Lohr (1996). Men who claimed to be heterosexual were given a test to determine their levels of homophobia and, based on their responses, were categorized as homophobic or nonhomophobic. They were then shown a series of erotic videotapes. While watching the videos, they were attached to a penile plethysmograph, a device that measures sexual arousal. Although all the men became aroused while watching videos depicting heterosexual or lesbian sex, only the homophobic men became aroused while watching videos of gay male sex. This study suggests that unconscious feelings of attraction toward other men may, in fact, act as a catalyst to homophobia and, perhaps, antigay hate crimes.

It is by now almost a truism in social psychology that our opinions and behavior are strongly influenced by such factors as the attitudes of others around or near to us, the norms of our group, and the relationships between our group and others. (p. 31)

With respect to hate crimes, the data support this view. As we have already discussed, most such crimes are committed by young people (who are especially susceptible to social influences) with a group of friends. Furthermore, most of these offenders are not hard-core, committed members of hate groups; rather, they are youths who are looking for excitement and wanting to fit in with what they perceive to be the norms of their social group. In Dunbar's (in press) recent study, only one quarter of the convicted hate crime offenders had clear signifiers of bias motivation, such as possession of hate paraphernalia or affiliation with a hate group, and only one quarter of the offenders had acted by themselves when they committed their crimes.

Indeed, recent research suggests that most hate crime offending may be a function not so much of the personality of the offender, but rather of the situation in which he finds himself. In this section we will explore some of the situational factors that might influence hate crime offending. It is worth noting again that no two offenders are alike and that these factors almost certainly play differing roles for each individual. This is supported by research that indicates that there are several typologies of hate crime offenders (Dunbar, in press; McDevitt et al., 2002).

The Influence of the Group: Conformity, Obedience, and Groupthink

As mentioned previously, almost all of the research shows that hate crimes are not usually committed by individuals acting alone, or by members of organized hate organizations, but rather by small groups of friends or associates (Craig, 2002; Dunbar, in press; B. Levin, 1993). This fact strongly suggests that social groups play an important role in explaining hate crimes. The fact that few offenders have any previous history of bias-related crimes particularly implies that in most cases, hate crimes may be due not so much to the offenders' own levels of prejudice as to the impact of group dynamics.

Much research has been conducted to try to explain why people sometimes do things when acting in groups or when influenced by others that they would not ordinarily do by themselves. Some of the early studies in this area were inspired by what could, arguably, be called the biggest hate crime of all: the extermination of 12 million Jews, Gypsies, homosexuals, handicapped people, and others during the Holocaust. How can we explain why most Germans and millions of other Europeans cooperated in genocide or at least failed to take any actions to prevent it? The answer seems to lie in the related constructs of conformity, obedience, and groupthink.

Solomon Asch (1956) conducted the classic set of experiments on conformity. In these experiments, a group of people was gathered, and each individual was asked to state which of several lines was the same length as a comparison line. This was not a difficult task. However, in reality, all but one of the people present were actually confederates of the experimenter and purposely gave the same wrong answer. Then it was the turn of the remaining person, the true subject of the experiment. Would he give the answer that his eyes told him was obviously correct, or would he give in to group pressure and give the same wrong answer as the others? Surprising even Asch, three quarters of the subjects responded incorrectly at least once, and more than one third of all the answers the subjects gave were incorrect.[3] These studies clearly indicated the power that social influence can have on individual behavior. However, even if a person is willing to give an obviously wrong answer on a perceptual task, does that mean that person will conform to pressure to take the much more serious step of committing acts of violence against others?

Stanley Milgram (1974) attempted to answer this question. In his experiments, subjects were told they were participating in a learning study and that their job was to "teach" a "learner" (actually a confederate of the researchers) a series of word pairs. The learner was taken into another room and, supposedly, hooked up to a device that would administer electric shocks to the arm. The teacher was placed in front of a large control panel and told that whenever the learner made a mistake, he was to deliver to the learner shocks of increasing intensity. The buttons that controlled the shocks varied from 15 volts (labeled "slight shock") to 450 (labeled "XXX"). The teacher could hear the learner but not see him.

At the beginning of the experiment, the teacher gradually increased the shock level with each of the learner's wrong answers. After a while, the learner began to moan or cry out with each shock. Then he demanded to be released. There was growing indication of pain and then a continuing, ominous silence. The subjects were not threatened or bribed to continue the experiment but, if they hesitated, were simply told by a white-coated official that it was important for the experiment to continue.

Milgram had asked 40 psychiatrists what they thought the subjects' responses would be. Those mental health professionals predicted that most would quit when the learner first asked to be let out, and only 1% would continue up to 450 volts (Aronson, 1999). In actuality, a stunning 62% of subjects continued to administer shocks right up to the end, even though they obviously believed they might be giving a lethal dose (Milgram, 1974)! The implication of these studies is that even ordinary people might be willing to commit serious acts of violence merely because they were asked to do so by an authority figure.

Another study on social influence, the famous Stanford Prison Experiment, was conducted by Zimbardo, Haney, Banks, and Jaffe (1974). In this experiment, they created a mock prison in the basement of the psychology department at Stanford University. Volunteer subjects, who were all male college students, were randomly assigned to be either guards or prisoners. The researchers then attempted to replicate, as closely as possible, the conditions of a real prison. Although the experiment was supposed to last 2 weeks, it had to be stopped after only 6 days because the "guards" had become increasingly cruel and brutal, whereas the "prisoners" had become fearful and servile, and some were experiencing severe psychological distress. The "guards" in this study were not ordinarily vicious people, nor were they encouraged to act that way by the researchers, but they had nonetheless become so because of their own perceptions of what their new social role required.

Taken together, these three studies, as well as numerous others they have inspired, indicate that human beings are deeply influenced by their social situations. More specifically, they show us that we are often willing to do things we would not otherwise have done, even to the extent of inflicting harm on others, if such behaviors seem to be socially desirable. Many hate crime offenders may be led to commit their crimes not primarily out of bias, but rather out of pressure or a desire to follow their group. Again, then, the term *hate crime* may be somewhat of a misnomer.

Why should the influence of groups be so great? A number of processes likely account for this. One of these is *deindividuation,* which may occur when a person feels anonymous. It has been shown that deindividuated people are more likely to act aggressively. For example, in one study (Zimbardo, 1969), student subjects who were dressed in loose-fitting robes and large hoods (not terribly unlike the traditional Klan regalia) gave longer and more severe electric shocks (as in the Milgram experiment, the shocks were not real, but the subjects did not know that) to another person than did subjects who were easily identifiable. As Aronson (1999, p. 214) concludes, "When a person is part of a crowd, he or she is 'faceless' and, therefore, takes less responsibility for his or her actions." Wearing uniforms or other identical gear (as do not only Klan members, but also neo-Nazis and other extremists) further decreases the differences among members of a group, and therefore increases deindividuation.

A second process that can occur in groups is *identification.* This occurs when a person finds a group attractive and wants to be like its members. As a result, that person assumes the group's attitudes and behaviors, although that person does not actually believe in those attitudes very strongly (Aronson, 1999). The strength of this process becomes obvious when one observes adolescents enthusiastically adopting their peers' style of dress, musical tastes, slang expressions, and so on. They also may adopt less benign conduct, such as smoking, alcohol and drug abuse, delinquency, and bigotry.

Third, a person can internalize the beliefs of others. This tends to happen when the others are not only admired but are also considered particularly credible. The others' attitudes become a permanent, important part of the person's belief system and one that is extremely resistant to change (Aronson, 1999). This may help explain why some people join hate groups: They initially engage in moderately racist thoughts or actions because doing so gains them acceptance from their peers, but when exposed to the teachings of a charismatic leader, they come to embrace racism as a core value.

Internalization can also occur because of cognitive dissonance, which we discussed briefly in Chapter 3. The primary tenet of cognitive dissonance theory is that people are driven to avoid inconsistency (dissonance) in their ideas and beliefs (cognitions). When they become aware of inconsistencies, they must either change one of the cognitions or create new ones to deal with the contradiction (Aronson, 1999). A corollary to this theory, which has been demonstrated in research on human aggression (see, e.g., Glass, 1964), is that when people inflict harm on others, they often self-justify their actions by denigrating their victims.[4] This may be particularly true when their harmful actions were done publicly, in view of others. Imagine, then, a young person who has engaged in verbal slurs, condemnation, or harassment of members of another group because he thinks doing so will gain him popularity or acceptance from his friends. Although he might not have been strongly biased to begin with, he may justify his behaviors by convincing himself that his targets *deserved* what he did to them. Those initially mild prejudices may become reinforced and, eventually, solidified. The study by Byers and Crider (2002) confirms this hypothesis. Young men who had committed hate crimes against the Amish told the researchers that for one reason or another, their victims had deserved what had happened to them.

A final process that can occur with groups is known as "groupthink," a term coined by Janis (1972). Under the groupthink model, groups sometimes strive so much to achieve or maintain consensus that dissent and critical thinking are strongly discouraged. As a result, groups of people often make poor decisions, even though the individuals involved may be aware that they are poor decisions and may not have reached the same decisions on their own. Although Janis (1989) has primarily used this theory to explain errors in judgment by policymakers and business leaders, it may apply equally well to the actions of hate crime offenders.

The research on social influence and group dynamics provides reasonable explanations for why some people commit hate crimes. To date, unfortunately, very little research has directly addressed these issues specifically with respect to group influence and hate crime offending. Franklin (2000) found that young adults who had committed antigay hate crimes were frequently motivated by their friends' antigay ideology. Byers, Crider, and Biggers (1999), in their research on young men who had committed hate crimes against the Amish, concluded that the offenders' families and peers often held anti-Amish beliefs. Again, this is another area of research that deserves more attention.

The Influence of Economics

Although little research has been done on the links between social influence and hate crimes, a fair amount of research has been conducted on the influence of another factor: economics. Specifically, a number of studies have tested the assumption that hate crimes are closely related to economic difficulties on the part of the offenders.

Conflict theory predicts that intergroup hostility arises when groups are in conflict for scarce resources (Campbell, 1965). For example, in their famous Robber's Cave experiment, Sherif, Harvey, White, Hood, and Sherif (1961) were able to successfully create a great deal of hostility between two previously collegial groups of Boy Scouts by setting the two groups in competition with one another for prizes and other desirable things.[5]

Similarly, scapegoat theory predicts that, when things are difficult, people will strike out at a convenient target. According to Allport (1954), it is perennially tempting for us to shift the burdens of our own misfortunes onto the backs of another. Allport also pointed out that, historically speaking, certain groups have made particular favorites as scapegoats, including Jews and blacks. An example of scapegoating occurred 2 days after the September 11 attacks when Jerry Falwell blamed abortionists, pagans, feminists, gays, and the American Civil Liberties Union (he later apologized; Gerstenfeld, 2002). And soon after the terrorist attacks, David Duke's Web site sported a large pop-up banner blaming the Jews.

The conflict and scapegoat theories help explain an interesting phenomenon: A significant number of hate crimes occur not between poor whites and poor blacks, but rather between two minority groups. Examples of this include the conflict between blacks and Jews in Crown Heights (discussed earlier in this chapter); between blacks, Hispanics, and Asian Americans in Los Angeles, New York, and Washington, D.C.; and between people of color and gays throughout the United States (Perry, 2001). Often, these groups are in competition for the second-lowest rung on the social and economic ladder. It makes sense that attacks by minorities might be targeted not at whites, whose position on that ladder is several steps ahead and perhaps viewed as unattainable, but instead at the attackers' closest rivals. In addition, as Perry (2001) points out, members of minority groups, too, have "internalized the dominant aspects of white masculine supremacy" (p. 134) and thus might choose white culture's traditional scapegoats.

Apart from the stereotype of racists as poor rednecks, there is some anecdotal evidence to support economics as an explanation for hate crimes. Money matters almost certainly played a part in the Klan persecution of Vietnamese shrimpers in Galveston Bay (discussed earlier in this chapter) and the well-publicized murder of Chinese American Vincent Chin by two unemployed autoworkers in Detroit in 1982 (the killers mistook their victim for Japanese American; Perry, 2001). And, in the Narrative Portrait at the end of this chapter, you will read about C. P. Ellis, who grew up in a desperately poor family during the Depression, just over the tracks from the black neighborhood. As an adult, Ellis, like his father before him, joined the Klan.

Ezekiel's (1995) work among young neo-Nazis also illustrates the potential link between poverty and hate crimes. All of the young men came from families that were working class at best, extremely poor at worst. Most of the men were unemployed, although a few had occasional low-paying jobs. They lived in Detroit, a city in which industrial employment, the mainstay of men of their class, was rapidly shrinking. They all lived in neighborhoods that were either largely or primarily African American, giving them plenty of opportunity to find handy targets for their anger.

Craig (2002) cites several examples outside the United States in which ethnic violence has been tied to difficult economic times in recent years. In Germany, China, France, Italy, and several former Soviet bloc countries, there have been high rates of unemployment and other monetary stresses that have frequently been blamed on immigrants, Jews, and others. During a recent trip to Hungary, which is still facing great difficulties in the transition from communism, I heard frequent resentment voiced against the Roma (Gypsies), who actually make up only a small proportion of the population. People told me that the Roma were to blame for Hungary's crime problems and that they put a strain on the government by having large families that they were unable to support. Roma teenagers at a high school I visited in the Czech Republic told stories of having been harassed and otherwise marginalized by their fellow citizens.

Economic themes frequently appear in hate literature and propaganda. In a content analysis my colleagues and I recently completed of extremist Web sites, half the sites mentioned economic issues (Gerstenfeld, Grant, & Chiang, 2003). Some of the claims were rational, if incorrect (e.g., "they" are stealing our jobs, "they" are costing us too much in welfare, etc.), whereas others were extremely far-fetched (e.g., kosher food certification constitutes an unfair tax on non-Jews, etc.). However, just because money issues are often discussed does not necessarily mean they are the root of hatred.

Perhaps the earliest controlled research on the link between hate crimes and economics was conducted by Hovland and Sears (1940), who found an inverse correlation between the price of cotton and the number of lynchings of African Americans in the South between 1882 and 1930. More recently, Medoff (1999) reported that hate crime activity decreases when market wage rates increase. Gale, Heath, and Ressler (1999) also found a complicated connection between economics and hate crime.[6]

Despite evidence to the contrary, it is not clear that bad economic times are, in fact, a major stimulus of hate crimes. Certainly, not all extremists are poor. For example, Vincent Bertollini, a Silicon Valley millionaire, was one of the biggest supporters of Richard Butler, the founder of Aryan Nations. When Butler lost his property in Idaho in a lawsuit, Bertollini bought him a new home. Bertollini and fellow high-tech millionaire Carl Story also founded a Christian Identity ministry called the 11th Hour Remnant Messenger. Other prominent white supremacists have also come from the higher socioeconomic strata: Richard Butler himself was once an aerospace engineer; the late William Pierce (leader of the National Alliance and author of the 1978 book *The Turner Diaries*) was a physics professor; and Arthur R. Butz, who is active in Holocaust denial, teaches electrical and computer engineering at Northwestern University. Among the leaders of the white supremacist organizations that Daniels (1997) studied, five of eight had college degrees, and two had graduate degrees. Aho (1990), in his study of Christian Patriots in Idaho, found that his subjects were actually slightly better educated than the average members of their communities.

Like Ezekiel (1995), Hamm (1993) also conducted extensive research among young members of organized hate groups in America. Hamm's (1993) research among skinheads found that, whereas most of them came from working-class backgrounds, they did not report particularly traumatic childhoods. Most of them were either currently enrolled in high school or college, or held blue-collar jobs. Thus, the stereotype of hate crime offenders as poverty-stricken, isolated, disaffected youth proved erroneous.

Recent research by Green and his colleagues also failed to find empirical support for a direct link between economics and hate crime. Green, Glaser, and Rich (1998) reevaluated Hovland and Sears's (1940) data on cotton prices and lynchings and concluded that there was no relationship between the two variables. In a second study, Green, Strolovitch, and Wong (1998) found that hate crimes in New York City neighborhoods were unrelated to the economic status of those neighborhoods. And when Green, Abelson, and Garnett (1999) examined responses to a survey given to residents of North Carolina, they found that members of extremist groups and hate crime offenders had only slightly more negative views pertaining to economics than did the other respondents. These studies led Ezekiel (2002) to question his own earlier (Ezekiel, 1995) conclusion that economic pressure is a major factor influencing whether youths become white supremacists.

The link between the economy and hate crimes is yet another area that needs additional research. Although there is some credible evidence that such a link exists, recent studies raise serious doubt. At the very least, the relationship appears to be complex, and we should question assumptions that hate crimes are a function of poverty or economic conflict.

The Influence of the Social Milieu

A variety of sociological theories have been proposed as explanations for hate crimes. Instead of focusing on individual differences, as do psychological theories, or exclusively on the offender's economic status, sociological theories concentrate on the entire social environment that surrounds offenders. Although the research in this area is enlightening, it suffers from a significant shortcoming in that most of it has been conducted among members of white supremacist groups. As we have already seen, only a small minority of hate crimes are actually committed by people affiliated with these groups. It is reasonable to propose that significant differences exist between avowed white supremacists and the "casual" or situational hate crime offender, so the generalizability of much of this research is questionable. Furthermore, there are important differences between extremist groups as well; this means that, for example, the data on Ku Klux Klan members is likely to be quite different from those on skinheads.

One theory that has frequently been used to explain hate crimes is strain theory. According to this theory, anomie occurs when people feel unable to reach a society's desired goals in a socially approved manner (Merton, 1957). Deviant behavior results. With respect to hate crimes, this concept is closely related to theories concerning the effects of economic difficulties but is broader. Strain theory does not require that the offender actually be living in poverty but rather that he have the subjective view that the offender or others like him or her are powerless and unable to fulfill the American Dream.

Again, there is a great deal of circumstantial evidence to support Merton's theory. For one thing, hate movements often have gained momentum during times of social and economic strife, even though the participants themselves are not necessarily directly affected by this strife. For example, Christian Identity, "patriot," tax protest, and militia groups (these groups are discussed in greater detail in Chapter 5) experienced large increases in membership in the American Midwest during the farm crisis of the 1980s (see, e.g., Corcoran, 1990).

Another piece of evidence to support strain theory has already been mentioned: Much extremist literature and propaganda centers on economic and related themes. David Duke's Web site, for example, claims that immigration costs Americans $30 billion a year, and one of the central goals of his "EURO" (European-American Unity and Rights Organization) is to fight affirmative action and similar policies.

A third piece of evidence is that many extremists, in addition to being opposed to ethnic and racial minorities, are also against anyone else they view as interfering with the privileges that should be accorded to white, Christian men. Misogyny is common; for example, the Father's Manifesto site (www.christianparty.net) contains, in addition to anti-Semitic material, exhortations to repeal the Nineteenth Amendment and to blacklist feminists. Homophobic invective is also widespread, often decrying gays' and lesbians' attempts to obtain "special privileges."

Furthermore, much of the literature on hate crime offenders has focused on their feelings of alienation and powerlessness. Dunbar (in press), Korem (1995), and Ezekiel (1995) all found that hate crime offenders tended to come from families that had histories of dysfunction and violence. Ezekiel (2002), especially, concluded that feelings of social dislocation and social isolation were two major factors influencing white supremacist behaviors.

Perry (2001), however, counters that strain theory does not adequately explain hate crimes. For one thing, who should feel strain more than women and minorities, who remain relatively disempowered and disadvantaged? Yet the majority of hate crime perpetrators are male and white. Moreover, hate crime offenders do not seem to feel powerless at all; rather, they appear to be interested in exercising and maintaining their power over others (Perry, 2002). In fact, maintenance of power is a major theme of white supremacists, who sport "White Power!" tattoos and recite David Lane's 14 Words: "We must secure the existence of our people and a future for white children."[7]

Perry (2001) also points out that hate crimes are frequently committed not by members of the underclass but rather by those who are privileged and socially powerful, such as college students and police officers. Offenders generally do not reject the authority of the status quo, but instead are hyperconformists who embrace it with extreme vigor (Hamm, 1993; Perry, 2001). Consider, for example, the words of former skinhead Thomas James Leyden, reflecting on why he allowed himself to be recruited into the movement: "I thought these were good guys, that I was being patriotic. I believed we were cleaning up America by drinking and fighting" (www.wiesenthal.com).

Two closely related sociological theories that might serve as more complete explanations for hate crimes are differential association and differential identification. According to differential association theory, people act in deviant ways because they were socialized to do so by a deviant subculture (Sutherland & Cressey, 1970). A modification of this concept, differential identification theory, states that people need not interact personally with those on whom they model their behavior but instead may choose models presented to them by the mass media. Furthermore, after they have chosen these models, they may be given rewards for doing so, such as the approval of others, notoriety, and material goods such as T-shirts or music CDs (Hamm, 1993).

In the context of hate crimes, then, offenders may learn to be bigoted from family, friends, and others around them, who socialize them to believe that prejudice and action against outgroups is acceptable behavior. Alternatively, these offenders might be influenced by television, music, movies, the Internet, and so on.

Although very little research has been conducted specifically to test this hypothesis, once again there is circumstantial evidence that supports it. As we have already seen, several decades of social psychological research tends to show a relationship between biased parental values and prejudice on the part of children. We have also seen that most hate crimes occur while the offenders are among groups of friends, and these friends might very well be a major source of biased attitudes. J. Levin and McDevitt (1993), for example, argue that many cases of hate crime involve small groups of people who were initially relatively nonbigoted, following a highly prejudiced leader.

Some research also suggests the power of friends in persuading people to join hate groups. Ezekiel (1995) concluded that most of the neo-Nazis he studied became involved because their friends recruited them. Similarly, nearly two thirds of the white supremacists in McCurrie's (1998) sample were recruited into hate groups; this cause of joining differs from that of more traditional gangs, in which youths tend to seek gang membership for protection or to make money. Several authors have emphasized the fact that extremist groups rely heavily on their recruitment efforts, especially among young people, to maintain memberships (Aho, 1988; Blazak, 2001). Finally, Turpin-Petrosino (2002) found that among secondary and college students, word-of-mouth contacts with white supremacists were more common than any other type of contact and were associated with stronger support for the groups than other types of contact.

Hero worship of a particular sort certainly seems to be an important facet of the white supremacist movement. Meeting places and gatherings frequently are adorned with depictions of racist luminaries such as Adolf Hitler and Adolf Eichmann, and white supremacists of many ilks wear Nazi uniforms and insignia. It is also common for them to adulate other champions and martyrs of white supremacy, such as Randy Weaver (whose wife and son were killed by federal agents in Idaho in 1992), William Pierce, George Lincoln Rockwell (founder of the American Nazi Party, who was murdered in 1967), and Bob Mathews (founder of The Order, who died during a standoff with the FBI).

It is not difficult for would-be bigots to find role models within the mass media. Ezekiel (1995), for example, reported that the neo-Nazis he studied gained most of their knowledge about nazism from watching movies about World War II. Nearly a century ago, the Klan enjoyed its largest membership, in part due to the popularity of the film *Birth of a Nation* (Griffith, 1915), the first full-length motion picture. In the movie's depiction of a white woman whose virtue had been compromised by a black man, it glorified the Klan.

In the 21st century, hate on the Internet is commonplace (Gerstenfeld et al., 2003). Estimates of the number of hate-related Web sites range from several hundred (Franklin, 2001) to more than 2,000 (Simon Wiesenthal Center, 2002). There are also extremist electronic mailing lists, chat groups, and the like (see Box 4.6). The extent to which these Internet sources appeal to people who are not already hard-core bigots is uncertain. In a survey by Turpin-Petrosino (2002), only 10 of 567 high school and college students reported online contact with white supremacist groups, but this may be an underestimate because Leets (2001) found that students did not necessarily recognize the true nature of extremist Web sites. One need not be a fan of old war movies or seek out extremist Web sites to be exposed to bigotry, however; as I discuss in the next section, biases of all types are a significant segment of mainstream culture.

Box 4.6   Hate Online

Hate groups have long used computers as a method of communicating with current and potential members. In the early 1980s, long before most Americans had heard of the Internet, groups such as White Aryan Resistance had created electronic bulletin boards (Hamm, 1993). As the Internet became a popular part of American society, extremist groups made frequent and enthusiastic use of it (B. Levin, 2002).

Hate groups can use Web sites and other online features to communicate with members, to recruit members, to spread their doctrines to a wider audience, and to sell merchandise. Moreover, the Internet has advantages over more traditional methods such as fliers and telephone hotlines. It is fast and inexpensive; it enables connections with a very wide audience, including people in other countries; it permits groups to attract young people with multimedia content and, sometimes, intentionally misleading information; and it allows groups careful control over their own images (Gerstenfeld et al., 2003).

By some counts, there may be more than 2,000 extremist Web sites. Getting a count and an accurate picture of their content is difficult because of the fluidity of the Internet in general and because these types of sites have a particular propensity to quickly appear and disappear.

Some people have called for these sites to be shut down, and some major internet service providers, such as Yahoo and Geocities, refuse to host sites that they consider unacceptable. Other people argue, however, that extremists have the constitutional right to express their views. Some Web hosting services refuse to censor the content of the pages they host. A few, such as Stormfront, which is run by former Ku Klux Klan leader Don Black, were actually created specifically for the purpose of hosting and promoting white supremacist sites.

A few antiracist organizations have taken a creative approach to combating hate online: They engage in "cybersquatting." Under this practice, they acquire a domain name that is similar to the name of a hate group or that is otherwise associated with racism. For example, a visitor to any of the following sites will find antiracist content:

www.aryan-nations.com

www.whitepride.org

www.14words.org

www.WCOTC.net (World Church of the Creator)

The idea behind this practice is that it allows an antihatred message to be spread while still avoiding censorship and protecting First Amendment rights.

As predicted by Hamm (1993), involvement in hate crimes and related activity may bring social benefits. I have already mentioned the potential impact of approval by friends and peers, which is an important reward at all ages, especially during the teenage years. Offenders may also gain greater notoriety: Appearances by white supremacists on talk shows were popular for a while, and some offenders, such as Timothy McVeigh and the Williams brothers, seem to have deliberately set out to make examples or martyrs of themselves. More tangible benefits are possible as well. In our survey of extremist Web sites (Gerstenfeld et al., 2003), nearly half of the sites contained multimedia content, such as free music downloads. More than half of the sites sold some kind of merchandise, such as T-shirts, stickers, video games, and CDs.

If the sociological theories just discussed are accurate, it might be the case that many hate crime offenders simply happened to associate with people who were already strongly prejudiced. Hate crime offending, then, might be more a matter of luck than personal, economic, or social circumstances. This might have some important policy implications when it comes to hate crime prevention.

There is a major weakness with using differential association and differential identification to explain hate crimes, apart from the lack of direct empirical evidence. Neither theory explains why some youths choose to identify with bigoted models, whereas most do not. Most young people today have access to the same array of role models in the mass media. Why do some find Nazis and other racists so appealing? The answer to that question might lie in the factors we discussed in the previous section; individual differences in personality may play a part.

The Influence of Culture

In her thoughtful critique of existing theories of hate crimes, Perry (2001) makes an important point: It is a mistake to think of hate crime offenders as deviants, because in our culture (and in many others) bigotry is anything but deviant. She argues, "Hate crime is not abnormal; rather it is a normal (albeit extreme) expression of the biases that are diffused throughout the culture and history in which it is embedded" (Perry, 2001, p. 37).

Does this seem extreme? American history is replete with examples of government-sponsored or -sanctioned violence and discrimination on the basis of race, religion, gender, sexual orientation, and so on. Within the last 2 centuries our country enslaved African Americans and committed genocide against millions of Native Americans.[8] Too far in the past? Women were denied the vote until 1920 and were not allowed to serve on juries for several decades more. Immigration, land ownership, and citizenship were limited on the basis of national origin. Japanese Americans had their property confiscated and were interned in camps during World War II. Legalized segregation continued well into the lifetimes of most living Americans. Still too distant? This very minute, homosexuals are not allowed to serve in the military if they are open about their sexual preference, and sodomy remains illegal in several states. Shortly after the September 11 attacks, in several separate incidents, FBI agents interrupted prayer services at mosques and recorded the license plate numbers of those who were attending Muslim religious services (Ibish, 2001b). One member of Congress said during a radio interview, "Someone who comes in that's got a diaper on his head and a fan belt wrapped around that diaper on his head, that guy needs to be pulled over" (Ibish, 2001a).

Even more common than government-sponsored bias, however, is that depicted in the mainstream mass media (see Box 4.7). Of course, hard-core white supremacists can listen to hate-filled music by such bands as Skrewdriver, Angry Aryans, and Nordic Thunder. However, very few teenagers are likely to buy the CDs of such groups. On the other hand, Eminem's latest offering is currently number 50 in sales rank at Amazon. One song ("Drips") contains these lyrics: "Now I don't wanna hit no woman, but this chick's got it comin', someone better get this bitch, before she gets kicked in the stomach/and she's pregnant, buts she's eggin' me on, beggin' me to throw her off the steps of this porch/my only weapon is force and I don't wanna resort to any violence of any sort." Besides their frequent misogyny, Eminem's songs are also infamous for their homophobic content.

Box 4.7   Bias in the Media

Many of our perceptions and beliefs about others are influenced by what we have seen on television, in movies, and in other forms of mass media. Unfortunately, research strongly suggests that these sources tend to perpetuate false stereotypes. Stereotypes in the media are probably as old as the media themselves. The first full-length motion picture, *Birth of a Nation* (Griffith, 1915), glorified the Klan, and the first talking movie, *The Jazz Singer* (Crosland, 1927), featured Al Jolson in black-face. To some extent, the situation has improved in recent years. Few modern film-makers would cast a servant character like those played by Stepin Fetchit, Aunt Jemima has been given a modern makeover, and Native Americans are now gener-ally treated more sensitively than in the cowboy movies of the 1950s. However, bias is still alive and well in the media.

A number of studies have been conducted on the depiction of crime in the news and on crime dramas. Overall, these studies agree that minorities are disproportion-ately shown as the perpetrators of crime, especially violent crime. Moreover, whites are disproportionately portrayed as crime victims (see, e.g., Dixon & Linz, 2000; Dorfman & Schiraldi, 2001; Mastro & Robinson, 2000; Oliver, 1994; Weiss & Chermak, 1998).

Although African Americans, and African American men especially, are frequently associated with crime and violence, this is not the only group that is portrayed nega-tively. People of Middle Eastern or Arab descent are often shown as terrorists or the like (Madani, 2000). In fact, after the bombing of the federal building in Oklahoma City, members of the media almost immediately speculated that the perpetrators were Arabs, and law enforcement agents began stopping and questioning people in the area who appeared to be Middle Eastern (Gerstenfeld, 2002). Of course, as we know now, this mass murder was committed by white, Christian Americans.

Racial and ethnic stereotypes are not the only stereotypes that appear in the media. Women are overrepresented in commercials for domestic products and underrepre-sented in others (Bartsch, Burnett, Diller, & Rankin-Williams, 2000; Hurtz & Durkin, 1997). A recent orange juice commercial is typical. A woman is shown frantically trying to feed her children breakfast and help them get ready for school while also talking on the phone. Her husband, meanwhile, sits at the table, hidden behind a newspaper and seemingly oblivious to the domestic crises. Another recent commer-cial even makes light of this pattern, showing two men discussing toilet cleaning products while they watch sports on television; the humor of this ad lies in the presumed improbability of men engaging in such a discussion.

Of course, much of the bias to which we are exposed is subtle. Sports fans might watch a game involving the Chiefs or the Redskins. Those who feel they have gotten a bad deal might complain about getting "gypped" or "Jewed down." A high school student wishing to express that something is undesirable might say, "It's so gay!" We might turn on the television almost any time to see young black men depicted as gangsters, Middle Eastern people as terrorists, and gay men as effeminate. In advertising, women are shown as sex objects if they are young and, if they are a little older, they become domestic drudges, responsible for cleaning toilets, doing laundry, and cooking meals. All of these things and many more reinforce our stereotypes, often without us even being aware of it. They also lead to a situation in which bigots believe that their beliefs are mainstream and acceptable.

Arguably, in fact, the beliefs of bigots really are not so very far from the ordinary. Consider, for example, the official platform of one Klan group, the Texas Knights of the Ku Klux Klan (www.texasamericanknights.org/platform.html). The Texas Knights call for returning to traditional Christian values, getting tough on crime, outlawing abortion and gun control, limiting welfare spending, reforming education, allowing prayer in schools, limiting the rights of homosexuals, restricting immigration, making English the official (and only) language, eliminating affirmative action, lowering taxes, and limiting foreign aid. This sounds remarkably similar to the platforms of mainstream political parties, especially those on the more conservative side of the political spectrum (see, e.g., the official Republican Party platform at www.gop.org/gopinfo/platform).

Despite the fact that many people assume we are living in an enlightened era, bigotry is still alive and well in the minds and hearts of many Americans. As discussed earlier in this chapter, we are often less willing to admit to it openly today, so it has become harder to measure, but it has certainly not disappeared. In a poll commissioned in the spring of 2002 by the Anti-Defamation League (ADL), 17% of respondents were strongly anti-Semitic, and an additional 35% were moderately so (ADL, 2002a). Anti-Semitic attitudes were also correlated with intolerant attitudes in general. Feelings against other groups run even stronger: Herek (2002b) reports that in 1998, 54% of Americans surveyed felt that homosexual behavior is always wrong and, in 1999, 44% of Americans surveyed considered homosexuality an unacceptable lifestyle. In Franklin's (2000) study, 10% of high school and college students reported having physically assaulted or threatened homosexuals, and another 23% reported having called them names. Of the remaining students, nearly one quarter had witnessed verbal or physical abuse committed by friends. Morris (1991) found strong evidence of anti-minority sentiment among college students, especially those living in fraternities and sororities. For example, 24% of those living in Greek housing responded that interracial marriage was clearly wrong, and an additional 20% had mixed feelings (Morris, 1991).

Many commentators have noted the relationship between prejudice in society at large and the commission of hate crimes. Petrosino (1999) points out that extremist ideology has gained growing acceptance in modern America. White supremacists have run in elections at local, state, and national levels and have sometimes been successful (the most obvious example is David Duke, who was elected to the Louisiana House of Representatives in 1989). Pat Buchanan's 2001 book is called *The Death of the West: How Dying Populations and Immigrant Invasions Imperil Our Country and Civilization.* In its arguments that Western (i.e., white, Christian) civilization is in peril due to immigration and high birthrates of non-Westerners, it closely reflects the dominant themes of most major white supremacist rhetoric.

Garland (2001) links the murders of gay serviceman Allen Schindler and gay textile worker Billy Jack Gaither[9] to antigay policies in the military and in state and federal governments. He concludes,

> Throughout the United States, harassment, punishment, and other forms of intrusion into the love lives of gay and lesbian people are well-rehearsed as a matter of culture and policy. . . . For those Americans who spend life along the culture's most violent and primitive margins, brute force must feel like an effective way to participate in the culture's promotion of death of its gay citizenry. (Garland, 2001, p. 91)

Speaking of a somewhat broader context, Wang (2002) argues that most of those who commit hate crimes do so mostly out of the desire to fit in or obtain other personal benefits.

As we have seen, this argument is well supported by empirical evidence. The hate crime perpetrator, she states, will choose those social groups whom the social environment has marked as acceptable targets.

Perhaps Craig (2002) best summed up the role of societal prejudices in fostering hate crimes. She wrote, "Patterns of prejudice are normative within this culture, and linked to social and historical processes. For the cynic then, the question 'Why do hate crimes occur?' reasonably becomes 'Why don't hate crimes occur more frequently?'" (Craig, 2002, p. 92).

WHY DO PEOPLE JOIN ORGANIZED HATE GROUPS?

It seems that most people who commit hate crimes are "casual" offenders. Their biases are perhaps not that much stronger than most people's, and their offending is primarily a function of the situations in which they find themselves. But what about the exceptions to this rule, the hard-core bigots? What leads a person to be so firm in his or her prejudices as to join an organized hate group?

There is no clear answer to this question. Certainly, no single factor stands out as a clear basis for hate group membership. Even previous family membership in such a group is probably not a very good predictor; many current members were the first in their family to join, and the children of current members do not inevitably end up joining themselves. McCurrie (1998) found that among the white supremacists he studied, one quarter had a father who encouraged them to join, and one fifth had a mother who provided such encouragement.

As we have already seen, there is no clear profile of extremist group members. On the whole, compared with other people, they do not seem to have come from particularly dysfunctional families or economically disadvantaged households. In one study of 82 white supremacist gang members, McCurrie (1998) found that most had been involved with drug dealing, most came from single-parent homes, one third were bullied in school and two thirds were bullies, and a fifth had been forced against their will to have sex at some time in their lives. However, McCurrie's sample drew very heavily from prison gang and motorcycle gang members, and may not be representative of extremists in general. Drug dealing, for example, although typical of prison and motorcycle gangs, is not usual for most other extremist groups. In fact, because some of the groups are strongly affiliated with conservative Christian ideology, drug use is strongly discouraged.

What does stand out as one reads the research on hate group membership is the role of recruitment, as mentioned previously. People join these groups because they know someone else who already belongs. As Aho (1990) concluded in his work with Christian Patriots,

> Respondents rarely joined the movement because they saw it initially as compatible with their political interests. Indeed, many confess at first to have been revolted by the Identity message or by Constitutionalism. Rather, they "joined with" others already in the movement, and only later began articulating its dogma. (p. 187)

In other words, virulent racism does not cause hate group affiliation, but hate group affiliation causes virulent racism.

What appears likely is that particular individuals are especially vulnerable to hate group recruitment. Perhaps this is due to their personal circumstances—feelings of frustration or

NARRATIVE PORTRAIT

GROWING UP RACIST, II

Clairborne Paul ("C. P.") Ellis, like Larry L. King, grew up during the Depression. He lived in Durham, North Carolina, in the poor white section of town, just across the railroad tracks from the black neighborhood. His father worked at a cotton mill at night and painted houses by day. C. P. quit school in the eighth grade, worked a series of low-paying jobs, owned a service station, and eventually worked as a maintenance worker at Duke University. During the civil rights era, he joined the Ku Klux Klan.

When the schools in Durham were to be desegregated, C. P. was named to the desegregation committee, as was Ann Atwater, a local black activist. Surprisingly, the two found common ground and gradually became close friends. In 1996, Osha Gray Davidson wrote a book chronicling C. P.'s and Ann's pasts and ongoing friendship. In this excerpt from that book, Davidson describes C. P.'s childhood:

Paul Ellis [C. P.'s father] was descended from yeoman farmers forced into tenancy and then off the land altogether when tobacco prices plummeted, thrusting him and thousands like him into the new industrial society that was transforming the South. C. P.'s mother, Maude, was an emotionally distant figure, already defeated by hard work and poverty by the time C. P. was born. . . . Paul Ellis drank himself into oblivion every weekend, often turning violent when intoxicated. Years later, C. P. was still haunted by the memory of his father, drunk and wielding an ice pick on the porch of their house, while he huddled inside with his mother behind the locked door.

C. P. and his one sister were raised in chaos and poverty, as their parents had been, and the future held for them nothing more than it had for those who had gone before: a few years of schooling and then the mills. If they didn't die there, amidst the chattering machinery and the cotton dust, they could look forward to a brief and exhausted "retirement" before returning to the red Piedmont soil, their lives having slipped away, trivial and unnoticed. . . .

When he had been drinking, Paul Ellis would sometimes sit with his son on the porch telling the boy stories about the glory days before the Civil War. C. P. relished the tales about Jefferson Davis, Jeb Stuart, Stonewall Jackson, and all the other warriors who had fought to preserve the "Southern way of life" and who, said his father, had been defeated by outsiders and traitorous blacks. But the Lost Cause was not lost forever. Paul Ellis told his son that there was a secret society of Southern men who would rise up someday and restore Dixie to her former greatness. During one of those drunken sessions, Paul Ellis whispered to the boy that years ago he himself had been a member of this group. Did he sense a glimmer of doubt in his son's eyes? Perhaps. Who was Paul Ellis after all but a nearly illiterate linthead who barely provided for his own family and who drank himself to sleep almost every weekend night.

Ellis, though, had proof. He climbed unsteadily to his feet and went inside, returning with a memento of his more illustrious past. In his outstretched hands, he held the white robes and conical hat of the Ku Klux Klan. "Only the Klan looks out for the white man," Paul Ellis told his son, with the profound solemnity of inebriation. "You just watch. The Klan'll save this country."

SOURCE: Davidson (1996, pp. 63-69).

alienation, economic stresses, religious or political views, and the like—or perhaps they simply have a need to feel like part of a group. Many of us have this need; it is why we join clubs and fraternities and the like. Hate groups, in particular, are often very good at making new members feel welcome. And it takes no special skills for a member of a hate group to be told that he is valued, powerful, and superior. All it takes is for him to be the right race, ethnicity, religion, and sexual orientation.

Extremist groups tend to produce a large amount of propaganda: pamphlets, books, Web sites, music, cartoons, and so on. It is likely that the major consumers of all of this material are people who have already joined the groups. Why spend so much energy preaching to the choir? Because it is after people have joined that they need to be persuaded to buy into the extremist ideology, and then they need to be constantly reminded to adhere to it.

CONCLUSION

A moral to be drawn from all of the research and theories presented in this chapter is that there is really nothing abnormal about those who commit hate crimes. It is tempting to think of them as deviant or evil. In fact, doing so reinforces our own self-images as good people. Certainly, those who do commit these acts are responsible for the choices they make and should be held legally and morally accountable. However, we should not become too comfortable in our own ethical superiority because, by most accounts, the type of people who commit hate crimes are not "them"—they're "us."

DISCUSSION QUESTIONS

1. Is it surprising to you that hate crimes are a serious problem on college campuses? Why do you think this is the case? Are you aware of hate crimes on your own campus or others? How prevalent are noncriminal acts of harassment and prejudice on your campus? What can and should be done about this, and by whom?

2. Consider the typology of hate crime offenders created by McDevitt, Levin, and Bennett (2002). What categories do the hate crimes you have heard about fall into? Do you think that other categories might exist as well? How different do you think the overall breakdown of types might be in areas other than Boston—for example, more rural locales? If you were a social scientist, what research concerning hate crime types would you like to conduct?

3. In what ways do stereotypes affect how we perceive other people? The research suggests that the processes that underlie stereotypes are automatic but can be reversed through conscious effort on the part of the perceiver. Under what circumstances do you believe this effort will and will not be made? If you were a teacher, how would you design a curriculum that would encourage children to make this effort?

4. Consider the stories of Larry L. King and C. P. Ellis. In what ways was racism evident in their childhoods? What roles did their parents play? King ended up educated and a liberal, whereas Ellis dropped out of school and joined the Klan. What factors do you think might explain these differences, considering that there are a lot of similarities in their childhoods? Why do you think King later disavowed racism?

5. A major controversy in research on prejudice is where the cause of prejudice lies: in the individual personality or in the social factors that affect that person. Describe the evidence for each view. With which side do you find yourself more in agreement?

6. Many people assume that hate crimes are committed by poor, uneducated people. How well is this assumption supported by research? Why do you think this assumption persists?

7. One thing that stands out in nearly all the research on hate crimes, both those committed by members of hate organizations and those committed by those who are unaffiliated with such organizations, is the role of the group. Hate crimes, more than any other kind of offense, appear to be a group activity. Why do you think that is? At least one state (Oregon) further enhances penalties for hate crimes committed by more than one individual. In light of what the research suggests, would you support such a law in your own state?

8. Some of the research suggests that whether a particular person becomes a hate crime offender is largely a function of whom he or she knows—that is, those who happen to have friends or family who are inclined to commit such crimes may become inclined to do so themselves. If this is the case, what policies or programs does it suggest might be useful in preventing hate crimes?

9. Describe some of the subtle ways in which our culture depicts and reinforces biases. Pay careful attention to the ways in which advertising, mass media, and everyday speech portray members of different groups. Do you feel that, in the past few years, this situation has gotten worse, gotten better, or remained the same? What are some strategies you would suggest to reduce bias within the larger cultural context?

10. Why do people join hate groups? What is attractive to them about these groups? In what ways do you think hate group members differ from "ordinary" people? What are the policy implications of this?

INTERNET EXERCISES

1. The Implicit Association Test (IAT), available online, offers an opportunity for you to explore your own biases. You can take tests concerning age, race, and gender bias at https://implicit.harvard.edu/implicit/. At the Southern Poverty Law Center's site (www.tolerance.org/hidden_bias/02.html), there are additional tests on biases about sexual orientation, Native Americans, and other topics. Take a few of these tests. Are you surprised at the results? To learn more about the tests, read this page: www.tolerance.org/hidden_bias/tutorials/05.html. The following page contains a bibliography about the IAT: https://implicit.harvard.edu/implicit/demo/bibliographyarticles.html.

2. The U.S. Department of Education has a comprehensive guide to protecting students from harassment and hate crimes. You can download it at www.ed.gov/PDFDocs/harassment.pdf. Partners Against Hate has a Web site with extensive information on hate crime prevention among school-age children and youths: www.partnersagainsthate.org/. The Center for Substance Abuse and Prevention also has information for educators on dealing with hate crimes: www.health.org/govpubs/ms716/.

3. You can find out more about Zimbardo's famous Stanford Prison Experiment at www.prisonexp.org. Among other things, you can view a slide show of the experiment or read discussion questions related to it.

4. For an interesting exploration of racist depictions of African Americans in the media, visit the Jim Crow Museum of Racist Memorabilia, which has a Web site at www.ferris.edu/

news/jimcrow/menu.htm. Michigan State University's museum has an exhibit on images of immigrants. Some of the exhibit can be viewed online at http://museum.cl.msu.edu/RC/collection/appel/exhibit/exhibit_home.html. A Web site that explores media images of Native Americans is www.bluecorncomics.com/stertype.htm.

NOTES

1. The Southern Poverty Law Center ended up representing the Vietnamese Fisherman's Association in a lawsuit against the Klan. After a bizarre trial, during which Klan leader Louis Beam apparently attempted to perform rites of exorcism against Southern Poverty Law Center (SPLC) attorney Morris Dees, a federal judge enjoined the Klan from continuing its activities in Galveston (Stanton, 1992).

2. In November 2002, Benjamin Williams committed suicide in his jail cell.

3. When subjects were placed in a room alone and asked to complete the same task, there were virtually no errors.

4. Social psychologists study the tendency to denigrate victims as a part of cognitive dissonance theory. Sociologists also have studied this phenomenon, and it is one of the central tenets of Sykes and Matza's (1957) neutralization theory. Neutralization theory holds that one way offenders justify their behavior is by denying the harm that their actions cause.

5. Significantly, the researchers were later able to eliminate this hostility and restore friendship by creating situations in which the two groups were forced to cooperate to repair the camp's water supply and restart a stalled bus.

6. These researchers proposed a rather complex mathematical model for hate crimes. Among other things, they found that measures of envy and altruism were helpful in predicting hate crime rates. For a different perspective, see Dharmapala and Garoupa (2001); these researchers use an economic benefits analysis to explore when penalty enhancements are likely to serve as effective deterrents to hate crimes. See also Dharmapala and McAdams (2001); using economic analysis, they conclude that hate crime offenders, unlike other criminals, stand mostly to gain esteem. They also discuss the relationship between hate speech and hate crimes.

7. David Lane was a founding member of The Order, a terrorist organization that, among other things, murdered Denver talk show host Alan Berg and robbed an armored car in California. Lane was subsequently convicted of several federal charges and sentenced to 190 years in prison.

8. For an excellent analysis of historical events as hate crimes, see Petrosino (1999).

9. Twenty-two-year-old Allen Schindler was beaten to death in 1992 by two of his shipmates because he was gay. Billy Jack Gaither, 39, was beaten to death in 1999 in Alabama by an acquaintance and the acquaintance's friend.

Chapter 5

ORGANIZED HATE

I went to college in Portland, Oregon, in the mid-1980s. When I went downtown, I would frequently see small groups of people about my age—mostly men—idling on street corners and in parks. Their heads were closely shaven, and they wore Doc Martens boots, jeans, white T-shirts, red suspenders, and black bomber jackets. I knew that they called themselves *skinheads,* and I had heard that they were violent at times. They did look vaguely menacing as they watched passersby, but no more so than the punk groups with their dyed mohawks and face piercings, or the grunge kids in their dirty flannels and watch caps.

A few months after I left Portland to attend graduate school, three young men were dropped off at the Portland apartment of one of the men, Mulugeta Seraw, by a friend after having attended a party. Seraw's apartment was not far from where mine had been. Seraw, an Ethiopian native, had come to Portland to attend university. He also worked full-time and sent money home to support his family in Ethiopia. Suddenly, a car pulled alongside the curb. Out of the car came several skinheads—perhaps some of the same people I had seen downtown. For no reason other than the color of their victims' skin, the skinheads attacked Seraw and his friends. They bludgeoned Seraw's head with a baseball bat; he died several hours later (Dees & Fiffer, 1993).

Before this murder, when I thought of hate groups at all, the only things that came to mind were clichés: Southern good ol' boys in white robes and pointy hats, or Hitler wannabes in fake SS uniforms. But Seraw's killers fit neither of these stereotypes. And the killing had occurred not in some place like Mississippi or Georgia, nor deep in the inner city, but in a working-class neighborhood in Portland, my hometown. Clearly, there was much more to organized hate than I had imagined. In fact, the Southern Poverty Law Center (SPLC) estimated that in 2001, there were 676 active hate groups in the United States, with groups in every state except Maine and Vermont (SPLC, 2002a).

In this chapter, we will explore the world of organized hate groups. There exists a large variety of different groups, each with its own history and practices. However, the links between these groups are many and complex, and they tend to share many of the same beliefs. We will also look at the changing face of extremism, as these groups take advantage of new developments and technologies, and as many of them strive to recraft their public images. Specifically, I attempt to define the term *hate group*, explore several major American hate groups, and explain common hate group typologies and ideologies. The chapter then discusses possible reasons people join and leave hate groups, the activities hate groups engage in, and women's roles in these groups.

WHAT IS A HATE GROUP?

Problems in Defining Hate Groups

There is no simple way to define a hate group, and surprisingly few scholars have tried. Although many of the groups I discuss in this chapter share certain common ideologies, and there are intricate ties between many of them, these groups also differ in many ways. Whereas some, like Tom Metzger's White Aryan Resistance, are overt and even boastful of their bigotry, many others cloak their real views with code words and appeals to the mainstream. Many of the groups claim not to be hate groups at all. For example, Duke's European-American Unity and Rights Association claims to be dedicated to equal rights for all (and special rights for none), and many organizations maintain that they are simply devoted to white solidarity and white culture. "We don't hate anyone," they proclaim. "We simply love white people."

Whether a particular group is to be classified as a hate group is sometimes in the eye of the beholder. For example, the Anti-Defamation League, a Jewish organization, considers Louis Farrakhan's Nation of Islam to be a hate group. Furthermore, some militant white supremacists such as Tom Metzger have met with Farrakhan, agree with him on several points,[1] and sometimes even distribute Nation of Islam literature to their own followers. The Nation of Islam itself, however, maintains that it is devoted to peace and the brotherhood of mankind. Some white supremacists assert that organizations such as the Anti-Defamation League are the real hate groups.[2] Furthermore, some scholars argue that there are significant differences between white supremacist groups and groups such as the Nation of Islam or the New Black Panthers. Whereas white supremacists seek to protect traditional power hierarchies, the latter groups are composed of members of traditionally disenfranchised groups.

Another problem with identifying hate groups is that some organizations have certain factions that are clearly bigoted although other factions are not. A good example of this can be found in the antigovernment movement of the 1990s. Many within this movement opposed the U.S. government because they believed it to be a front for Jewish/Israeli interests (and thus they referred to it as the Zionist Occupational Government), and some claimed the supremacy of state or local power, including even the invalidity of the Thirteenth, Fourteenth, and Nineteenth Amendments.[3] Others within this movement, however, held a variety of different beliefs, including beliefs in racial and gender equality, and some leaders were even members of minority groups themselves (Berlet & Lyons, 2000).

It can be very difficult to distinguish hate groups from other extremist groups, such as terrorists. In fact, some authors assert that no such distinction exists—that hate groups are terrorist groups, and that hate crimes are terrorist acts (Hamm, 1993, 1998; Petrosino, 1999; Smith & Damphousse, 1998). The line between them, if it exists at all, is certainly very thin. Timothy McVeigh and Terry Nichols and the September 11 hijackers all destroyed symbolic buildings, with terrible attendant loss of lives. McVeigh and Nichols were loosely affiliated with neo-Nazi and antigovernment groups, and they were inspired by the white supremacist novel *The Turner Diaries* (Pierce, 1978), whereas the hijackers were affiliated with an Islamic extremist organization. Aside from the fact that one group was domestic and one foreign, was there any meaningful difference? Moreover, some U.S. extremist organizations have established links with foreign terrorist organizations, certainly to share information, and also perhaps to share funds and weapons (Gerstenfeld, Grant, & Chiang, 2003; Petrosino, 1999).

It may also be difficult to distinguish hate groups from street gangs. Anderson, Mangels, and Dyson (2001) argue that hate groups have much in common with traditional gangs in that both types of groups present domestic threats, strike fear in citizens, form a surrogate family, have established leadership patterns, use distinctive symbols, and have identifiable territories. McCurrie (1998), too, assumes that white supremacist groups are gangs. This may be accurate for racist prison and motorcycle gangs, such as the Nazi Low Riders and the Aryan Brotherhood, and, to some extent, for some skinhead groups as well. However, other groups, such as the Klan, Christian Identity organizations, and others, seem quite distinct from street gangs in several ways, including the activities in which they engage.

Finally, at times it may be difficult to distinguish hate groups from legitimate political bodies as well as political dissenters and revolutionaries. In the United States, for example, Petrosino (1999) makes a compelling argument that many historical acts committed by the government or under its authority would qualify as hate crimes. Examples include the intentional genocide of Native Americans, slavery, and the Japanese internment camps. Internationally, as we will discuss in Chapter 8, it would be problematical to distinguish hate groups from organizations such as the Irish Republican Army, Hamas, Kahane Chai, the Shining Path, and the Khmer Rouge.

It must also be emphasized that, although all white supremacist groups are hate groups, not all hate groups are white supremacist groups (see Box 5.1). As already mentioned, some would argue that the Nation of Islam and the New Black Panthers, both African American groups, are hate groups. The Nation of Aztlan is a Mexican American group whose message is largely anti-Semitic. Fred Phelp's Westboro Baptist Church is viciously homophobic. And the Father's Manifesto, although it espouses anti-Semitic and antiblack dogma, is primarily misogynist in its message.

Box 5.1   Nonwhite Supremacist Hate Groups

Although white supremacists make up the bulk of extremist groups, they do not have a monopoly on hate. Here is a sampler of other kinds of extremist groups.

Jewish Defense League—The JDL is a militant, right-wing, pro-Jewish organization head-quartered in Los Angeles. Although it denies that it is a hate group, it has allegedly threatened those whom it considers anti-Semitic. In late 2001, two JDL members were arrested for allegedly plotting to bomb a mosque in the Los Angeles area and Congressman Darrell Issa's office.

Kahane Chai—This militant Jewish supremacist group is opposed to Christians and, especially, Arabs. It has been declared a terrorist organization by the Israeli government. One member, Dr. Baruch Goldstein, murdered at least 40 people in a Palestinian mosque in 1994. Members have also been accused of several attacks on Palestinians in the West Bank. Many members are U.S. citizens.

Muslims of the Americas—This radical Islamic group has offices in the United States and Canada, and ties to a terrorist organization. It is anti-Semitic, anti-Christian, and homophobic.

Nation of Aztlan—This California-based group is pro-Mexican American. It calls for the return of the Southwestern United States to people of Mexican descent. The contents of its Web site, http://aztlan.net, are primarily anti-Semitic.

Nation of Islam—This African American group was founded in 1935 by Elijah Muhammad and is now led by Louis Farrakhan. Members of the group have spoken out publicly against whites in general and against Jews especially.

Radio Islam—This group is apparently headed by Ahmed Rami, a Moroccan native now living in Sweden. Radio Islam operates both a radio station and a large Web site. Its motto is "No hate. No Violence. Races? Only one Human race."; however, Radio Islam is extremely anti-Semitic. Its Web site, http://abbc.com, contains anti-Jewish rhetoric in a number of languages, and U.S. citizens are among its supporters.

Characteristics Shared by All Hate Groups

Despite all the problems inherent in identifying hate groups, there are some characteristics that they do have in common. First, despite whatever image they attempt to present, their viewpoints are bigoted. They are opposed to some group (or, much more commonly, groups) on the basis of race, religion, ethnicity, gender, sexual orientation, and so on. At least one of their primary goals is to advance their own interests at the expense of those they oppose.

Second, they are organized. The level of organization varies a great deal. Skinheads, for example, tend to form small groups with informal organization. Other groups, such as some Klan units, may have very formalized and complex structures. In any case, there is an acknowledged leader or leaders. Unlike the unorganized groups of friends and associates who commit most hate crimes, these groups have names and, usually, common symbols, uniforms, or traditions. They often even have official membership criteria, membership applications, and membership cards. They frequently have newsletters or other types of communication, membership lists, planned gatherings, and literature. They may engage in fundraising activities. They are like many other kinds of organizations—churches, clubs, fraternities, and so on—except that the primary common interest their members share is bigotry.

Characteristics Shared by Most Hate Groups

Although all hate groups, by definition, are bigoted and organized, there are also some characteristics shared by most, but not all, of these groups. The majority are white supremacist, and they tend to define *white* very narrowly. Jews, Arabs, Asian Indians, and Hispanics, for example, are not included within this designation, and frequently people of northern European ancestry are considered superior to people with ancestors from southern or eastern Europe. The majority of these groups also primarily espouse views from the very far right of the political spectrum, although occasionally more leftist positions are incorporated as well, such as environmentalism, feminism, and support of organized labor. For many, although not all, of these groups, religion plays a central role (see Box 5.2 for an example). This is often a particular brand of fundamentalist Christianity known as Christian Identity. On the other hand, some groups, such the World Church of the Creator, are pagan, a few are Jewish or Muslim, and some have no particular religious slant at all (other than, usually, anti-Semitism).

Box 5.2   The Westboro Baptist Church

The Westboro Baptist Church is based in Topeka, Kansas, and headed by the Reverend Fred Phelps. Its membership is small, consisting mostly of members of Phelps's large family, but it has received much media attention. Phelps has made statements against Jews, blacks, and other Christians, but the church's primary antipathy is toward gays and lesbians. It runs a Web site called "God Hates Fags" (www.godhatesfags.com). Included on its Web site is a graphic indicating the number of days that the church claims Matthew Shepard and Diane Whipple have been burning in hell. Shepard, of course, was the gay University of Wyoming student who was murdered in 1998, and Whipple, who was a lesbian, was mauled to death by her neighbors' dogs in 2001 in San Francisco.

The church has picketed numerous events and locations across the country. Its signs are often vicious, including "Thank God for AIDS." Among other things, members picketed at Matthew Shepard's funeral.

HATE GROUP TYPOLOGIES

It is sometimes useful to classify extremist organizations into discrete categories, and several attempts have been made to do so. Dobratz and Shanks-Meile (1995) discuss two main divisions within white supremacism: the right-wing racial movements, a category that includes some Klan groups and Christian Identity adherents, and the white resistance movement, which encompasses other Klan groups, skinheads, neo-Nazis, and others. Groups in the second category tend to be more militant, whereas those in the first category tend to focus on political change.

Kleg (1993) describes five kinds of white supremacist groups: neo-Nazis, skinheads, the Ku Klux Klan, the Identity Church, and the Posse Comitatus. These categories, however, are not truly discrete. Many skinheads are neo-Nazis, and Identity Church doctrine is very common among Klan and Posse groups. Kleg (1993) also overlooks some groups, including Holocaust deniers (B. Levin, 2001a) and racist militias (Cook & Kelly, 1999; Pitcavage, 2001). In a study my colleagues and I conducted of extremist Web sites (Gerstenfeld et al., 2003), we found that even expanding the number of categories to seven was insufficient, and we proposed that some white supremacist organizations fall under the general rubric of nationalist or white separatist groups. Other extremist organizations fall into none of these categories, such as the neo-Confederate and ultra-right-wing groups (e.g., the Council of Conservative Citizens and the John Birch Society). Of course, some extremist groups are not white supremacist at all, such as the Westboro Baptist Church and the Jewish Defense League.

There are a number of reasons why it is difficult to categorize hate groups. Frequently, membership and even leadership overlap, and separate groups may be affiliated with one another. For example, the skinheads who murdered Mulugeta Seraw had existed as a distinct group, East Side White Pride. Tom Metzger, who leads White Aryan Resistance (WAR), based in southern California, and his son John, who headed WAR's youth arm (called Aryan Youth Movement) during the 1980s, decided that the Pacific Northwest was fertile ground for their views. They sent two of their followers to Portland to try to organize the skinheads. While in Portland, they distributed WAR literature to skinheads and encouraged them to act violently. The WAR emissaries had been in direct contact with the East Side White Pride members who killed Seraw. Consequently, the Metzgers and WAR were sued for indirectly inciting the murder, and Seraw's family received a $12.5 million judgment.

Leaders within hate groups often bicker with one another about ideological or financial matters, and some break off from established groups to form their own. Tom Metzger once worked with the Knights of the Ku Klux Klan under David Duke, founded the California Knights, and then, disenchanted with the Klan's stance, which he felt was not hard-line enough, founded WAR. Duke himself later disengaged from the Klan to found the National Association for the Advancement of White People and, later still, the European-American Unity and Rights Organization. He has also been affiliated with both the Populist Party and the Republican Party. William Pierce once belonged to the American Nazi Party and, when its leader was killed, formed the National Alliance. Bob Mathews, with ties to both the National Alliance and Richard Butler's Aryan Nations, created a group called The Order. As Ridgeway (1995) points out, the interconnections between these groups are often complex and confusing.

Not only do many hate group leaders have multiple or successive affiliations, but a single name may encompass several disparate groups. The major example of this is the Ku Klux Klan. In the United States today, there is no single Klan; rather, there is a variety of separate organizations that all use some variation of the Klan name. Although all

Klans are racist, individual organizations may take very different approaches, with some being more aggressive and others maintaining images that are more polished and professional (Dobratz & Shanks-Meile, 1995). These groups often argue over which is the "real" Klan, and rancor between leaders of different hate groups can be high.

Categorizing and delineating hate groups is sometimes made even more difficult because of a practice known as *leaderless resistance* (B. Levin, 2002; Perry, 2000). This practice was first advocated by Louis Beam, a former Klan leader who later joined the Aryan Nations. Beam argued that there should be small "phantom cells," without formal leaders or ties to other groups. Such a plan would make it harder for the government or other enemies of organized racism to damage or eliminate the hate movement because any particular legal action would affect only a few individual people. This structure also makes it more difficult for government agents to infiltrate hate groups. Christian Identity leader Pete Peters has also been a frequent advocate of leaderless resistance (Blazak, 2001), as has racist activist Alex Curtis.

Even when groups are not formally affiliated with one another, and even though the leaders sometimes do not get along, basic ideologies are often the same (see Box 5.3 for examples of acronyms and phrases shared by many hate groups). Groups will sometimes share meetings and literature, and they do communicate with one another in various ways. In our study of extremist Web sites (Gerstenfeld et al., 2003), we found that more than 80% of the Web sites contained external links, and about half of the sites had at least one link to a site in a category other than its own. Links between groups may also cross international and ethnic borders. More than half of the sites in our study linked to international organizations. Quite a few white supremacist groups' sites linked to Islamic extremist sites, such as Radio Islam and Hamas, and Radio Islam's site included links to, among others, the sites of David Duke and Louis Farrakhan. The Nation of Islam site includes an anti-ADL text by (white) Holocaust denier Michael Hoffman.

Box 5.3   Hate Acronyms and Numbers

A variety of words, acronyms, and mottos pop up frequently among white supremacist groups. Some of these can be subtle. Target stores recently pulled from the shelves a line of clothing featuring the term "88"; store management had been unaware of its meaning. Here are a few of the more common ones:

14 Words: Coined by white supremacist David Lane, the 14 words are "We must secure the existence of our people and a future for white children."

4/20: Hitler's birthday

88: This stands for Heil Hitler, because H is the 8th letter of the alphabet.

RAHOWA: Racial Holy War

SWP: Supreme White Power

WPWW: White Pride Worldwide

ZOG: Zionist Occupied Government

Clearly, it is difficult and often confusing to talk about discrete hate groups. Nevertheless, different groups do exist, and there are interesting and important historical, ideological, and functional differences between them. As one author (Weinberg, 1998) points out, "For those interested in the politics of contemporary right-wing extremism, American soil is the equivalent of the Brazilian rain forest for botanists. One hardly knows where to look first, so great is the profusion" (pp. 17-18). The next section will introduce some of the major hate groups in the United States, beginning with the first organized group: the Ku Klux Klan.

MAJOR AMERICAN HATE GROUPS

The Ku Klux Klan

Hate groups exist only when there is something to oppose; if the official government policy is itself sufficiently racist or bigoted, there is no real need for other forms of organized hate. This was the case before the American Civil War, when slavery was permitted in much of the country and endorsed or supported by most of the rest.[4] With the end of the war, however, came the end of slavery. Blacks, especially in the South, began to make social, economic, and political gains. White Southerners felt threatened: The wealthy feared losing their source of cheap labor, the poor feared being dislodged from the second-lowest rung of the socioeconomic ladder, and all feared losing the power they had enjoyed over other people.

Berlet and Lyons (2000) point out that not only were whites anxious about losing control over blacks, but white men feared losing control of white women. American society in general, and perhaps Southern society especially, was highly patriarchal, with only men being permitted land ownership and voting rights. Women had very little opportunity to exert independent authority. White men worried that black men would "steal" white women away, even further depriving white men of their supremacy. This belief is evidenced by the overtly sexual themes of much antiblack rhetoric and behavior. After the Civil War, although white supremacists frequently raped black women (further asserting their dominance), they also frequently sexually mutilated black men. Lynchings were frequently justified on the grounds that the victims had had some kind of sexual contact, even minimally, with a white woman (Ferber, 1998). Almost 100 years after the Civil War, in 1955, 14-year-old African American Emmett Till was murdered in Mississippi for flirting with a white woman.[5]

Less than a year after Robert E. Lee surrendered the Confederacy in 1865, the Ku Klux Klan was born. Because the Klan is a secret society, many of the details of its history are disputed. The first meeting took place in a judge's office in Pulaski, Tennessee, probably in December 1865 (Quarles, 1999). The six founders had been officers in the Confederate Army. The name came from the Greek word *kuklos,* meaning circle, and its creators were most likely inspired by Greek fraternities; at least three of the founders had belonged to fraternities while in college (Quarles, 1999).

Some authors claim that initially, the Klan was just a social club that engaged in pranks and minor mischief. Chalmers (1965), for example, wrote,

> Their problem was idleness, their purpose amusement. . . . They met in secret places, put on disguises, and had great fun galloping about town after dark. They engaged in much horseplay, for which purpose the secret initiation was the focal point of their activities. (pp. 8-9)

It is unclear whether simple entertainment was really the original intention, because frightening blacks was always a goal. Klan members would cover themselves and their horses in white sheets, ostensibly because the superstitious blacks would mistake them for ghosts, but more likely because it preserved their anonymity and presented a more fearsome appearance than would men in ordinary clothes. Not incidentally, as discussed in Chapter 4, anonymity such as that provided by their disguises leads to the deindividuation effect, which results in an increase in the propensity for aggression.

Whatever its original purposes, the Klan very soon grew in both violence and popularity. Reconstruction began in the South, and military rule was imposed. The backlash among Southerners was severe. Flogging, mutilation, beating, and murder of blacks (and, occasionally, Northern *carpetbaggers* and Southern *scalawags*[6]) was common. It has been estimated that between 1868 and 1871, more than 1,500 black people were killed in Georgia alone (Berlet & Lyons, 2000). The Klan became a terrorist organization overtly dedicated to white supremacy, its activities similar to but more violent than the slave patrols before the war.

Klan units quickly arose throughout the South. There was nominally a complicated organizational structure, and in 1867 the many Klan groups had a meeting in Nashville, at which former Confederate General Nathan Bedford Forrest was named Grand Wizard. In reality, however, the command was mostly decentralized, and local units acted independently. The Klan claimed membership to be as high as 500,000 (Quarles, 1999). The leaders tended to be prominent men in their communities, but the members came from all walks of life.

It should be noted that although the Klan at this time was mostly a Southern institution, its beliefs were not—that is, bigotry against blacks and other minorities was common throughout the country. During the Civil War, for example, working-class whites in New York City took part in the Draft Riot, murdering perhaps hundreds of blacks (Berlet & Lyons, 2000). Working-class people in the North and West considered blacks to be economic competitors and refused them inclusion in labor unions. Blacks were barred from residency in California in that state's first constitution in 1849 (Berlet & Lyons, 2000).

In 1869, Grand Wizard Forrest issued a proclamation officially disbanding the Klan. Ostensibly this was because he felt that it was "becoming injurious instead of subservient to the public peace and public safety for which it was intended" (Quarles, 1999, p. 49). In reality, Forrest may simply have been attempting to absolve himself of any responsibility for Klan members' actions. In any case, Forrest and the central Klan headquarters had no real control over local units, and Klan violence only increased. Some state governments used militias to try to control the Klan, with mixed success. Congress held hearings and, in 1870 and 1871, passed a series of laws aimed at curtailing Klan activities (some of these laws are discussed in Chapter 2).

By 1872, the Klan died out. It is possible that this was due to governmental efforts. However, by then Reconstruction had ended, and the federal government had ceased to control Southern states. Blacks were swiftly ousted from positions of power, and the first in a series of laws was passed throughout the South to disenfranchise, disempower, and generally demean blacks. Thus, the Klan was no longer needed, because government-sanctioned racism was once again fully in place.

In 1905, a North Carolinian named Thomas Dixon, Jr., wrote a book called *The Clansman.* The book was made into a motion picture in 1915 by D. W. Griffith; titled *The Birth of a Nation,* it was the first full-length movie, complete with a score played by a 32-piece orchestra. It was hugely successful and was even given a special screening at the White House for the President, members of Congress, and the Supreme Court. By 1927, more than 50,000,000 people had paid $2 or more each to see the film (Quarles, 1999).

The Birth of a Nation (Griffith, 1915) depicted the South shortly after the Civil War. It romanticized Southern chivalry and demonized carpetbaggers and blacks. The Ku Klux Klan was shown heroically protecting the white race, especially white women. The same year the film came out, 1915, the Klan was reborn at Stone Mountain, Georgia. At Stone Mountain, a cross was burned. The original Klan had not burned crosses, but some other organizations had, and Griffith's film portrayed a cross-burning.

Five thousand members had joined by 1920, when Colonel William Simmons paid a publicity firm to recruit new members (each of whom, naturally, paid dues to Simmons). By the 1920s, the Klan boasted millions of members in all 48 states. President Warren G. Harding was sworn in as a member in a White House ceremony in 1921 (Berlet & Lyons, 2000).

Aside from the Griffith film, there were a number of factors that accounted for the Klan's great popularity at this time. One of these was the First World War, which, as wars often do, caused general xenophobia and rampant patriotism. At the end of the war, communism came to power in Russia, and violently anticommunist sentiment arose in the United States. Among other things, the Klan was strongly anticommunist. A second factor was the huge influx of immigrants to the United States. Unlike previous immigrants, these came from Southern and Eastern Europe and from Asia; many of the Eastern Europeans were Jewish. Anti-immigrant sentiment was strong throughout the country, especially in the Northeast and West.

A third factor contributing to the Klan's popularity were the new sciences of genetics and anthropology. Some Americans and Europeans began to argue that human beings were divided into a number of biologically distinct races, that these races determined individual characteristics such as intelligence and industriousness, and that some races were inherently (and inescapably) superior to others. These men claimed that the Northern European, or Nordic, race was the highest quality. Furthermore, interbreeding of races was to be especially avoided because it would lead to the degradation of the superior race. Jews, who were considered to constitute a separate race, were to be particularly feared, perhaps because their identity was not always obvious: A Jew could easily pass as "white" and could assimilate into white culture, but such a hidden menace would endanger the gene pool.

The activities of the U.S. government also encouraged Klan membership. Racist policies were common after the First World War, such as laws in California that prohibited Japanese Americans from owning land (Almaguer, 1994); the killings of hundreds of Mexican Americans in the Southwest;[7] and the forced sterilization of prisoners, the mentally ill, and the handicapped. The federal government supported groups that were opposed to "radicals" and "reds." In the Palmer Raids of 1919 and 1920, federal and local authorities arrested thousands of people who had been born outside the United States and who were suspected of being radicals. They were held without formal charges for long periods of time, forbidden from communicating with family or attorneys, and were sometimes threatened, beaten, and tortured.

Several other sources contributed to the Klan's huge surge in membership as well. These included the labor movement (much of which was led by immigrants and non-whites) and the emergence of religious fundamentalism. In addition, the new Klan often emphasized local concerns, such as education, law enforcement, and civic reform.

The new Klan was considerably different from the original. For one thing, it was no longer confined to the South. In fact, it was strongest in the Midwest and also very popular in parts of the West. It was more an urban than a rural phenomenon, and its members came from a broad range of socioeconomic classes. Although it was frequently violent, many Klan groups sought change through political and social means instead. The Klan was influential in electing politicians in many states, including Oregon, California, and Wisconsin (Chalmers, 1965).

The focus of the Klan had also changed—it was no longer primarily antiblack. Although it still hated blacks, it now also hated Jews, immigrants, Catholics, communists, people of color in general, and anyone else who was not considered 100% American. Klan groups devoted themselves to upholding what they considered to be appropriate codes of morality and were willing to attack anyone who violated those codes, regardless of race or religion. The Klan favored temperance, religious devotion, and "family values."

By 1925, the Klan began to crumble. This was due, in part, to anti-Klan activities and anti-Klan legislation (such as laws against wearing masks in public places; see Chapter 2). Some members became dissatisfied with the Klan's more violent behavior. There were also leadership conflicts, financial difficulties, and several scandals involving prominent Klansmen (Berlet & Lyons, 2000). Although a thousand or so members remained, primarily in the South, the Klan would never again enjoy such prominence.

The remaining Klan members continued some activities during the 1930s and 1940s, primarily focusing on their anticommunist efforts. They also forged some ties with German and German American pro-Hitler groups, with whom they shared much ideology. In 1944, the IRS presented the Klan with a bill for back taxes in excess of $600,000. Again the Klan was officially disbanded, although individual units persisted. For the most part during this period, however, Americans were preoccupied with first the Depression and then World War II, and little public support for the Klan remained.

The Klan's third incarnation began in 1954, when the Supreme Court held in *Brown v. Board of Education* that segregated public schools were unconstitutional. Once again, black Americans striving for civil rights inspired thousands of white Americans to lash back. Anticommunism was also widespread at this time, and this era saw the emergence of other ultra–right-wing organizations such as the John Birch Society and the Liberty Lobby.

Klan members engaged in protests, marches, and riots throughout the South in the late 1950s and 1960s. They also violently attacked blacks and civil rights activists. In Birmingham, Alabama, they bombed an African American church, killing four young girls. One Klan leader in Florida responded to the girls' deaths at a rally by stating, "They weren't children. Children are little people, little human beings, and that means white people. . . . They're just little niggers" (Ridgeway, 1995, p. 70). In Mississippi, Klan members murdered three civil rights workers. When local law enforcement refused to respond (and, in fact, was probably involved in the murders), the FBI sent in more than 150 agents,[8] and Congress passed the 1964 Civil Rights Act. One author (Quarles, 1999) states that between 1954 and 1968, acts of violence by Klansmen in the South occurred almost every week. At least 44 black churches were burned in the summer of 1964 alone (Nelson, 1993).

Although opposition to civil rights was a major Klan theme during this period, so was anti-Semitism. In fact, some Klan leaders taught that the civil rights movement was a plot run by communist Jews and that blacks were just the Jews' simple puppets. Of course, part of the implication was that blacks were too unintelligent to organize themselves. In Mississippi, the White Knights of the Ku Klux Klan waged a terrorism campaign against Jews, bombing synagogues and the homes of prominent local Jewish people (Nelson, 1993).

By the end of the 1960s, the Klan once again faded into near obscurity. This might have been prompted by wider acceptance of blacks' civil rights among the general American public. In addition, other social movements were at hand: the antiwar movement, the women's movement, the beginnings of the gay rights movement, and so on. There were also Congressional hearings on the Klan, as there had been a century before.

In the 1980s, together with a revival of conservatism in America, the Klan again experienced an upsurge in membership, but this one was relatively small. Racists still existed, of course, but now there were a wide variety of newer racist groups to join, including neo-Nazis and militias. Some racists, such as David Duke, rejected the Klan because they wished for a more polished, socially acceptable image than that associated with the Klan. Others, such as Tom Metzger, felt that the Klan was not militant enough.

Today there is no single Klan, but rather a number of groups that are unaffiliated with each other. In fact, as stated previously, some of these groups actively oppose one another and argue over which is the "real" Klan. One of the largest Klans today is the Knights of the Ku Klux Klan, run by Thomas Robb in Harrison, Arkansas. Robb strives for a clean-cut image, claiming that his Klan is a love organization, not a hate group. Although most Klans today have abandoned their anti-Catholic message, they remain opposed to people of color and Jews, and they are also antigay. Many Klan groups today are also strongly affiliated with the branch of racist, fundamentalist Christianity known as the Identity Church; I will discuss Christian Identity in "The Racist Militia Movement" subsection later in this section. See Box 5.4 for a discussion of some other major white supremacist groups.

Box 5.4   Some Other Major White Supremacist Groups

There are many white supremacist groups in the United States. Here are a few of the more prominent ones:

Aryan Nations—This group was founded by Richard Butler in the 1970s. It encompasses both neo-Nazi and Christian Identity teachings. It originally had a compound in rural Idaho but lost it in a lawsuit. It is currently headquartered in Pennsylvania, but followers have also created their own groups in Ohio and Idaho.

Council of Conservative Citizens—This group is headquartered in St. Louis and proclaims itself "The True Voice of the American Right." It has had associations with several prominent mainstream conservative politicians.

European-American Unity and Rights Organization—EURO was founded by David Duke. It claims to defend white interests in the same way that the National Association for the Advancement of Colored People (NAACP) defends black interests. It is antiblack and anti-Semitic.

National Alliance—William Pierce founded the National Alliance in 1974. Pierce died in July 2002, but Erich Gliebe succeeded him. This group is neo-Nazi and is especially anti-Semitic. It owns Resistance Records, the Internet's largest seller of white power music.

NSDAP/AO—This neo-Nazi group, based in Lincoln, Nebraska, is run by Gary ("Gerhard") Lauck. Lauck was imprisoned in Germany for distributing Nazi propaganda there, which is illegal.

The Order—This group, founded by Bob Mathews, was an offshoot of the National Alliance and Aryan Nations. Its purpose was to precipitate a race war in the United States, similar to that depicted in the *The Turner Diaries* (Pierce, 1978). Members murdered a Jewish radio talk show host and robbed several banks and armored cars. Mathews was killed in an FBI standoff in 1984, and other members of the group were imprisoned.

Third Position—This is not a single group but a philosophy shared by several—a neofascist viewpoint that advocates racial and ethnic separatism and anti-Semitism. It also adopts certain leftist ideals, including antielitism.

White Aryan Resistance—WAR is run by Tom Metzger from his home in Fallbrook, California. It is anti-immigrant, antiminority, and anti-Semitic. It espouses some Third Position views. WAR has made active use of various media and was responsible for attempts to organize skinheads in the United States.

Racist Skinheads

The history of the skinheads starts a century after the Klan's and on another continent. Unlike the Klan, the skinheads began as an unorganized youth cultural movement in England. Although it is misleading to think of the Klan as a single unified group, it would be even more erroneous to think of the skinheads in this way. White supremacy was not the skinheads' original goal. Over the years, however, skinheads became increasingly violent, and eventually the movement traveled across the Atlantic to the United States, where it developed ties with other white supremacists.

Although the precise origin of the skinheads is unknown, it is clear that it lies in the youth scene of 1960s London. At that time, London teenagers took part in a large variety of fashion and cultural cliques, including the teddy boys, the mods, the rockers, and the hippies. Each had a distinctive style of dress and music. Ridgeway (1995) and Moore (1993) state that the skinheads splintered from the mods, who dressed conservatively, listened to American black soul music, and cultivated coolness. Hamm (1993), on the other hand, implies that the skinheads developed more or less separately.

Whatever their precise origins, the skinheads deliberately invoked a working-class persona. Ironically, they were also heavily influenced by black Jamaican gangs known as *rude boys*. What both the white British working class and the black Jamaican delinquents had in common was an image of rebellion against powerful authority.

By the early 1970s, working-class people were suffering from economic difficulties as Britain's heavy industry faltered. Unemployment was high and wages were low. As Moore (1993) puts it, "Many English kids must have felt they were sailing on a rotten, sinking ship below whose deck they were performing—if they were working at all—drudge jobs" (p. 20). At the same time, new waves of immigrants were arriving, primarily from the West Indies and Pakistan. As in the United States, the newcomers became easy scapegoats to blame for financial woes. Race riots became common.

The original skinheads were not affiliated with Nazism. They were, as Hamm (1993) says, "proletarian, puritanical, chauvinistic, clean-cut, and aggressive" (p. 25). They valued their neighborhoods, beer-drinking, slam-dancing, and football (i.e., soccer). They wore their hair cut short or shaved, both in challenge to the hippies' long hair, and also as a practical asset in street fighting (longer hair can be grabbed during a fight). They also wore working-class clothes: heavy boots, jeans or workpants, and plaid shirts.

One way a group can define itself is in opposition to others. The early skinheads particularly disdained the hippies and frequently got into fights with them. Later, they turned on Pakistanis (this was known as Paki bashing) and gays (Dobratz & Shanks-Meile, 1997). Thus, the skinheads became antielitist, nationalistic, and ultramasculine. Gradually, the skinhead and Jamaican cultures evolved away from each other, and the skinheads became influenced by the punk movement. The punks, in turn, had a fascination with Nazism; Hamm (1993) states that there was even a British punk band called Elvis Hitler. The skinheads developed their own kind of music, Oi. A mixture of heavy metal and punk, Oi music often had overtly racist and white nationalist lyrics. One of the earliest and most popular Oi bands was Skrewdriver.

During the early 1980s, skinheads spread across Europe and became extremely violent. Hamm (1993) states that in England there were reportedly 70,000 attacks a year, including numerous murders. During this period, the skinhead movement migrated to the United States.

As was the case in Britain, the early 1980s were difficult years for the American blue-collar workers. Traditional heavy industry and factory jobs were swiftly disappearing and being replaced by low-paying, low-status service jobs. Thanks to Reaganomics, however, those on the upper rungs of the socioeconomic ladder were enjoying unparalleled prosperity. Many members of the working class found themselves surrounded by a culture of conspicuous consumerism and yet perceived no realistic way to achieve that for themselves.

The first two skinhead gangs in the United States, Romantic Violence and the American Front, were created in Chicago and San Francisco, respectively. They were both led by young American men with histories of violence. Neither group was directly affiliated with British skinheads, but they copied the British skinheads' style of dress and taste in music as well as their neo-Nazi and bigoted beliefs. Both groups disbanded rather quickly, but their former members went on to form groups of their own.

As I have already discussed, the Klan was in severe decline in this period. However, hate was not, as evidenced, perhaps, by the fact that this is the time in which many states began passing hate crime legislation. The most influential hate groups of the time were extremely militant: The Order committed at least two murders, bombed a synagogue, stockpiled weapons, and stole millions of dollars during armed robberies (George & Wilcox, 1996). This was also the time in which Tom Metzger founded WAR, in which the Aryan Nations peaked in popularity, and in which Christian Patriot organizations became popular (Aho, 1990).

Many of these other groups saw the potential for organizing skinheads into a powerful ally. Tom Metzger made particular efforts in this regard (Hamm, 1993), as did the Aryan Nations and the Nazi organization NSDAP/AO[9] (Dobratz & Shanks-Meile, 1997). Because of their youth and their violent tendencies, skinheads were portrayed as the foot soldiers of white supremacy. For the most part, however, skinheads remained only loosely organized. There were a few umbrella organizations, such as the Hammerskin Nation, but most skinhead groups were (and are) autonomous.

Compared to the Klan, the skinheads are more urban. Whereas the Klan today is most often in the Southeast and Midwest, skinheads are more often in the West, Northeast, and Upper Midwest. Skinheads still promote a working-class image, although many individual members come from solidly middle-class backgrounds. Many of them do not shave their heads or don traditional skinhead garb, perhaps so they will be less easily targeted by law enforcement (Dobratz & Shanks-Meile, 1997).

The media have played a major role in increasing the skinheads' notoriety. In the late 1980s, skinheads appeared on talk shows such as *Oprah* and *Geraldo* (in one famous incident, a skinhead broke Geraldo Rivera's nose), and in national magazines like *Rolling Stone* and *Time* (Hamm, 1993). The skins have also created mass media forms of spreading their word, including magazines, Web pages, and music. (See Box 5.5 for a summary of the major "classics" of white supremacist literature.) The Anti-Defamation League (ADL) estimates that Resistance Records, one of the primary sellers of hate music (it is owned by the National Alliance), will have $1 million in sales this year (ADL, 2002c).

According to the Southern Poverty Law Center (SPLC), in 2001 there were 43 skinhead groups in the United States (SPLC, 2002a), significantly fewer than the 144 counted in 1991, but more than the 30 groups in 1995 (Dobratz & Shanks-Meile, 1997). Thus, it appears that skinheads may be heading toward a second peak in popularity. The number of individual members is unknown.

Box 5.5   "Classics" of Extremist Literature

No matter what their specific affiliation is, extremist groups frequently cite, reprint, and sell several well-known writings. Among them are the following:

The International Jew: The World's Foremost Problem—This is actually a series of essays originally written in the early 1920s by Henry Ford (founder of Ford Motors). It discusses the threat of a worldwide Jewish conspiracy. Ford later apologized for his writings but still accepted a medal of honor from the German Nazi government in 1938. This book is now in the public domain, so it appears in its entirety on numerous Web sites, and can be reprinted by anyone who chooses.

Mein Kampf—The title means "My Struggle." It was dictated by Adolph Hitler in the early 1920s while he was in prison, and it details his thoughts on Aryan supremacy and his plans for Germany. Also now in the public domain, it remains a mainstay of neo-Nazis.

The Protocols of the Learned Elders of Zion—The Protocols, first published in Russian in 1897, was translated into English and brought to the United States in 1920. Translations now exist in numerous languages. It purports to be the proceedings from a secret meeting of rabbis in which the Jews plot to take over the world. Although it was proven as early as 1920 to be a hoax, anti-Semitic groups still publish it and treat it as authentic.

The Turner Diaries—This piece of fiction was written in 1978 by William Pierce, founder of the National Alliance. It tells the story of a guerilla army that starts a white revolution, and it depicts killings of race traitors and destruction of government buildings. It inspired the members of The Order, as well as Oklahoma City bomber Timothy McVeigh.

Dobratz and Shanks-Miele (1997) conclude that not enough is known about the skinheads. Some of what has been published is contradictory. For example, although Hamm (1993) states that skinheads are opposed to illegal drug use, Moore (1993) disagrees. Moore and Hamm also disagree on whether skinheads are antiauthoritarian and on their degree of alienation.

It is important to note that not all skinheads are racists. As mentioned previously, in its earliest days, the skinhead movement was heavily influenced by black Jamaican culture. The racist skinheads subsequently moved away from this, but the antiracists did not. The antiracist skinheads (sometimes known by the acronym SHARP, for Skinheads Against Racial Prejudice) actively promote tolerance and oppose bigotry. Members are of a variety of races and ethnicities. Like the racist skinheads, they tend to be working-class and antielitist, but they often adopt a far-left, rather than far-right, perspective. They listen to Oi music as well, but without the racist lyrics. Violence between racist and antiracist skinheads is common.

The Racist Militia Movement

If the skinheads were the racist media favorites of the 1980s, the militias were clearly the favorites of the 1990s. It was the bombing of the federal building in Oklahoma City on

April 19, 1995, that made most Americans aware of this movement, but its origins had existed long before that.

Right-wing paramilitary organizations have a long history in the United States. The early incarnations of the Klan were to some degree quasi-military, there were fascist groups prior to World War II, and, during the Cold War, there were anticommunist groups (Pitcavage, 2001). Modern racist militias had their strongest roots, however, in several convergent groups from the 1970s and 1980s, especially the Posse Comitatus and the Christian Patriots.

The Posse Comitatus and the Christian Patriots

The Posse Comitatus was founded in 1969. Literally, the name means "power of the county," and it refers to the group's core belief that all government should be centered at the local level, with the greatest power vested in the county sheriff. Thus, the Posse rejected the authority of state and, especially, federal governments. The Posse was originally prevalent in Oregon, California, and Idaho. However, during the farm crisis of the mid-1980s, Posse adherents teamed up with Christian Identity ministers to spread their word in the Great Plains states and Upper Midwest (Bennett, 1995). Among other things, Posse members believed that they did not have to pay taxes, and they were involved in several armed conflicts with federal agents.

Like the Posse, Christian Patriots distrust the federal government. They believe that the federal government is, in reality, part of a satanic conspiracy run by Jews to destroy white Christians. They staunchly defend their own interpretation of the U.S. Constitution; they believe in only the main text and the Bill of Rights, and reject all subsequent amendments. As implied by their name, Christian Patriots subscribe to fundamentalist Christian theology. The Patriots have been primarily located in the Pacific Northwest (Aho, 1990).

Neither the Posse nor the Patriots had any central authority or organization, but individual groups sometimes had contacts or affiliations with other like-minded organizations, such as the Aryan Nations. Posse and Patriot members were extremely ardent opponents of gun control and tended to embrace survivalism.

Contributors to the Rise in the Militia Movement's Popularity

By the early 1990s, several factors had combined to contribute to the sudden rise in popularity of the militia movement. There were two well-publicized, violent incidents in which federal agents were widely perceived to have overstepped the boundaries of acceptable behavior (George & Wilcox, 1996):

- In August 1992, federal law enforcement agents engaged in an armed standoff at the rural Idaho home of Randy Weaver, whom they accused of weapons violations. Weaver's 14-year-old son and his wife were killed by federal snipers. Weaver and a friend were also wounded, and a U.S. marshal was killed.
- In 1993, the Bureau of Alcohol, Tobacco, and Firearms surrounded the Waco, Texas, headquarters of a small religious sect known as the Branch Davidians. There was a gun battle, with fatalities on both sides, followed by a 51-day standoff. On April 19, federal authorities assaulted the building, which burned to the ground, killing 104 members of the group, many of whom were children.[10]

Although the racist right had had previous martyrs, these two events catalyzed many Americans in an unprecedented manner. In fact, both the Weaver homestead and the site of the Branch Davidian compound have become shrines of sorts (Ronson, 2002). Many people who had formerly entertained no particularly radical views began to severely question the government and its means and goals. Antigovernment feeling affected even the establishment itself with, ironically, many mainstream politicians running on an antigovernment (or at least new government) platform.

Another major factor in the proliferation of militias was the dual rise of political conservatism and religious fundamentalism. In 1994, trumpeting its Contract With America, the Republican Party won both houses of Congress. Shrinking the federal government (especially its social welfare programs) and protecting "family values" became major goals (Bennett, 1995). Meanwhile, members of the Christian Coalition were elected in numerous local races.

Militia membership was also spurred by the passage in 1993 of the Brady Bill, which banned some kinds of weapons and established a waiting period for others. Those who already distrusted the government saw this not only as a violation of the Second Amendment but also as a ploy to enslave Americans (B. Levin, 1998).

Finally, conspiracy theories in general were becoming popular in mainstream American culture (Pitcavage, 2001). Many Americans were willing to believe in elaborate government coverups about UFOs, "black helicopters," and the like. Many individuals who seemed otherwise perfectly sane accepted as true tales that sounded like paranoid delusions.

Characteristics of Modern Militia Groups

Like their predecessors, the militias that formed in the 1990s had no central authority but were instead run primarily at the very local level. They were quite diverse, and not all of them were racist (B. Levin, 1998). In fact, some members of some militia groups were members of minority groups themselves. What all the militias had in common was a deep distrust of the government. Conspiracy theories, especially those involving the government, abounded. Militias were also opposed to globalism, which they perceived as an attempt to achieve a "New World Order."

Most militia members had a particular aversion to any attempts at gun control. Most also distinguished between that which is legal (i.e., permitted by the current laws of the state and nation), and that which is lawful (i.e., permitted by certain natural or moral laws). It was the latter to which they were dedicated.

Militias often formed ties with other groups that shared similar views. For example, militia members sometimes had ties to antiabortion extremists (Clarkson, 1998), tax protesters, and common-law courts adherents (ADL, 1997). In some cases, these associated groups were racist. Right-wing racism appealed to some militia members because its belief in a government and economy controlled by a vast Jewish conspiracy fit in well with militias' own love of conspiracy theories and distrust of the government. Thus, they shared a hatred for the federal government, which they referred to as ZOG, the Zionist Occupational Government. Because many militia members were conservative Christians, Christian Identity doctrine was also attractive to them.

The Peak and Decline of Modern Militias

Militia membership peaked in 1996. Although it was impossible to get a reliable count of groups or individual members, some estimated that as many as 50,000 people belonged. Groups were thought to exist in almost every state (Pitcavage, 2001). The Southern Poverty Law Center estimated that in 1996, there were 858 militia groups (SPLC, 2002b).

NARRATIVE PORTRAIT

AT A KLAN RALLY, I

In 1976, Patsy Simms attended a Klan rally in Charleston, West Virginia. She later described the rally in a book titled *The Klan*. One of the speakers was 22-year-old Grand Dragon Bill Miller. The following is an excerpt from the book, which includes part of Miller's speech and part of the speech given by Miller's wife, Linda Miller.

"Since I can ever count Jews have corrupted, destroyed, and defaced ever'thing that they've ever brought their hands on!" he shouted. "And anybody that tells you Jews are God's chosen speaker or Christ was a Jew, I wanta meet 'em." . . .

A small circle of men and women hovered around the podium like moths attracted to a bare light bulb. At first they listened intently, and one robed woman leaning against a budding tree interspersed a hearty *"Aaaaa-men, brother!"* Encouraged, Bill Miller shifted into high gear. "Now let's go to the niggers!" He leaned over the railing, toward the increasingly restless audience. "Many of you people have daughters. Now what if your daughter'd come home one evenin' with a nice big black buck? Right now so many people scared of the niggers anyway, what difference does it make? You'd accept it. You'd have mulatto children. And it comes back again to the greatest creation—the white man—and the bloodstream is polluted. Who gives anyone this right? No one! Not even God himself!"

There were scattered claps and the lone *"Aaaaa-men!"* and Miller continued his racial hopscotch, finally bringing the Klan to the rescue. "Protest days are over," he warned. "They are a thing of the past. We get out and march, what do we get? We get nothin'! We are committed to do one thing, and that's force and force only!"

"Aaaaa-men, brother!"

Miller briefly aimed his wrath at bickering among brothers and sisters of the Klan before returning to his appeal for members. "If you feel as we do, come forward an' let us teach you how we feel an' how the Klan is goin' to ride to save America!" . . .

Bill Miller returned to the podium to introduce his wife, Linda, a porcelain-skinned woman with red hair and a red robe, who apologized that this was her first time to speak in public. "We need more women," she started in an angelic tone. "We need the men, but we also need the women behind the men, to do little things like we've done this evenin'. Stand over there an' cook an' take your money."

Her blue-gray eyes made a timid sweep of the audience, then focused fiercely ahead. "So let's talk about another thing. What about the niggers? What about our little chil'ren that's goin' to school? The things they have to put up with?" Her saccharine voice turned sassy. "Niggers are *not* like us. We're goin' somewhere, where they're not. We've got the brains. They have nothin'."

"Aaaaa-men, brother—I mean sister!"

SOURCE: Simms (1996).

However, the movement stalled and then quickly declined by 1997, in part due to a series of arrests of militia members. The more moderate members quit because they were disenchanted with the radical behavior of some militia members, and the more extremist members quit because they felt the actions were not radical enough. The movement has never completely died, though, and the Southern Poverty Law Center estimated that in 2001, there were 73 active militia groups in the United States (SPLC, 2002a). Although the focus of attention has moved away from militias in recent years, Pitcavage (2001) argues that policymakers should still be aware of their presence. He concludes,

> Although not matching the hardcore white supremacist movement in terms of the number of major acts of criminal extremism over the past 6 years, the militia movement has emerged as probably the second most dangerous threat for domestic terrorist acts in this country. (Pitcavage, 2001, p. 976)

HATE GROUP IDEOLOGIES

The previous section serves as an introduction to only a few of the major hate groups. In reality, these groups exist in a bewildering variety, with many permutations and combinations. Individual groups frequently squabble over specific beliefs, practices, and tactics; indeed, these squabbles frequently result in the genesis of new groups, as disgruntled members desert an established group to form their own. There are, however, certain basic principles that constitute the core ideology of almost all hate groups.

Power

The first of these principles and, arguably, the most important, is power. White supremacist groups, which constitute the majority of hate groups, are concerned with whites (especially white men) losing their traditional control over others. Other hate groups are usually dedicated to gaining some of that power, generally at the expense of other groups.

Interestingly, the theme of power is rarely overtly discussed within hate group literature and propaganda. It does, however, form the rallying cry for the white supremacist movement: "White *power!*" And its importance is strongly implied both by the circumstances that have led to hate group formation and by the indirect messages those groups convey.

As we discussed in the previous section, hate groups are created, and membership increases, when the dominant group feels that there is a threat to its dominance. In the Reconstruction era, Southern whites founded the Ku Klux Klan when the slaves were freed and when blacks began to exercise independent political and economic authority. When Reconstruction ended and that authority was once again denied to blacks, the Klan disbanded. It reemerged around the time of the First World War and again during the civil rights era, both times when there were realistic threats to white domination. Similarly, the skinhead movement arose, both in Britain and in the United States, among youth who felt that their own paths to social and economic success were blocked by others. The skinheads purposely adopted an intimidating image, and membership in that group became a way for these young men to assert influence over their own neighborhood at least. Even the antiracist skinheads oppose those they consider to be elitist and authority figures. The militias gained popularity during a time of great distrust of the government. Militia members felt that government was abusing the powers it held and was impermissibly appropriating rights to which individual citizens are lawfully entitled.

The theme of power is also implicit in much hate group doctrine. White supremacists teach that race exists as a biological phenomenon, that race determines individual character traits, and that some races are superior to others. Consider, for example, the following statement:

> The reason why Whites and East Asians have wider hips than Blacks, and so make poorer runners is because they give birth to larger brained babies. During evolution, increasing cranial size meant women had to have a wider pelvis. Further, the hormones that give Blacks an edge at sports make them restless in school and prone to crime. (Rushton, 2000, p. 14)[11]

In a similar vein, here is part of the "General Principles" of the National Alliance:

> Our world is hierarchical. Each of us is a member of the Aryan (or European) race, which, like the other races, developed its special characteristics over many thousands of years during which natural selection not only adapted it to its environment but also advanced it along its evolutionary path. Those races which evolved in the more demanding environment of the North, where surviving a winter required planning and self-discipline, advanced more rapidly in the development of the higher mental faculties—including the abilities to conceptualize, to solve problems, to plan for the future, and to postpone gratification—than those which remained in the relatively unvarying climate of the tropics. Consequently, the races vary today in their capabilities to build and to sustain a civilized society and, more generally, in their abilities to lend a conscious hand to Nature in the task of evolution. (National Alliance, n.d.)

A hierarchy of races strongly suggests that superior races should be in a position of governance and control over those races that are less capable. (See Box 5.6 for a summary of the biology of race.)

Box 5.6   The Biology of Race

The concept of race is obviously very important to extremists, and it is an idea with which all Americans are familiar. It might surprise you to learn that, from a biological standpoint, there is no such thing as race.

The theory that the human species could be divided into several separate races, based on skin color and geographic origin, was proposed in the 18th century. The number of races proposed by these early scientists (all of whom were European) varied from three to five (Daniels, 1997; Ferber, 1998). The idea of race was quickly adopted by Americans who favored slavery because the theory states that human groups are biologically different, implying that some groups are superior; thus, slavery is justified on "scientific" grounds.

In the early and mid-20th century, the idea of race was widely embraced by many Americans and Europeans, who proposed that there are many races, equivalent to subspecies. The German Nazis believed that, in general, white races were superior

(Continued)

> **Box 5.6 (Continued)**
>
> to nonwhite (Jews were considered to be a separate race), and that the *Aryan* race
> was superior to other whites. White supremacists today still accept this.
>
> Most modern scientists, however, have rejected the biological reality of race
> (Ferrante & Brown, 1999). The American Anthropological Association issued a
> statement in 1998 officially denying the existence of race as a biological entity. The
> "races" we are used to thinking of are simply one arbitrary method of categorizing
> groups of humans, who might just as well be classified by tooth shape or blood type.
> There is much more genetic variety within an individual "race" than between races,
> and an individual's "race" tells us nothing about his or her qualities or abilities.

White supremacists believe the power that is rightfully theirs, by virtue of their
superior race, is being stolen by others. They believe Jewish bankers and media moguls
conspire with a corrupt government to take that power. They believe blacks take it by force
through their criminal behavior. They believe nonwhites in general take it by immigrating
to the United States in great masses and by reproducing when they get here, thereby eventually taking power through sheer force of numbers.[12] They believe affirmative action
takes it by removing white males from their jobs and replacing them with less qualified
minorities. They believe race traitors take it by interbreeding with nonwhites, thus con-
taminating the superior Aryan gene pool. Over and over in white supremacist writings
appear these words: "We must take our country back!"

The theme of power often appears in the art and symbols employed by extremist
groups. Images such as fists, swords, and other weapons are common, as are illustrations
of soldiers and warriors.

Racial Separatism

Power may be the dominant underlying theme of hate group ideologies, but it is not
the only recurring subject. Racial separatism is also frequently discussed. The Aryan
Nations, for example, advocates for an Aryan homeland. Previously, this group wanted
to claim the Pacific Northwest as its own (Ridgeway, 1995). Perhaps now that Aryan
Nations headquarters have moved from Idaho to Pennsylvania (after it lost its Idaho
property in a lawsuit), this group wants the Northeast instead. Only whites of pure Aryan
(i.e., according to www.aryan-nations.org, "Anglo-Saxon, Germanic, Nordic, Basque,
Lombard, Celtic and Slavic") origin will be allowed. The National Alliance, on the other
hand, claims all of Europe, Australia, the southern tip of Africa, and the temperate zones
of the Americas. All nonwhite people and culture will be banned (which means, the
National Alliance Web site at www.natvan.com announces, that young inhabitants will
dance to polkas and waltzes instead of rock music). Generously, however, the National
Alliance will allow whites of a variety of cultures, including Celtic, Germanic, Slavic,
and Baltic.

It is not only the white supremacist groups that advocate racial separatism. Berlet
and Lyons (2000) point out that Tom Metzger and the followers of Lyndon LaRouche
have had ties with Nation of Islam leader Louis Farrakhan. All three of these parties
are united in their opposition to the ADL, and Metzger and Farrakhan, at least, would

apparently be pleased with independent white and black nations. One former skinhead admitted,

> I used to sell Nation of Islam books because of their philosophy. They want the same thing. We used to take their literature and it would say "Black" and we'd put "White." If it said "Nation of Islam," we'd take that off and put our name on it. The philosophy is the same. It's racial separatism. Blacks here, whites here. (Simon Wiesenthal Center, 2002b)

Not all hate groups advocate complete separatism. Some would permit a limited number of nonwhites in their country as long as those nonwhites conformed to white culture, stayed in their place, and did not intermarry with whites. Members of these groups believe they have the right to define what constitutes an American, and residency in their country can be permitted only on their terms. The Knights of the Ku Klux Klan, for instance, advocates that everyone "not satisfied with White Christian rules of conduct" should be repatriated to the lands of their ancestors.

Religion

A third theme that frequently plays an important role with hate groups is religion. Not all groups have a religious basis. For traditional National Socialists (Nazis), for example, religion might even be discouraged. Still, for many groups, religion is the bedrock of their belief system. In some cases, this religion is fundamentalist Christianity (e.g., the Westboro Baptist Church), radical Judaism (e.g., the Jewish Defense League), or Islam (e.g., Nation of Islam). Two religious sects are of particular importance to hate groups: Christian Identity and paganism.

Christian Identity

Christian Identity has its roots in a 19th-century doctrine known as British Israelism. This doctrine, which was first articulated by a man named John Wilson and was then popularized by Edward Hines, taught that the people who call themselves Jews are not the true Jews. Instead, God's Chosen People were the residents of Great Britain. Those who were known as the Jews were actually Khazars, people of Mongolian-Turkish origin (Coates, 1987).

British Israelism came to the United States in the early 20th century. In the mid-1940s, Wesley Swift founded the Church of Jesus Christ Christian in California. This church blended British Israelist ideas with zealous anti-Semitism and political extremism (Barkun, 1994). Its doctrine, which came to be known as Christian Identity, was adopted by several prominent racists of the mid-20th century, including Richard Butler, one-time director of the Christian Defense League and eventual founder of Aryan Nations, and William Potter Gale, cofounder of the Posse Comitatus. Today, Christian Identity theology is widely embraced by a large number of white supremacist groups, including many Klans. Randy Weaver was an Identity believer, as were members of The Order. Even some neo-Nazi organizations such as the National Alliance have accepted Identity beliefs to some extent.

Christian Identity is attractive to white supremacists because it puts a theological seal of approval on their ideas. One author has referred to it as the "theological glue" of white supremacism (Ostendorf, 2001). Identity teaches that not only are Aryans the Chosen People but that people of color, whom they term "mud people," are not even fully human. The precursor of these "mud people" was created by God before Adam; Adam's descendents (the Adamites) are white people. Another main tenet of Identity is that Eve had a sexual relationship with the serpent and that Cain was the offspring of this liaison. Jews are the descendents of Cain, and so are the literal spawn of Satan; they hold dominion over the mud people and use them to their nefarious ends. Abel was the result of Eve's relationship with Adam, and Abel's progeny are the true Israelites (Barkun, 1994).

Besides the fact that it provides rationale for their racist views, many white supremacists also like Christian Identity because it teaches that Adam's descendants are engaged in an apocalyptic struggle with Satan's seed. Thus, the white supremacist movement becomes not simply an effort for cultural and racial domination but a battle between Good and Evil. The movement's followers are literally doing God's work.

Non-Christian Religions

Christian Identity is the most prominent religious movement among modern white supremacists. It is not accepted by all, however, and some reject Christianity altogether. Some are attracted to Celtic or Norse pagan beliefs (Dobratz & Shanks-Meile, 1997). Not only do these beliefs fit in with their reverence for Aryan peoples but, because the warrior mythos permeates these beliefs, it reinforces white supremacists' self-image as soldiers for their race.

Another non-Christian religion tied to white supremacism is the World Church of the Creator. This church was founded by Ben Klassen in 1973; its leader since 1996 has been Matthew Hale. According to Creativity, the white race is the true Creator. The church's primary objective is "The Survival, Expansion and Advancement of the White Race," and its motto is RAHOWA (Racial Holy War). Creativity embraces much of the doctrine of Nazism and opposes people of color and Jews.

Although the World Church of the Creator claims not to advocate violence, several of its followers have committed violent hate crimes. In January 1999, Benjamin Smith was named a "Creator of the Year" for his efforts in distributing Church materials; in July of that year he went on a shooting spree against Jews and people of color, killing two and wounding nine. Benjamin and James Williams, the California brothers who burned three synagogues in Sacramento and murdered a gay couple, were influenced by Church literature and believed their actions would help precipitate a holy war. In 1993, eight people with ties to the Church and to WAR were arrested for plotting to bomb a black church and to assassinate Rodney King. Another member took part in firebombing an NAACP office in Tacoma (ADL, 2002g).

Common Antipathy for the Same Groups

Jews

Another factor that unifies hate groups is a common antipathy for the same people. Foremost among the hated groups are Jews (see Box 5.7 for a discussion of Holocaust denial, which is common among extremists). Chapter 6 discusses the roots and consequences of anti-Semitism in more detail. Briefly put, whether or not individual members of hate groups subscribe to the Identity tenet that Jews are Satan's spawn, they all agree that Jews are the architects of a vast conspiracy against whites. Jews, they say, control the government and the media. They are responsible for globalism and the excesses of capitalism, but also for communism. Other minorities, although dangerous in their own right, are menaces primarily because they are the pawns of the sinister Jews.

Box 5.7   Holocaust Denial

Numerous groups and individuals today are involved in a specific form of hatred known as Holocaust denial. They call themselves *revisionists* and, though they claim not to hate anyone, they also maintain that the Holocaust never happened.

There are different strains of this movement. Some admit that millions of Jews died but argue that this was a result of disease and malnutrition rather than purposeful genocide on the part of the Nazis. Others assert that the entire thing is a hoax by the Jewish people to obtain money and sympathy in furtherance of the worldwide Jewish conspiracy. They all deny that the gas chambers existed. Many Holocaust deniers portray themselves as impartial scholars and maintain that they have scientific evidence to support their claims.

Holocaust deniers have a number of Web sites and publications. Although most Holocaust deniers are American or European, very recently their message has been taken up by some Islamic extremists as well.

An excellent documentary on the work of some of these Holocaust deniers is *Mr. Death: The Rise and Fall of Fred A. Leuchter, Jr.* (Morris, 1999).

Although anti-Semitism is an ancient theme among bigots, recent events have increased its popularity or have at least given those bigots fresh ammunition. The events of September 11 were alleged by some hate groups to have been part of a Jewish/Israeli plot. The more rational groups, however, merely claim that September 11 is the direct result of Jewish-led American policies in the Middle East.

Anti-Semitism is espoused not only by white supremacists but often by other hate groups as well. As discussed previously in this chapter, it sometimes unifies groups that are seemingly extremely disparate. As the Aryan Nations Web site (www.aryan-nations.org) trumpets (above a link to its Ministry of Islamic Liaison), "War makes strange bedfellows." Aryan Nations (which is a Christian Identity group) declares that Muslims are Aryans' cousins. The National Alliance tries to have it both ways: It condemns the Jews' racial agenda, which leads to the aggressive bombing and killing of Muslims overseas, while, in the same paragraph, it decries immigration policies that allow Middle Easterners ("colonies of the belligerents themselves") to live in the United States.

People of Color

Of course, Jews are not the only people who are hated. These groups hate people of color in general, with special invective being aimed at blacks. The antiblack rhetoric tends to have three foci: (a) blacks are genetically inferior and so are less intelligent, less civilized, less industrious, and less capable than whites; (b) blacks are economic drains on white Americans, both because Affirmative Action allows them to steal whites' jobs and because they have an overreliance on welfare and other public aid; and (c) blacks are dangerous because they commit a disproportionate share of crime, especially violent crime, much of which targets white victims.

There is frequently also a sexual premise to antiblack writings that is mostly absent from anti-Semitic materials. Blacks, it is claimed, will steal your daughters, and the result will be mongrel children. Some groups even claim that if a white woman has a baby with a black man, any subsequent babies she has will be tainted and part black, even if they are fathered by white men! In other words, once a white woman is polluted through her contact with black men, that taint can never be expunged.

Interestingly, sexual relationships between white men and black women are rarely mentioned in antiblack writings (Daniels, 1997; Ferber, 1998). There are several possible explanations for this: (a) these relationships are not as important, because they will not contaminate the white gene pool; (b) it is assumed that white men, unlike white women, can be trusted to control their sexual impulses; (c) it is assumed that, whereas white women are the epitome of attractiveness and sexual desirability, black women are uniformly repellent; and (d) sexual relationships between white men and black women, unlike those between black men and white women, do not threaten white male dominance over all women.

Rhetoric against people of color other than blacks is fairly rare, although it does exist. The themes are similar: They are taking our jobs, committing crimes, and generally endangering white culture. WAR's Web site (www.resist.com), for example, has a section with cartoons about Mexican Americans, who are portrayed as illegal aliens who reproduce rapidly, join gangs, and rely on government handouts.

Immigrants

Anti-immigrant sentiment is extremely common among hate groups, who assume that immigrants are nonwhite. Some groups, such as the Knights of the Ku Klux Klan, focus primarily on illegal immigration, whereas others wish to stop or severely limit any kind of immigration. These groups claim that immigrants put a financial strain on Americans and that they endanger white culture. Some organizations are also concerned with enacting English-only laws and eliminating bilingual education. According to some watchdog organizations, the racist right not only spearheaded the anti-immigration movement in America but is still strongly involved (SPLC, 2002c).

Gays and Lesbians

Gays and lesbians are also frequently targeted by hate groups. Many groups, especially those espousing Christian Identity or other types of fundamentalism, claim that homosexuality is morally wrong. Others claim that it is simply unnatural. Sometimes homophobia is tied to AIDS fears. Groups are divided on what should be done to gays and lesbians. The most extreme groups call for executing them; others call for simply imprisoning them. The least extreme simply want to close the closet doors and ensure that gays and lesbians have no influence on public policy. It is interesting that, although antigay dogma is usually only a secondary theme for most hate groups (the Westboro Baptist Church being an exception), many of the individuals who commit hate crimes target gays. The official doctrine of a hate group is not always directly reflected in the actions of its members nor in the actions of unaffiliated people. Again, we will discuss antigay hate crimes in more detail in Chapter 6.

Common Antipathy for Particular Beliefs and Actions

In addition to hating the groups of people just described, hate groups are also often united in hating particular beliefs and actions. Most are vehemently opposed to abortion, and some violent abortion opponents have had ties to white supremacism (Clarkson, 1998). Eric Rudolph, who is suspected of a 1998 abortion clinic bombing that killed a police officer, also is believed to be involved in the bombing of a lesbian bar in 1997 and the 1996 Olympics bombing in Atlanta. Rudolph, who remains at large (he is currently on the FBI's list of Ten Most Wanted Fugitives), has become a hero to some extremist, survivalist, and conspiracy-theory groups.[13] Some hate groups, such as WAR, oppose abortion only in the case of whites and actually encourage it for nonwhites.

Other beliefs shared by many groups include a hatred for communism (which is often considered part of the Jewish conspiracy) and political liberalism. Some, but not all, are also opposed to feminism. We will discuss women's roles in hate groups in the "Women in Organized Hate" section later in this chapter.

HATE GROUP RECRUITMENT AND DEFECTION

There is a large body of literature on what makes people bigoted, and a number of people have conducted research on why individuals commit hate crimes. Surprisingly, however, little has been studied about how hate groups recruit members or why most people quit these groups fairly quickly.

Recruitment

Most hate groups produce a fair amount of propaganda such as books, fliers, magazines, Web pages, and so on. At first glance, it might seem that these are used to attract new members into the movement. As it turns out, however, this is probably not the case. As discussed in Chapter 4, and as pointed out succinctly by Bjørgo (1998), "Young people usually do not join racist groups because they are racists, but they gradually adopt racist views because they have become part of a racist group" (p. 234). The propaganda is not used to lure new members, then, but to indoctrinate existing ones. If it is not these groups' ideologies that draw members, what is it?

Hamm (1993) discusses at length the methods used by Tom Metzger to organize and enlist skinheads in his cause. Metzger's efforts were aimed at people who were already skinheads and thus were already involved in organized (or semi-organized, anyway) hate groups. What remains largely unanswered in this study is why these young people joined the skinheads in the first place. However, Hamm (1993) does provide case studies of four skinheads (three male and one female). All four were recruited into the groups via personal contact with someone who already belonged. These findings comport with Ezekiel's (1995) research on neo-Nazis in Detroit: Each of the people he studied had also been persuaded to join by a friend who was a member. Similarly, McCurrie (1998) concluded that most members of racist groups were recruited rather than taking the initiative to join on their own.

What sort of person is targeted for recruitment in a hate group? Perhaps one of the most complete studies to address this question was conducted by Blazak (2001) among skinheads in several U.S. cities, as well as in some European countries. He found that skinhead recruiters focused on schools and neighborhoods where there were already some skinheads in existence or where there was some perceived threat to straight, white males. Such places might include areas where there had been large layoffs, where white students had been victimized by minority gangs, or where multiculturalism had recently been introduced into the curriculum.

After a suitable location was found, Blazak (2001) reported, recruiters would direct their attention to those individuals who were experiencing strain as a result of feeling left out, frustrated, or harassed. The skinheads could serve as champions or surrogate big brothers for these disaffected youths. They offered structure to those who seemed to need it and a proposed program to solve their problems in life. Blazak (2001) concludes that skinhead recruitment is very similar to the strategies employed by cults.

Aho's (1990) study of Idaho Christian Patriots is consistent with Blazak's findings. There were no particular demographic features that distinguished Patriots from their nonmember neighbors, and the enlistment techniques used by the group were similar to those used by other social groups and movements. Aho (1990) concluded that social mobilization theory accounted for affiliation with these groups: Prospective members rationally determined that joining would bring them collective benefits, such as the preservation of their culture, as well as individual benefits, such as the opportunity to socialize (the group would offer potential members enticements like recreational and entertainment opportunities).

Aho's (1990) study also confirms the importance of personal ties in hate group recruitment. Most people were drawn into the group not through mass appeals like pamphlets and leaflets but through personal contact with individual members. Moreover, Aho (1990, p. 191) concludes that the members' primary interest was not being "dutiful Christian patriots" but rather "preserving, establishing, and expanding love relationships and family ties, friendships, and positive work environments."

Bjørgo (1998) also agrees that young people are usually recruited into racist groups by friends or family members. These recruits are searching for substitute families, for friendship, and for status and self-identity. Of course, a great many people seek out these things in their teenage years; hate groups seem to simply focus on those individuals who have not yet attained them through other means.

TJ Leyden, a former skinhead who now works with the Simon Wiesenthal Center, has spoken about the techniques he used to attract new members. His experiences are consistent with the empirical research just mentioned. First, he says, his group would intentionally stir up racial tensions in a school by distributing white power literature. Then he would find a particular kind of student, as he describes here:

> We would always look for the kid who was ditching school because he was hanging out in the street. We knew that would be an easy kid to get. We had kids on campus who would find a kid who was hanging out reading a book all by himself who was a loner. Then they'd go over and start talking to him. (Simon Wiesenthal Center, 2002b)

These students would then be gradually introduced to white supremacist dogma, beginning first with fairly mild rhetoric and moving, step by step, into rhetoric that was more hard-core.

The small amount of literature on hate group recruitment suggests that the techniques used by hate groups are basically the same as those used by other types of groups in the following examples:

- I live in a brand-new subdivision where many of the residents are new to the community. Recently, a woman who lives nearby went door to door in our neighborhood, inviting people to join her church. She actually said very little about the religious aspects but instead gushed about what a friendly group of people the congregation was and how the church offered benefits like concerts and free childcare. Presumably, had I decided to attend a service, I would have then been introduced to their theological viewpoints and practices.
- On my campus, soon after classes start every fall, a Club Fair is held in the quad. The timing is clearly aimed at freshmen and other new students. Fraternities,

sororities, academic clubs, and other campus organizations host booths. Signs and fliers advertise the groups' activities and benefits. Those who stop by the booths are frequently offered small gifts and are invited to attend the groups' next meetings.

These techniques, which focus on those who are looking for friendship or a sense of belonging, are used by thousands of nonracist organizations. They offer potential members social ties as well as more tangible benefits and then, after the prospects have taken the initial steps of trying out the new group, they begin to initiate the recruits into their beliefs and culture.

Aside from the studies mentioned here, there has been very little research on hate group recruiting. That is unfortunate, because a greater understanding of this issue might help policymakers better craft strategies against hate crimes. It would be especially enlightening to see more studies conducted among those who joined hate groups as older adults, rather than as youths, and among those who joined groups other than the skinheads (see Box 5.8 for a list of some major hate group leaders). No research at all has been conducted on members of hate groups that are not white supremacist.

Box 5.8   Hate Group Leaders

The following is an incomplete list of some of those who have recently been most prominent in the white supremacist movement in the United States:

Louis Beam: Klan and Aryan Nations leader

Don Black: Former Klansman, runs Internet site Stormfront

Richard Butler: Founder of Aryan Nations

Alex Curtis: Encouraged "lone wolf" white activism

David Duke: Former Klan leader, politician, and founder of National Association for the Advancement of White People (NAAWP) and EURO

Matthew Hale: Head of the World Church of the Creator

Tom Metzger: Founder and head of WAR

Pete Peters: Christian Identity minister

William Pierce: Founder of National Alliance and author of *The Turner Diaries* (1978); died in July 2002

Thom Robb: Christian Identity minister and Klan leader

Bradley Smith: Founder of Committee for Open Debate on the Holocaust (CODOH), a Holocaust denial group

---

Defection

Another area that has received very little research is the study of why people leave hate groups. Several people have noted that the turnover rate among these groups is high (see, e.g., Blazak, 2001; Ezekiel, 1995). One possible reason people leave hate groups is suggested by the literature on joining: People join these groups not for ideological reasons but because of the social and other benefits they believe these groups can give them. If they fail to receive these benefits, then, they may soon leave. As Aho (1988) puts it, "The act of joining a hate group is like disembarking at a railway station, the passengers departing when they sense their interests are better served elsewhere" (p. 166).

Aho (1988) conducted case studies of six individuals who had left the hate movement. Some of these people had held leadership positions within their group, and some had committed violent acts on their group's behalf. Aho found that two factors combined for each of his subjects to lead to their departure from the group. First, they received some "social push" away from hate. That is, there was something that made continuing involvement difficult, such as isolation from the group via imprisonment, living too far away from the group's activities, or concerns about their personal safety. Second, there was also a "social pull" away from hate, a force that led them in some other direction. In two cases it was a girlfriend who disapproved of their racist affiliation, in one case it was a daughter who disapproved, and in the remaining three it was a religious experience or contact not related to racism.

Aho (1994) told of another specific example of a person leaving a hate group. Larry Trapp was an ex-Klansman and neo-Nazi living in Lincoln, Nebraska. Trapp was facing criminal charges for actions related to his racism. Michael Weisser, the cantor of a Lincoln synagogue, first left messages on Trapp's answering machine, reminding him that he would have to answer to God for his actions. Weisser then offered him help, including rides to the grocery store (Trapp was legally blind and a double amputee). Eventually, Weisser and his wife invited Trapp to their home, and the couple befriended him. Trapp renounced his bigoted views.

Other reports tell similar stories. C. P. Ellis, who was discussed in Chapter 4, left the Klan for two reasons: his children were being given a hard time in school because of their father's affiliation, and he developed a friendship with an African American civil rights activist (Davidson, 1996). The neo-Nazis in Ezekiel's (1995) study left when they found nonracist girlfriends, obtained steady jobs, or, in some cases, had children. TJ Leyden left when he saw his two small children begin copying his racial slurs and began to fear for their future (Simon Wiesenthal Center, 2002b).

Bjørgo (1998, pp. 241-244) identifies nine fairly specific reasons why people consider leaving hate groups:

- They may receive negative social sanctions, such as family criticism or criminal prosecution.
- They may feel that things have gone too far, particularly regarding the group's violent behavior.
- They may become disillusioned with the group's activities, finding them too frivolous, the members lacking in loyalty, or the leaders too manipulative.
- They may lose their status within the group.
- They may become emotionally and physically exhausted.
- They may decide that being a racist is jeopardizing their future career prospects.
- They may establish new family responsibilities.
- They may feel that they are simply too old to continue.
- They may lose faith in the group's ideology.

Bjørgo (1998) states, however, that just as most people do not embrace racist beliefs until after they have joined the group, most do not reject those beliefs until after they have left. This hypothesis would be supported by cognitive dissonance theory (discussed in

Chapter 4), which holds that behavior usually precedes attitude change, rather than the other way around.

Again, defection from hate groups is a topic badly in need of more research. If more were known about why people leave these groups, perhaps more could be done to encourage separation.

HATE GROUP ACTIVITIES

What do hate groups do? Of course, in some cases they commit hate crimes. But not all members commit hate crimes, and even those who do must surely spend a limited portion of their days actually planning and perpetrating illegal acts. Surprisingly, virtually no studies have specifically focused on this question, and so information must be gained through indirect methods.

Meetings

Hate groups sometimes have meetings or rallies for their members. Often, these gatherings may be attended by several different groups. The documentary film *Blood in the Face* (Bohlen, Rafferty, & Ridgeway, 1991), for example, depicts a get-together at the Michigan farm of Pastor Bob Miles, a former Klan leader. The attendees appear to be a motley mixture of Klansmen, neo-Nazis, Identity adherents, and assorted other white supremacists. A variety of activities take place at these meetings, including speeches, cross (and sometimes swastika) lightings, and sales of various paraphernalia (books, T-shirts, CDs, videotapes, etc.). Music is usually present, and white supremacists sometimes have white power music concerts (see Box 5.9 for more information on white power music). Although illegal drugs are usually discouraged at these gatherings, beer drinking is usual.

Box 5.9   White Power Music

An important force within the white supremacist movement today is white power music. This music may be a particularly effective method of attracting and retaining young hate group members. Concerts are common at rallies and meetings, Web sites offer free music downloads, and annual CD sales are probably in the millions.

There are a variety of types of white power music. One type is Oi, a combination of punk and heavy metal. There are also, among others, white power country, thrash metal, and industrial rock groups.

Several companies specialize in selling white power music. The biggest, Resistance Records, is owned by the National Alliance. The ADL lists major white power bands at www.adl.org/learn/Ext_US/music_country.asp.

Rallies

Besides private events for members, hate groups sometimes engage in public rallies and marches. These may involve a single group or multiple organizations. Actual attendance by the group members at these events is usually quite small. One of the largest recent gatherings occurred in August 2002, when about 475 assorted white supremacists marched in Washington, D.C. Generally, only a handful of racists appear, and they are typically far outnumbered by counterprotesters. Hate groups seem to have three goals for these events: First, they get the groups a lot of media publicity; second, they help reinforce to their membership that the group is actually doing something; and third, because of the antagonistic presence of counterdemonstrators, they offer the opportunity for violence. The white supremacists may even, if they are lucky, appear on television as the victims of others' aggression.

Propaganda

Hate groups also engage in creating and distributing propaganda of various sorts. On the low-tech side, this might include fliers, pamphlets, newsletters, and so on. These can be given to existing members or distributed to the public at large. In the 9 years I have been teaching, faculty at my university have received anonymous, racist mass mailings at least three separate times. A book on the Klan that I checked out of my campus library had been defaced with stickers advertising a local Klan group. Propaganda efforts can also be more high-tech. Many groups create videotapes and audiotapes, Tom Metzger used to have a public cable television show called *Race and Reason,* and several extremist leaders host radio shows.

Internet

As already discussed, hate groups make wide use of the Internet as well. Before most people had ever heard of the Internet, a few white supremacists such as Tom Metzger sponsored computer bulletin boards, which allowed members to communicate with one another electronically. Estimates of the current number of Web sites sponsored by extremists run from several hundred to several thousand (Gerstenfeld et al., 2003). The content and complexity of these sites vary. Some are quite slick and sophisticated, and some contain large amounts of propaganda and other materials.[14]

Because of the first amendment, it is nearly impossible to censor the content of extremist Web sites in the United States. There are some indications that hate groups are using the Internet as a means not only to expand membership domestically but also to increase their ties with groups and interested parties abroad. Because many other countries ban materials that are constitutionally protected within the U.S., this has led to some extremely complex legal difficulties. We will explore these problems in more detail in Chapter 8.

Organized Political Activity

Organized political activity may also take place among hate groups. Ezekiel (2002) asserts that the primary motivation for most hate group leaders is not racism but power. Often, they attempt to exert that power not only over their members but also over the community at large via the democratic process. Thom Robb's Knights of the Ku Klux Klan styles itself as a political party, as do several other white supremacist organizations. The Populist Party has offered several racist candidates. Others choose to affiliate with more mainstream politics. Tom Metzger ran for the California legislature as a Democrat, and David Duke won a seat in the Louisiana House of Representatives on the Republican ticket. Later, he came very close to becoming governor, also as a Republican.

Hate groups' political activities have been joined by a recent tendency for many of these groups to wish to appear more socially acceptable. Perry (2000, 2001) refers to this as "button-down terror." Members have packed away their white robes or combat boots or let their hair grow out. In doing so, they have emphasized the resemblance of their own views to those of the mainstream right. Not only have hate group leaders been accepted by conventional political parties, but individuals with a strong mainstream presence have been fairly open in their ties to extremist groups. Former Republican Senate leader Trent Lott, who eventually had to resign from his leadership post because of comments he made that appeared to support slavery, for instance, was a frequent speaker for the Council of Conservative Citizens, which the ADL characterizes as a white supremacist organization.

Socializing

Even combined, however, all of these activities probably still take up little time for most hate groups and their members. What information there is suggests that they spend the majority of their time together simply socializing. The neo-Nazis in Ezekiel's (1995) study, for example, mostly hung out at members' homes. This fits in with the model of hate group affiliation we have already discussed—namely, that its most attractive feature to most members is the opportunity for companionship.

WOMEN IN ORGANIZED HATE

Most research on extremists ignores women's roles entirely, and very little work has focused on female bigots. Many scholars who do discuss the issue assume that the hate movement is monolithically antifeminist and that women play a secondary, subsidiary role to men.

For example, concerning the place of women in neo-Nazism, Ezekiel (2002) writes,

> This is a men's movement. Some women are around but always in quite traditional supportive roles. They are the girlfriends or wives of members, and at gatherings they can serve the food they have cooked. In my 7 years around the movement, I heard many speeches, but never one by a woman. I never saw a woman in a leadership role. Women were servants and nurturers. (pp. 54-55)

Ezekiel (2002) also notes that many of the drawings of women created by neo-Nazis are highly sexualized, like junior high school fantasies. Having interviewed white supremacist leaders, he concludes,

> As I recall the stories the leaders told me and the things they said to their followers, everything that comes to mind is masculine: The actors in the stories are masculine; the stories are about combat, domination, and subjugation; the stories are not about nurturance or about cooperative effort that adds new elements, not about creativity or tenderness. (p. 57)

Although these observations are undoubtedly accurate, Ezekiel himself is guilty here of stereotyping about what it means to be female. To fight is male, he assumes, whereas nurturing and tenderness are women's roles. Presumably, if women ran the hate movement, it would be kinder and gentler.

Other commentators have also noted the sexism inherent in the neo-Nazi movement. Coates (1987) wrote,

The neo-Nazi movement is utterly sexist as well as racist. Women do not get to wear its uniforms, and they file down the altar behind their men. Their role is to serve as the planting ground for the "white seed." (p. 78)

Bushart, Craig, and Barnes (1998) agree: "An Aryan woman is the mother of the future, a doe-eyed Eve of the new garden, the promise of racial survival" (p. 264). Ferber (1998) wrote, "White supremacist discourse is a masculine enterprise" (p. 60).

Some hate group rhetoric confirms these ideas. Most of the images of women contained in hate group literature and on their Web sites depict women as mothers (and, hence, the future of the white race), as damsels in distress who need to be saved by the Aryan warrior, or as overly voluptuous sexual figures (Daniels, 1997; Ferber, 1998). Women are frequently exhorted to stand by their men and to keep to traditional women's roles (ADL, 1998). Some hate group doctrine is explicitly antifeminist (Blee, 2002). A leader of the Covenant, the Sword and the Arm of the Lord, a militant Identity group, stated, "This nation was founded on racism and sexism" (Bennett, 1995, p. 351). Furthermore, virtually all of the well-known leaders of hate groups are men, and the majority of hate crimes are committed by males.

However, it is important to note that the role of women in extremist groups is somewhat contradictory. Women have always played an important role in extremism. During its height in the early 20th century, the Klan often had ties with the temperance and suffrage movements (Berlet & Lyons, 2000). Although it certainly romanticized motherhood and other traditional feminine roles, the Klan also championed some degree of feminism, perhaps as an enticement to women to get involved (Blee, 1991). During the early 1900s, there was an organization called the Women of the Ku Klux Klan. In some cases, these women merely performed traditional duties such as food preparation (Bennett, 1995), but others resisted male control and participated actively in Klan activities (Berlet & Lyons, 2000).

Women were not active only in the Klan. Jeansonne (1996) wrote of the Mothers' Movement, which arose in the United States in 1939. This movement of the far right was racist, anti-Semitic, and opposed to American involvement in World War II. At one time, it may have had more than 5 million members. A photo in Jeansonne's book depicts a middle-aged woman in a flowered hat, her hair in a bun, as she gives a Nazi salute after leaving a courtroom during a sedition trial (Jeansonne, 1996, p. 153).

Today, too, women have an active presence in hate groups. Moore (1993) states that skinhead women have the same attitudes as the men, are sometimes violently criminal, and occasionally take a leadership role in organizing skinheads. He concludes, "Women are definitely on the skinhead bus, though perhaps toward the back and rarely as drivers" (p. 80).

After interviews with numerous women involved in the hate movement, Blee (2002) drew several conclusions. Like their male associates, most did not hold particularly bigoted views before they joined the movement, but they did develop those views after they had joined. Most did not come from especially impoverished or dysfunctional backgrounds. Contrary to some assumptions, most were not drawn into the movement by their boyfriends or husbands. Most were opposed to feminism. Blee also predicted that women will likely not be central figures within the hate movement and that their interpersonal conflicts and disillusionment will lead them to defect and will weaken the movement as a whole.

Why do many hate groups seem to simultaneously repudiate and court women? On the one hand, white supremacists wish to maintain their existing privileges, and that includes power over women. Thus, feminists are demonized (often as part of the Jewish conspiracy), and women are relegated to secondary and supporting roles. On the other hand, hate group leaders realize that they need female members. Not only will women add to the groups' overall size but, as discussed in the previous section, male members of the movement tend to defect if their wives or girlfriends are not supportive.

The role of women in extremism is yet another issue that is badly in need of additional research. That research may be difficult to conduct because hate groups' views of women are complex and, it seems, often contradictory. Moreover, it may be difficult for researchers to see objectively beyond their own biases and expectations about women.

CONCLUSION

In Chapter 4, we saw that few hate crimes are actually committed by members of hate groups. So why devote an entire chapter to the topic of hate groups, as this book does? And why does this chapter make frequent pleas for more research on topics related to organized hate?

The answer to these questions is that hate groups have an influence far beyond what the actual numbers might indicate. At the individual level, they are dangerous because they appear to cause otherwise relatively unbigoted individuals to adopt strongly prejudiced viewpoints. Using seemingly ordinary techniques, they target seemingly ordinary people, and the result, often, is a hard-core hater. This in itself should be cause for concern.

At a higher level, hate groups affect our entire society. They are visible, and they are heard. Even if we do not belong, and even if we do not know anyone else who belongs, we cannot help but be exposed to their messages. In a very direct way, they contribute to an overall climate of intolerance that influences us all and that, in some instances, will lead to hate crimes.

DISCUSSION QUESTIONS

1. What is a hate group, and how can it be distinguished from terrorists, gangs, and other groups? What do hate groups have in common? Discuss the assertion that only groups associated with those in power (i.e., whites and Christians) qualify in this category, whereas groups associated with disempowered minorities do not.

2. On behalf of Mulugeta Seraw's family, the Southern Poverty Law Center sued Tom and John Metzger. Although the Metzgers were more than a thousand miles away at the time of the murder, the SPLC argued that they should be liable because they had encouraged Portland skinheads toward violence. The Metzgers' defense was that they were exercising their First Amendment rights. The SPLC obtained a $12.5 million judgment against the Metzgers, but many civil libertarians were concerned about the repercussions of this case. In your view, to what extent should extremist leaders like the Metzgers be held liable for the actions of their followers?

3. What circumstances led to birth, death, and repeated rebirths of the Klan? Berlet and Lyons (2000) argue that the reason the Klan, unlike other racist organizations, has been revived so often is that it promotes a "palingenetic myth": that white civilization will rise from the ashes after nearly being destroyed. According to these authors, this

NARRATIVE PORTRAIT

AT A KLAN RALLY, II

Earlier in this chapter, there was an excerpt from Patsy Simms's 1976 visit to a Klan rally. Twenty-five years later, Jon Ronson attended another Klan rally. This rally took place in Arkansas, and the speaker was Thom Robb, leader of the Knights of the Ku Klux Klan. Robb has sought to project a kinder, gentler image of the Klan.

Thom began, dramatically, by holding up a poster with the words GET OUT NIGGER! scrawled out in bold letters. It was an arresting moment.

Thom scanned the marquee.

"This is *stupid*," said Thom, waving the poster in the air. "This is stupid, stupid stuff."

Thom's son, Nathan, handed out photocopies of the GET OUT NIGGER! poster to the audience so they could scrutinize it further.

"When your grandmother sees this," continued Thom, "who is she going to support? Is she going to support you? No! So we don't call those black fellas the N-word, because the very people we're trying to reach, all they'll hear is the N-word. Right?" . . .

"The masses," said Thom, "are *feminine*. In the area of politics, the masses are *feminine*. OK? And the feminine masses look to what? They look to the *masculine* for protection. And who is the masculine? The government. So the feminine masses look to the government for protection. OK?"

There were nods.

"So, when the feminine masses see a Klansman on TV, or a militiaman running around with a gun, or a patriot wearing camouflage, what is she going to feel? She's going to feel that her *safety* is being what? She's going to feel that her *safety* is being *altered*. And she doesn't *want* to feel that her safety is being altered. So she's going to turn to who? She's going to turn to the *masculine* for protection."

Thom paused. He said, "I'm going to show you something else now."

Nathan handed out a photocopy of a leaflet that read, alarmingly, "You have been paid a friendly visit by the Knights of the Ku Klux Klan. Shall we pay you a *real* visit?"

"Do we want to go around threatening people?" said Thom, softly.

The audience shook their heads.

"*Come on!*" yelled Thom. "We're supposed to be the *knights* on the *white horses* who ride into town and save our people! We're supposed to be the *good guys! Shining armor!* Do we want to go around *threatening* people?"

A gust of wind blew the photocopies across the marquee. There was a short break while they were retrieved and secured onto the lectern with a rock. Then he resumed.

"Truth," he said "is what we *perceive*. To the feminine masses, what they *perceive* is the truth. OK? So we, as individuals, and we as a corporate body, are two different things."

Thom scanned the marquee. He looked at the individuals in the marquee.

"We as a corporate body," said Thom, "must have a corporate image. And that corporate image has to be projected to the feminine masses. . . ."

SOURCE: Ronson (2002).

myth is especially appealing to those who consider themselves beset by crisis. What is your reaction to this hypothesis?

4. How do the Palmer Raids of 1919 and 1920 compare with the Department of Justice's actions in 2001 and 2002 in response to the September 11 attacks?

5. Compare and contrast the Klan with the racist skinheads. What social factors have led to the creation of each group? How do the members compare demographically and behaviorally?

6. The relationships between racist skinheads and other white supremacist groups are complex. Some groups have tried (with varying degrees of success) to cultivate and organize skinheads, whereas others, such as some Klan groups, have rejected them. Why do you think the skinheads are attractive to some groups and not to others? What do you think accounts for the skinheads' popularity with the media?

7. Why did militias quickly gain and then lose popularity? Do you agree with Pitcavage's (2001) assessment that today they constitute the second biggest threat of domestic terrorism? What methods should be employed to counter illegal militia activity?

8. World Church of the Creator leader Matthew Hale has been denied admission to the Illinois Bar even though he graduated from law school and passed the bar exam. The reason he has been denied admission is that the Bar's Committee on Fitness and Conduct found that, because of his openly racist views, he is not of sufficiently good moral character (Sloane, 2002). In your opinion, should Hale be allowed to practice law?

9. What are the central tenets of most hate groups? Do you see any underlying themes in these beliefs? In what ways do these beliefs differ from mainstream views?

10. What is known about how hate groups recruit people and about why people leave? If you were a researcher, how would you design a study to explore these questions? What are some of the challenges involved in conducting such a study?

11. How do the activities of hate groups differ from those of nonracist organizations, such as clubs, fraternities, or religious groups you belong to?

12. In what ways are hate groups attempting to adopt a more mainstream public image? Why do you think they have adopted this strategy? How successful has it been? To what extent do you believe their clean-cut front is hiding a less savory reality? Do you think this tactic will help them lure new members, and, if so, how can it be thwarted?

13. In what ways is the role of women in the hate movement complex and confusing? To what extent are those who study this issue influenced by their own biases, and how can this be avoided? What do you think will be the place of women in hate groups in the future?

INTERNET EXERCISES

1. The Anti-Defamation League has a good summary of major American extremist groups and leaders: www.adl.org/learn/Ext_US/default.asp. The ADL also maintains a database of international terrorist organizations: www.adl.org/ict/default.asp. The Southern Poverty Law Center has a map of active hate groups: www.tolerance.org/maps/hate/index.html. Click on your state to view a close-up map, as well as a listing of groups by location.

2. Perhaps because of media attention, most people assume all skinheads are racist. However, there are also antiracist skinheads (who refer to the racist skinheads as *boneheads*). One site with information on antiracist skinheads is Rash United, at www.geocities.com/CapitolHill/Lobby/3475/#RASH.

NOTES

1. Specifically, these seemingly disparate groups often share anti-Semitic philosophies as well as a belief in racial separatism.

2. On his Web site (www.davidduke.org), for example, David Duke refers to the ADL as a Jewish supremacist organization.

3. The Thirteenth Amendment abolished slavery, and the Nineteenth Amendment secured voting rights for women.

4. Slavery was by no means the only form of large-scale, government-sanctioned bias. For example, a program of genocide was engaged against Native Americans, not only by the military but also via the government encouraging white settlers and encouraging the intentional eradication of bison, the economic mainstay of many Native Americans. Another example was legal discrimination against immigrants from China and, later, Japan. Both of these examples occurred more in the West than the South and continued well after the Civil War. Also continuing for many decades after the war, and throughout the country, was legalized discrimination against women.

5. Two brothers later confessed to the murder but were acquitted by an all-white jury. In 1962, Bob Dylan memorialized this murder in a song called *The Death of Emmett Till*. The song, optimistically, ends with this verse:

This song is just a reminder to remind your fellow man

That this kind of thing still lives today in that ghost-robed Ku Klux Klan.

But if all of us folks that thinks alike, if we give all we could give,

We'd make this great land of ours a greater place to live.

6. *Carpetbagger* was a derogatory term referring to Northerners who came south after the Civil War, supposedly for the purposes of financial or political exploitation. *Scalawags* were Southerners who supported Reconstruction.

7. The pretext for these killings was the Mexican Revolution, which lasted from about 1910 to 1920. Mexican Revolutionary forces had attacked some border towns and had also encouraged people of color in the United States to rise up and establish independent countries.

8. The movie *Mississippi Burning* (Zollo, Colesberry, & Parker, 1988) was made about this event. Ironically, it was this film that evidently inspired Todd Mitchell to encourage his friends to attack a white boy (see Chapter 3).

9. NSDAP/AO is an acronym for the German phrase "National Socialist German Workers Party–Overseas Organization."

10. It is unclear whether the fire was the result of law enforcement actions or whether the Branch Davidians set the fire themselves. The day of the fire, April 19, was the day Timothy McVeigh chose to bomb the federal building 2 years later.

11. The publisher of the book this quote is from, the Charles Darwin Research Institute, is apparently run by Rushton. A couple of years ago, I and most of my colleagues received an abridged edition of this book. The originator of this mass mailing was anonymous.

12. Many white supremacists exclaim with horror that soon whites will no longer constitute a majority in the United States. In fact, this is already true in California and Hawaii, although white non-Hispanics still make up the largest single racial or ethnic group in both

states. The implication is that once "they" outnumber "us," they will take advantage of their greater number to treat whites badly.

13. One of my favorite (due to its ridiculous nature) current conspiracy theory Web pages (www.conspiracyworld.com/web/Articles/mysterious_riddle_of_chandra_lev.htm) posits that Congressional intern Chandra Levy was murdered because, being a secret spy for the Israeli Mossad, she had advance notice of the impending September 11 attacks. She also, the Web page says, had seen documents linking Timothy McVeigh to a federal plot.

14. For a recent article on hate on the Internet, see K. B. Stern (2001).

Chapter 6

HATE CRIME VICTIMS

Matthew Shepard died on October 12, 1998, a few days after he was found beaten and tied to a fence outside Laramie, Wyoming. The 21-year-old college student died because he was gay. Alonzo Bailey's death on May 5, 2002, was no less horrible but went largely unnoticed and unremarked. Bailey, who was 33 years old, African American, and homeless, was chained to a fence in an Oklahoma City industrial park. Wooden pallets were piled on top of him and set on fire. By the time police discovered his burning body, he was already dead. Months later, it remained unclear whether Bailey's murder was a hate crime, but the suspected killer, Anthony Lee Tedford, did have white supremacist tattoos.

Fortunately, hate crime murders are rare. In 2000, for example, the Federal Bureau of Investigation (FBI) recorded only 19 of these crimes in the entire United States (FBI, 2001b). For comparison's sake, 666 people were murdered in Chicago alone in 2001 (FBI, 2002), and an average of 73 people a year in the United States are killed by being struck by lightning. However, even though very few people actually die because of a hate crime, this does not mean that hate crime victims are few. In 2000, the FBI counted nearly 10,000 hate crime victims nationwide; as discussed in Chapter 3, this is probably a severe underestimate, and the true number may be three or more times as high. Whatever the actual number, undeniably thousands of individuals and institutions become victims of crime each year because of their race, religion, sexual orientation, or other group affiliation.

In this chapter, I address the topic of hate crime victims. After touching on some of the problems inherent in trying to identify hate crime victims, I explore official hate crime data and advocacy group data to try to determine who the victims of these crimes are. Then I focus on three specific groups who are frequently targeted: African Americans, gays and lesbians, and Jews. I look at other groups as well, including Asians, women, and the disabled. Throughout the chapter, we will consider how and why certain groups are targeted.

PROBLEMS IN IDENTIFYING HATE CRIME VICTIMS

Difficulties in Reporting

A variety of factors make it difficult to determine who the true victims of hate crimes are. The most important of these is the severe underreporting of the crimes themselves.

Furthermore, this underreporting is not equal among different victim groups. Members of certain groups are particularly unlikely to report a hate crime because they have poor relations with the police, because they are inhibited by cultural or linguistic factors, or because they are among the people who are the most voiceless in our society.

Of course, crimes in general often go unreported. Consequently, researchers frequently rely on instruments such as the National Crime Victimization Survey, an annual poll that measures crimes that do not get reported to the police. Unfortunately, however, the survey does not ask respondents about the motives of the offenders, so it cannot be used to study hate crime victimization (Garofalo, 1997).

Difficulties in Recording

Even when hate crimes are reported, there are serious difficulties in interpreting the data. Agencies differ in how they record information, and accounting may be inconsistent. In addition, agencies may have trouble determining the group of the victim because human individuals do not always fit neatly into predetermined categories. The boundaries between race, ethnicity, and religion may be especially fuzzy. For example, if a black Hispanic person is attacked, which box should be checked? What about victims of mixed race or ethnicity?

A good example of the confusion that can occur within official hate crime data can be found in crimes against Muslims or Arabs. After September 11, 2001, there was a surge in hate crimes against those who were perceived to be Arab, Muslim, or Middle Eastern. California recorded most of these crimes under the "other ethnicity" category. The state government still retained a category under religion called "anti-Muslim," but few crimes were placed in that category. Illinois, on the other hand, created a special category under ethnicity—"Arab"—while also retaining "Muslim" as a religious category. Thus, crimes that resulted from the events of September 11 could be classified in a variety of different ways by different officers or by different states.

Another difficulty with official hate crime data lies in the fact that, although the majority of these crimes are committed against individuals, many are also committed against institutions such as synagogues, cemeteries, and community centers. In the case of an attack against an institution, usually that incident will be recorded as involving a single victim. In reality, all of the people who visit or belong to that institution could be considered victims of that crime, but this fact is not reflected in the official numbers.

In addition to official police data, some private interest groups also attempt to keep track of hate crimes. Examples include the Anti-Defamation League, the American Arab Anti-Discrimination Committee, and Lambda. Unfortunately, there are problems with the reliability of these groups' data as well. Most of them record not only hate crimes but also noncriminal hate incidents (such as distribution of fliers or usage of epithets). These associations tend to rely heavily on reports by victims to the organizations, and it is unclear what proportion of victims chooses to contact the organizations. Furthermore, these private interest groups do not represent the full range of hate crime victims; they tend to focus on a particular group, and some groups of victims have no organization tracking the crimes against them. Finally, because these are advocacy organizations, not neutral observers, they may have something to gain by under- or (more likely) overemphasizing victimization rates.

Difficulties With Self-Reports

Another source of information about hate crime victims comes from self-reports. Several studies have been done in which researchers attempted to determine hate crime victimization rates among a certain group. The most comprehensive work of this kind has been conducted among gays and lesbians. Again, however, there are problems. It is virtually impossible to obtain a representative sample to survey because some members of a group are much more likely than others to participate. Furthermore, those who do participate may have faulty or incomplete memories, or they may misunderstand exactly what constitutes a hate crime.

OFFICIAL HATE CRIME DATA

Although there are severe limitations with official hate crime data, there is still value in examining those data for patterns and trends. Table 6.1 presents recent data on hate crime victims from the FBI and from four individual states.[1]

Certain patterns can be discerned from these official data. Perhaps the most striking is that in every jurisdiction, the most common victims of hate crimes are African Americans. In the FBI data, more than one in three victims was selected because he or she was black. This rate of occurrence is far in excess of the actual proportion of Americans who are African American, which is estimated to be about 12.8% (United States [U.S.] Census, 2001). In every jurisdiction, no other group even comes close to blacks in frequency of hate crime victimization. These data are also consistent with those in other years and other jurisdictions (Gerstenfeld, 1998). Despite the cautions we must take in interpreting law enforcement data, it seems safe to conclude that blacks are the most likely victims of hate crimes in the United States.

Table 6.1 Official Data on Hate Crime Victims

Jurisdiction Year	FBI 2000	California 2001	Colorado 2001	Illinois 2001	Texas 2000
Total Victims	9924	2812	165	360	305
Victims According to Bias Type:					
Black	3535 (35.6%)	768 (27.3%)	38 (23.0%)	113 (31.4%)	132 (43.3%)
Jewish	1269 (12.8%)	209 (7.4%)	9 (3.6%)	27 (7.5%)	25 (8.2%)
White	1080 (10.9%)	155 (5.5%)	8 (4.8%)	40 (11.1%)	26 (8.5%)
Male homosexual	1060 (10.7%)	405 (14.4%)	13 (7.9%)	45 (12.5%)	30 (9.8%)
Hispanic	763 (7.7%)	282 (10.0%)	24 (14.5%)	15 (4.2%)	25 (8.2%)
Other ethnicity	453 (4.6%)	505 (18.0%)[a]	19 (11.5%)	17 (4.7%)	3 (1%)
Multiracial group	379 (3.8%)	114 (4.1%)	1 (.6%)	8 (2.2%)	9 (2.9%)
Asian/Pacific Islander	339 (3.4%)	114 (4.1%)	7 (4.2%)	6 (1.7%)	4 (1.3%)
Female homosexual	228 (2.3%)	70 (2.5%)	2 (1.2%)	10 (2.8%)	11 (3.6%)
Homosexual	226 (2.3%)	23 (.8%)	13 (7.9%)	5 (1.4%)	11 (3.6%)
Arab/Muslim[a]	36 (.4%)	87 (3.1%)	19 (11.5%)	49 (13.6%)	4 (1.3%)
Other	592 (6.0%)	80 (2.8%)	12 (7.3%)	25 (6.9%)	25 (8.2%)

SOURCES: California Attorney General (2002); Colorado Bureau of Investigation (2002); FBI (2001b); Illinois State Police (2002); Texas Department of Public Safety (2001).
NOTE: The data in this table came from official law enforcement reports. Victims are listed in the order of the frequency they occur within the FBI data.
a. This category includes anti-Arab and anti-Middle-Eastern crimes.

Box 6.1   The Effects of 9/11

Recent data indicate that September 11, 2001, resulted in a violet backlash against Americans who were presumed to be Muslim. Here are a few examples:

- According to the Council on American-Islamic Relations (www.cair-net.org), verified anti-Muslim hate crimes increased from 366 during the period of March 2000 through March 2001, to 1,125 during March 2001 through March 2002 (Bruner, 2002).
- The Massachusetts Governor's Task Force on Hate Crimes recorded 576 hate crimes in 2001 compared with 463 in 2000. In the first 8 months of 2001, there

---

Nationally, Jews are the second most common hate crime victims. Although the same does not hold true for the states in Table 6.1, it is the case that in every jurisdiction listed, Jews are still victimized at a disproportionately high rate. Only about 2% of the U.S. population is Jewish, and the percentage of the population that is Jewish in the states listed in Table 6.1 ranges from .6% (in Texas) to 3% (in California) (U.S. Census, 2001).

Whites are the third most common victim within the FBI data, and they are victimized at a fairly high frequency in most of the states. However, 71.3% of the U.S. population is white and non-Hispanic (U.S. Census, 2001). Therefore, unlike blacks and Jews, whites are victims of hate crimes at a disproportionately *low* rate. This directly contradicts the claims of some white supremacist groups that antiwhite hate crimes are the real problem in American society.

Gays and lesbians are also frequent victims of hate crimes. Within the federal data, one in five hate crimes was committed because of the victim's sexual orientation. The numbers vary from jurisdiction to jurisdiction, but generally the proportion is high. Crimes against gay men are considerably more common than those against lesbians or those against homosexuals in general. It is impossible to know exactly how many Americans are homosexual or bisexual, nor can we know what percentage of those people are open about their sexual orientation and thus likely to be targeted as hate crime victims. It is quite clear, however, that gays and lesbians are at considerably higher risk than heterosexuals for being the victims of hate crimes.

One additional pattern stands out within the data in Table 6.1: the effects of September 11, 2001, upon Americans who are Muslim, Arab, or Middle Eastern. Crimes against these people are relatively uncommon in the FBI and Texas reports, both of which come from 2000. They are significantly higher, however, in the California, Colorado, and Illinois reports, which are from 2001. In fact, total hate crimes in California increased 15% from 2000 to 2001, even though they would have decreased 5% if not for the attacks related to September 11, 2001. The same situation can be seen in Colorado, in which the total number of crimes increased by more than a third compared with 2000. No anti-Muslim attacks were recorded in Colorado in the first half of 2001, but 19 were recorded in the second half (Kelly & Robinson, 2002). In Texas, from which a complete report for 2001 is not yet available, the number of hate crimes increased by 50% from 2000 to 2001, and the number of crimes against Arab Americans went from 4 to 63 (Hughes, 2002). Obviously, the events of September 11 had a dramatic impact not only on Americans in

were five anti-Muslim or anti-Arab crimes; in the last 4 months, there were 86 (Szaniszlo, 2002).

- During 2000 and the first half of 2001, no anti-Islamic hate crimes were reported in Colorado. Following 9/11, there were 17 reported cases (Kelly & Robinson, 2002).
- The Canadian Islamic Congress reported 200 anti-Muslim hate crimes in the year after 9/11. The Canadian government recorded 100 such crimes, a 66% increase over the previous year (Bobak, 2002).
- In Florida, total reported hate crimes increased from 269 in 2000 to 335 in 2001, due to increases in crimes against Muslims and Arab Americans (Hegarty, 2002).

general but especially on those who were perceived to be Arabs or Muslims (Gerstenfeld, 2002) (see Box 6.1 for more on the effects of September 11, 2001).

ADVOCACY GROUP DATA

As mentioned previously in this chapter, several advocacy groups attempt to monitor hate crimes and related events. The data collected by these groups have several advantages over those collected by law enforcement agencies, the most important being that many victims who do not feel comfortable talking to police may be willing to report a crime to one of these groups. Therefore, advocacy groups are able to keep track of many hate crimes that fall through the cracks of the official reporting system.

There are disadvantages to advocacy group reporting as well. One downside is the fact that data collection usually relies on victims contacting the groups via telephone hotlines or online reporting systems. Many victims are probably unaware of these programs or choose not to use them. Also, there is generally no way to verify these reports, so an unknown number are likely false or mistaken. Furthermore, advocacy groups typically include not only hate crimes but also noncriminal incidents such as name-calling. Despite these shortcomings, however, these data can be used to corroborate certain trends and patterns that appear in the official data.

One thing that stands out in recent advocacy group reports, as it does in the official data, is the impact of September 11, 2001, on Americans of Middle Eastern and Asian ancestry. The American-Arab Anti-Discrimination Committee, for example, reported more than 600 violent incidents against Arab Americans in the 6 months following September 11 (American-Arab Anti-Discrimination Committee, 2002). The group South Asian American Leaders of Tomorrow, which used media resources to compile its statistics, reported 645 incidents against South Asians and Middle Easterners in the week after September 11 (South Asian American Leaders of Tomorrow, 2001). And the National Asian Pacific American Legal Consortium (NAPALC) counted 250 anti-Asian incidents in the last three months of 2001, compared with the 400 to 500 incidents the organization typically finds per year (NAPALC 2002).

The Orange County (California) Human Relations Commission relies on both law enforcement data and advocacy organization reports to compile its annual audit. It counted

69 hate crimes against Middle Eastern or Muslim Americans in 2001 compared with only 8 the previous year. There was a particular spike in overall hate crimes in September 2001. The number of crimes in previous months ranged from five to 11, but there were 66 in September (Orange County Human Relations Commission, 2002).

Another pattern that appears clear from the advocacy group data is that the official data seriously underreport hate crimes. For example, the number of Asian Americans victimized by hate crimes in 2000 was 40% higher in the NAPALC report (NAPALC, 2001) than in the FBI report (FBI, 2001b). Similarly, the National Coalition of Anti-Violence Programs (NCAVP) reported 2,210 victims of hate crimes in 2001 due to sexual orientation (NCAVP, 2002), whereas the FBI reported only 1,514 (FBI, 2001b). The actual discrepancy is actually much greater because the FBI data purport to include almost all of the United States,[2] whereas the NCAVP report included only a dozen reporting regions with a combined population of about 51 million. Of the 2,210 victims in the NCAVP report, only 762 had reported the crime to the police.

The NCAVP report also highlights one reason that hate crime victims often do not report hate crimes to the police. An astonishing 59% of the hate crime victims in that report claimed to have been subjected to verbal abuse and slurs *by police* pertaining to their sexual orientation (NCAVP, 2002). It is easy to see why many gays and lesbians might hesitate to call law enforcement if subjected to a hate crime.

Reports from advocacy organizations also indicate that the number of annual victims of hate crimes in general is not increasing; however, the number of victims affected by the 9/11 backlash is increasing. The NCAVP (2002) report showed a 12% decrease in sexual orientation crimes between 2000 and 2001, and the Anti-Defamation League reported nearly an 11% drop in anti-Semitic hate crimes (ADL, 2002f). Orange County had a 33% increase in overall hate crimes in that same period, but that was due to the precipitous rise in crimes against Middle Eastern and Muslim Americans (Orange County Human Relations Commission, 2002).

HATE CRIMES AGAINST AFRICAN AMERICANS

African Americans appear to be the most common victims of hate crimes, and they also appear to be victimized at a disproportionate rate. This probably comes as little surprise; blacks have, arguably, been subjected to the longest and most oppressive systems of discrimination and deprivation in the United States, and they were the primary targets of America's oldest organized hate groups. In fact, the very words *hate crime* tend to invoke images of black victims.

What is remarkable, however, is that extremely little research has focused on hate crimes against African Americans. A fair amount of scholarship has concentrated on hate crimes due to sexual orientation, and another body of literature exists on anti-Semitic crimes. It is also possible to find studies concerning other victims, such as Asian Americans, women, and the disabled. Aside from the specific problem of church burnings, however, virtually nobody has studied hate crimes against African Americans.

There are several possible explanations for this lack of research. To begin with, no advocacy organizations appear to monitor and focus on specifically antiblack crimes. Of course, this is not to say that there are no African American advocacy groups. But most of these, such as the NAACP (National Association for the Advancement of Colored People), concentrate their efforts on areas other than hate crime, such as discrimination in general, education, economic matters, and other legal issues. Other organizations, such as the Center for Democratic Renewal and the Southern Poverty Law Center, deal with hate crimes in general but do not focus specifically on crimes against African Americans. In contrast, groups that deal primarily with crimes against Jews, gays, Arab Americans, and Asian Americans do exist.

A second possible explanation has to do with who conducts hate crime research. By and large, it is people like me: professors. Understandably, many academics tend to be interested in topics that have some personal relevance. African Americans are badly underrepresented among the American professoriate, so hate crimes against blacks may be a topic that is not especially near and dear to most researchers' hearts.

Another possible reason so little research has been done on African American victims of hate crime is that it is simply taken for granted. Perhaps this kind of violence is so paradigmatic of hate crimes that it is less interesting for researchers, who might prefer to focus on slightly more exotic or controversial issues. Scholars may also feel (incorrectly) that everything worthwhile is already known about antiblack hate crimes and that they ought, therefore, to concentrate on those crimes about which we are less well informed.

Whatever the reason for the lack of research on antiblack hate crimes, the consequence is that very little is known about the subject. We are forced to rely almost entirely on either historical treatments, which do not give accurate depictions of modern-day problems, or official hate crime data, which, as we have already discussed, is of dubious validity. We do not know, then, how great a problem hate crimes truly are in African American communities. We do not know what number of blacks have suffered from these crimes, what forms the criminal acts take, or what proportion of black victims choose to report the crimes to the police. We also do not know if there are characteristics that distinguish antiblack offenses or offenders from other hate crime offenses or offenders. And we know virtually nothing about the subjective experiences of African American victims of hate crimes.

Hate Crimes Against African Americans, Yesterday

Petrosino (1999) has argued that the enslavement in America of millions of Africans and their descendants ought to be considered a hate crime. Although forced indenture was used against several racial and ethnic groups, only among blacks was it so widespread, and only among blacks was it so permanent that it exceeded the life of each individual and was passed on to his or her children. Thus, blacks were specifically selected for deprivation of liberty and life based on their race.

If slavery is to be considered a hate crime, then it ranks as one of America's earliest, dating to the early 17th century, well before the United States gained independence. Moreover, it also qualifies as perhaps the most massive hate crime in history, having been carried out against millions of people. And it has had the greatest impact, permanently changing the nature of this entire country and, arguably, having a legacy that continues nearly 140 years after its cessation.

Slavery was expressly sanctioned by the laws of its time, unlike the acts that today we call hate crimes, so it is debatable whether slavery should be classified as a hate crime. But another of Petrosino's (1999) claims is stronger: namely, that lynchings, another widespread historical act against African Americans, should be included within the classification of hate crimes. The case for classifying lynchings as hate crimes holds more merit because lynchings were not permitted by law (although laws against lynchings were rarely enforced when the victim was black) and because race was generally the reason lynching victims were selected. Blacks were hanged or burned, beaten, or shot to death for such minor offenses as being "saucy" to whites (Petrosino, 1999), trying to register to vote, participating in labor union activities (Turner, Stanton, Vahala, & Williams, 1982), or, like Emmett Till, having the temerity to flirt with a white woman. (See Box 6.2 for more about lynchings.)

Box 6.2   Strange Fruit

In 1937, a Jewish schoolteacher from New York named Abel Meeropol saw a photo of a lynching of two African Americans. He was inspired to write a poem titled "Strange Fruit." The poem was later turned into a song, which was a hit for Billie Holiday. The poem reads as follows:

Southern trees bear a strange fruit,
Blood on the leaves and blood at the root,
Black body swinging in the Southern breeze,
Strange fruit hanging from the poplar trees.

Pastoral scene of the gallant South,
The bulging eyes and the twisted mouth,
Scent of magnolia sweet and fresh,
And the sudden smell of burning flesh!

Here is a fruit for the crows to pluck,
For the rain to gather, for the wind to suck,
For the sun to rot, for a tree to drop,
Here is a strange and bitter crop.

It is impossible to know how many African Americans were lynched. Tolnay and Beck (1995) estimate that at least 2,500 blacks were lynched in the South between 1882 and 1930. Other estimates are higher; another author states that more than 3,400 blacks were lynched between 1882 and 1951 (Gibson, 2002), and others calculate that 2,050 blacks were lynched between 1882 and 1903 alone (Cutler, 1905). It was not unusual for photographs to be taken of lynchings, and these photos show smiling white mobs standing around the corpses; often, young children are present.

Whatever the precise number, lynchings of African Americans were clearly a serious problem in the late 19th and early 20th centuries. Local law enforcement was often complicit in these crimes or, at the very least, turned a blind eye to them. Prosecutions of whites for killing blacks were very rare, and convictions (by all-white juries) rarer still. Although the number of lynchings decreased considerably by the middle of the 20th century, they were occasionally recorded later. On March 21, 1981, a black teenager named Michael Donald was beaten and hanged by two members of the Klan in Mobile, Alabama. The Klan members were angry that a local jury had failed to find an African American man guilty of killing a white police officer, and they chose their victim solely because he was black (Stanton, 1992). One of the killers, James Knowles, was sentenced to life in prison. When the other man, Henry Hays, was executed in 1997, it was the first time a white man had been put to death for killing a black victim since 1913.

Hate Crimes Against African Americans, Today

Hate crimes continue against African Americans because prejudice continues. Although discrimination and segregation may no longer be legal, they are still commonplace, and many Americans harbor stereotypes of blacks as inferior, dangerous, or economically damaging to others. As one African American man stated, "Being black in America is like being forced to wear ill-fitting shoes. Some people adjust to it. It's always uncomfortable on your foot, but you've got to wear it because it's the only shoe you've got" (Terkel, 1992, p. 136).

There is certainly support for the proposition that in recent years, the situation has improved for African Americans legally, economically, and socially. Relatively few whites today openly maintain the superiority of whites or support segregation and discrimination (Sears, 1998). However, racism has certainly not disappeared. The subtlety of modern racism (see Chapter 4 for a discussion of this concept) makes it easy for whites to overlook it and, therefore, deny its existence. Thus, because whites today have a considerably rosier view of African American life than do blacks, whites typically resist programs and policies that would promote equality, such as affirmative action (Jones, 1999; Sears, 1998).

Negative stereotypes about African Americans are still common and continue to affect the ways in which blacks are perceived and treated (Operario & Fiske, 1998). Blacks experience inequality in employment (Bendick, Jackson, & Reinoso, 1999), in education (Pincus & Ehrlich, 1999), in health care (Byrd & Clayton, 2001), and in criminal justice (Walker, Delone, & Spohn, 1999), among other areas.

Given the continuing pervasiveness of antiblack attitudes and behaviors, it is not surprising that blacks are the primary targets of hate crimes. In his research among neo-Nazis, Ezekiel (1995) found that, although white supremacist leadership focused on Jews as the "real problem," antiblack feelings were much more important to rank-and-file members. These young men had little contact with and little knowledge of Jews, but "hatred and fear of black people [had] been a part of these Detroit kids' upbringing" (Ezekiel, 1995, p. 152).

One of the few academic explorations to focus on antiblack hate crimes was written by Torres (1999). Unfortunately, Torres relied heavily on FBI data, which, as we have seen and he acknowledged, are of questionable validity. The data do confirm that African Americans are consistently the most frequently targeted group.

A few studies have also been conducted on the wave of arsons of black churches that occurred in the 1990s. (See Table 6.2 for data on church arsons in the 1990s.) In 1996, an estimated 297 places of worship were bombed or set on fire in the United States. Between January 1, 1995, and August 15, 2000, there were 945 such attacks. About a third of the targets were churches with predominantly African American congregations; in the South, 44% of the targets were black churches. Most of the offenders were white (National Church Arson Task Force, 2000). In response to these acts, the Church Arson Prevention Act was passed by Congress in 1996 (Soule & Van Dyke, 1999), and President Clinton formed the Church Arson Prevention Task Force.

The destruction of black churches was especially distressing because the church plays a central role in many African American communities (Carter, 1999; Soule & Van Dyke, 1999). Furthermore, these kinds of crimes devastate not just a few individuals but often entire groups. The PBS documentary "Forgotten Fires" (Chandler, 1998), for example, documents the burning of two black churches in Manning, South Carolina, by several young Klan members. The arsonists were eventually given long prison sentences, one of the churches received a $21.5 million judgment against the local Klan group and its

Table 6.2   Church Arsons in the 1990s

Year	Total Churches Burned	Black Churches Burned
1995	52	25
1996	297	120
1997	209	54
1998	165	43
1999	140	37
2000[a]	82	31

SOURCE: National Church Arson Task Force (2000).
NOTE: a. Data through August 15, 2000, only.

leaders, and the churches were rebuilt. Such relatively happy endings are not common, however.

Although racially motivated destruction of black churches dates back to at least the Civil War, there have been several periods when the rate of such crimes has dramatically increased. The early to mid-1990s was one of these times. Soule and Van Dyke (1999) found that black church arsons were associated with increased levels of economic and political competition.

After 1996, the number of church burnings began to decrease sharply. In 1999, there were only 140 cases, less than half the number that had occurred three years before (National Church Arson Task Force, 2000). It is unclear whether this drop was due to the enactment of the federal law, to the prosecution efforts of the federal task force, to a strong national economy, to a shift in media focus to other topics, or to other factors.

Hate Crimes Against African Americans, Tomorrow

One thing that distinguishes most African Americans from many other potential hate crime targets is that they are immediately identifiable. Gays and lesbians and Jews, who are other common targets, often have particular places of activity, such as community centers. But unless they are gathered in these places, group affiliation is not often readily apparent among the members of these groups. I could wander into a strange neighborhood and the residents would not be able to discern that I am Jewish.[3] But when Michael Griffith and Yusuf Hawkins entered white neighborhoods in New York, their assailants could immediately tell that they were black. It would be interesting to determine if this distinction leads to a difference in the ways in which hate crimes are perpetrated against members of different groups. Are attacks against blacks more likely to be personal, whereas those against, say, Jews are more likely to be institutional? Do antigay and anti-Semitic offenders generally set out in search of their victims, whereas antiblack offenders happen across them by chance? These questions, which do have implications for hate crime prevention policies, are so far unanswered.

Very little public voice has been heard from black victims of hate crimes. A few books have been written about specific crimes—the murder of Mulugeta Seraw (Dees & Fiffer, 1993), the lynching of Michael Donald (Stanton, 1992), the cross-burning on the front lawn of the Jones family[4] (Cleary, 1994)—but all from the point of view of the lawyers involved. A recent exception to this rule is the case of James Byrd, who was dragged to death behind a pickup truck in Jasper, Texas. Two journalists published books about this case (King, 2002; Temple-Raston, 2002), and Texas's new hate crime law was named in Byrd's memory. There are a few other exceptions as well, but not many. Again, the reasons are unclear.

In many communities, African Americans have had a poor relationship with law enforcement. Los Angeles and New York are perhaps the most obvious examples of this, but they are by no means alone. Several black students have related stories to me about what they perceived to be harassment at the hands of police in the cities and towns in my area of Northern California. As a result, black victims of hate crimes might be reluctant to report those crimes to the police. We do not know, therefore, what proportion of blacks has actually experienced hate crimes, what forms those crimes took, or what impact they had on their victims. Perhaps research will be undertaken soon to answer these questions, similar to the recent efforts that have been made to understand hate crimes against gays and lesbians.

ANTI-SEMITIC HATE CRIMES

The persecution of Jewish people predates the creation of the United States by more than a millennium. Prager and Telushkin (1983) claim, "While hatred of other groups has always existed, no hatred has been as universal, as deep, or as permanent as anti-semitism" (p. 17). In the pre-Christian era, Jews were often reviled by the Egyptians, Greeks, and Romans. Christians accused Jews of having killed Christ and of engaging in the ritual murder of Christian children (this latter accusation is known as *blood libel*), and Jews who refused to convert to Christianity were massacred during the Crusades. Martin Luther, who founded Protestantism, advocated the persecution of Jews, and the Catholic church issued numerous decrees against the Jews. During the time of the bubonic plague, Jews were blamed for poisoning gentiles.

Historically, anti-Semitism took many forms. Jews were often confined to ghettoes and prohibited from owning land. As a result, many Jews earned their living as merchants, traders, and bankers; this helped fuel the stereotype of Jews as greedy and wealthy. In addition, when governments were in need of cash, they often seized the property of Jews, which was more readily marketable than land. Defamation of Jews was often promoted to justify these seizures. Of course, those who owed Jews money were often pleased to participate in getting rid of their creditors.

Jews constituted a minority in every country in which they lived. Not only was their religion different, but so were their customs, language, and manner of dress. These differences were significant in Europe, where ethnic identity was (and is) considered extremely important. Jews' loyalty to ruling governments was always questioned, and they were perpetually considered to be outsiders, no matter how long they had lived in a particular area (Smith, 1996). Jews were denied citizenship in their own countries (Yinger, 1964) and suffered from many forms of discrimination and restriction, such as the requirement, adopted during the 20th century by the Nazis, that they wear yellow stars on their clothing.

Actions against the Jews were sometimes even more severe. On many occasions, they were forced to convert to Christianity. They were expelled from many countries, such as Spain during the Inquisition, and refused entry to others. Sometimes their communities were destroyed, and the Jews themselves were exterminated. In Eastern Europe, for example, it is estimated that as many as 100,000 Jews were killed between 1648 and 1658 alone (Yinger, 1964). (For several examples of the historical persecution of Jews, see Box 6.3.)

Box 6.3   Historical Examples of the Persecution of Jews

- 66: 50,000 Jews killed in Alexandria, Egypt
- 325: Christian Church formulates anti-Jewish policy
- 527-564: Justinian Code removed Jews' civil rights
- 632: Forced conversion of Jews to Christianity in Spain
- 1190: 150 Jews massacred at York, England
- 1215: Catholic Fourth Lateran Council required Jews to wear special clothing
- 1298-1299: Thousands of Jews killed in southern and western Germany
- 1348-1350: Jews accused of causing Black Death by poisoning wells; massacres result in Spain, France, Germany, and Austria
- 1492-1497: Jews expelled from Spain, Sicily, Lithuania, and Portugal
- 1715: Pope Pius VI issues edict placing restrictions on Jews
- 1791: Pale of Settlements established in Russia; Jews allowed to live only within the Pale
- 1885: 100,000 Russian Jews expelled from Germany
- 1905: First public edition (in Russian) of the *Protocols of the Learned Elders of Zion*
- 1917-1921: Pogroms in Poland, Hungary, and the Ukraine
- 1925: *Mein Kampf* published

SOURCES: Dickerson (1997), Grobman (1990), Maddison (n.d.).

The pogroms against the Jews in the late 19th and early 20th centuries were a major factor in the migration of large numbers of Jews to the United States. Although anti-Semitism was common in the United States as well, with prominent men such as Henry Ford speaking openly against the Jews, the situation was considerably better than in Europe. In part this was because of the depoliticization of religion in this country, in part because America was already quite diverse, and in part because there were other ethnic groups, such as blacks and Native Americans, who seemed even more different and against whom hatreds were especially focused (Smith, 1996).

Next to the enslavement of blacks, perhaps the greatest event of modern centuries that could be considered a hate crime was the Holocaust. Millions of European Jews were stripped of citizenship, civil rights, and property; subjected to a large array of sanctions and discriminations; forced into ghettoes and then concentration camps; and, ultimately, deliberately murdered. An estimated 6 million Jews died as a result of Hitler's Final Solution. This was approximately two thirds of Europe's Jewish population; in some countries, such as Poland and Germany, 90% of the Jews were exterminated (B. Levin, 2001a).[5] Most of the Nazis' ideas about Jews were borrowed from earlier anti-Semites, and many of those ideas persist among modern American white supremacists.

Anti-Semitism in the United States Today

Jews make up only about 3% of the American population. Many Americans are not personally acquainted with any Jews and know little about them. It may be surprising, therefore, that anti-Semitism serves as the core of almost all white supremacist dogma. As

Ezekiel (1995) found, although the average member of an extremist group has more personal rancor against blacks and other people of color, the leadership is largely anti-Semitic in its teachings. Extremist literature blames Jews for economic woes, communism, disloyalty to America, and for inciting other minorities to miscegenation and other per-ceived usurpations of white power. At times, the literature even accuses Jews of being literally the spawn of Satan. It is claimed that a Jewish conspiracy runs the country's (and the world's) media, banks, businesses, and governments. Because blacks and other people of color, as well as women, are assumed to be too incompetent to organize their own strug-gles for equality, it is claimed that the civil rights and feminist movements are also led by Jews.

On one hand, the picture is quite positive for today's American Jews. As a group, they are well above average on measures of socioeconomic status and are represented at a disproportionately high rate among members of Congress and among high-status pro-fessions such as physicians and attorneys. In general, surveys have showed significant decreases in anti-Jewish feelings and stereotypes in recent years as well (Dinnerstein, 1994; Smith, 1996). In addition, both the Catholic and the Lutheran churches have offi-cially repudiated anti-Jewish doctrine and have apologized for wrongs committed by their forebears.

On the other hand, anti-Semitism is hardly extinct. According to the Anti-Defamation League, in 2002, 17% of Americans who were polled held hard-core anti-Semitic beliefs, and an additional 35% were moderately anti-Semitic (ADL, 2002a).[6] Other surveys have shown that significant numbers of people continue to believe that Jews have too much power, that they are greedy, and that they are disloyal to America (Dinnerstein, 1994; Smith, 1996). Although anti-Semitism had been generally in decline since World War II, it showed increases in 2002, both as measured by surveys and by the numbers of anti-Jewish hate crimes and hate incidents that occurred (ADL, 2002a). This was quite likely due both to the events of September 11, 2001 (which were blamed, by some, on Jewish-led American foreign policies), and the crisis in the Middle East.

Furthermore, the Jews' very success can create problems. It lends credence to allega-tions that Jews are too powerful and yield too much influence upon the government. It also may stir up resentment among ethnic minorities whose socioeconomic positions are lower. Several researchers have found, for example, that anti-Semitism is more common among African Americans and Hispanics than among whites (ADL, 2002a; Dinnerstein, 1994; Smith, 1996). This many seem odd because Jews were active participants in the civil rights movement. Dinnerstein (1994) offers an explanation:

One means of identifying with the majority, and therefore gaining greater confidence and esteem, has been to take on the prejudices of the majority group. To some extent, black antisemitism has been a psychological, perhaps an unconscious one at that, response to their positions in American society. By adopting a majority prejudice, members of the group can feel superior to the group that they too despise. Moreover, like members of the dominant culture, a number of African Americans have been com-fortable scapegoating Jews for their difficulties. (p. 248)

Perry (2001) points out other possible reasons for black-Jewish antagonism, includ-ing religious beliefs (i.e., the continuing belief that Jews killed Christ), the conviction that Jews are responsible for maintaining a hierarchy that subordinates blacks, and resentment of Jews because they claim shared oppression. (See Box 6.4 for a discussion of reparations, which have been used to address historical wrongs against blacks, Jews, and others.)

Box 6.4   Reconciling With the Past

How should people with a history of being persecuted be compensated? How should companies with a history of participating in persecution be treated? These questions have arisen lately in several contexts: Jewish survivors of the Holocaust, Japanese American survivors of internment camps, and African Americans whose families were enslaved or otherwise maltreated. Here are a few examples of recent attempts to help reconcile past wrongs:

- In 1994, survivors of the village of Rosewood, Florida, received a $2 million settlement from the Florida legislature. In 1923, in response to a white woman claiming she'd been attacked by a black man, at least eight African American residents of Rosewood were murdered, and the entire town was burned to the ground.
- In 1988, President Reagan signed a law permitting payments to Japanese American survivors of internment camps.
- In 1998, Volkswagen admitted using slave labor during World War II and set up a fund to compensate survivors. A year later, the German government created a compensation fund for Nazi victims.
- In 2001, a commission recommended that Tulsa, Oklahoma, pay reparations to the survivors of a 1921 race riot that killed 39 African Americans. As of this writing, this recommendation has not been followed.
- In 2002, an African American law student filed a multibillion-dollar lawsuit against several corporations that were alleged to have profited from slavery.
- In 2002, the city of Chicago passed an ordinance requiring companies that do business in Chicago to disclose profits made from slavery.

Dinnerstein (1994) asserts that no systematic study has been made of the reasons for anti-Semitism among people of color. This certainly seems to be a subject that deserves further attention. Researchers may also want to consider whether hate crime laws actually exacerbate these sorts of tensions between minority groups. Jacobs and Potter (1998) argue that this may be the case because these laws encourage identity politics.

Another troubling aspect of modern anti-Semitism is that it has served as a vehicle for the alliance of some very disparate interest groups. White, black, and Islamic extremist groups are united in their hatred of Jews. As mentioned in Chapter 5, these groups' Web sites now often provide links to one another's Web sites and, sometimes, they share literature as well. Many of these groups are also united in denying that the Holocaust occurred (B. Levin, 2001a); in reality, they claim, the Holocaust was a hoax intended to secure money, sympathy, and influence for Jews.

In addition, many organizations on both the far right and the far left of the political spectrum are currently opposed to American military action in the Middle East. For many, this opposition takes the form of anti-Israel viewpoints, which, for many, mean anti-Jewish viewpoints. These kinds of coalitions are potentially dangerous because they increase the voice and power of extremist groups and because they may incite violent actions.

Why the Jews?

Why have so many people hated Jews for so long? For some bigots, this very history is taken as evidence that Jews deserve ill treatment. The Aryan Nations Web site (www.aryan-nations.org), for example, contains an animation listing numerous historical examples of Jews having been "banned, deported, executed, expelled, persecuted, [and] slaughtered," with the suggestion that these practices ought to be used against the Jews worldwide in the 21st century. The implication seems to be that there must be something seriously wrong with a group of people that has been so reviled.

White supremacist rhetoric aside, how can the continuing abhorrence of Jews be explained? There is no simple answer to this question, but a number of factors probably contribute to the continuing popularity of anti-Semitism. One of these is its long history; anyone wishing to preach anti-Semitism can easily find a large amount of literature to back his or her arguments. In fact, extremist Web sites often contain anti-Jewish quotes (some genuine, some not) from a large variety of sources, from Cicero through Martin Luther and Thomas Jefferson to Adolph Hitler and beyond.

A second factor, and one that should not be underestimated, is religion. One author asserts, "Christian viewpoints underlie all American antisemitism" (Dinnerstein, 1994, p. ix). Later, the same author writes, "No idea in Christian teaching has been more solidly implanted among adherents of the faith, and more devastating to Christian-Jewish relations, than the accusation that the Jews had killed their Savior" (Dinnerstein, 1994, p. xxi). It is no accident, either, that the groups with the most strained relationships with Jews are also, from a religious perspective, Jews' closest relatives: Christians and Muslims. Like some sort of global sibling rivalry gone massively awry, a group may feel the need to distinguish itself from a similar group through prejudice and hatred.

A third reason that Jews may be so vilified is that, compared with people of color, they can much more easily assimilate; they can easily "pass." This represents a particular threat to white supremacy because Jews can infiltrate the white power structure in a way that people of color cannot. This may explain why some white supremacist rhetoric is concerned with proving that Jews are a separate, nonwhite race and also why some white supremacists take pains to "expose" certain powerful people as being Jewish. It may also explain anti-Jewish sentiment among nonwhites, who may resent the unique ability of this minority group to join the majority.

A fourth reason for anti-Semitism, as mentioned earlier, is that Jews in America have achieved great socioeconomic success. Again, this success leads to resentment by both whites and nonwhites who feel that they have been left behind. It may also lead to fear. Hate groups hate a great many people, but the group they have the greatest fear of is the Jews because people of that group are presumed to have the power to carry out their nefarious plans.

Fifth, most Americans actually have little exposure to Jewish people and Jewish beliefs. Thus, Jewish customs seem strange and can easily be misunderstood. A personal story illustrates this. When I was in graduate school in Nebraska, one of my professors invited a large group of people to his house for a Passover seder.[7] A crew from a local news station came to film part of the seder because they were doing a story on the holiday. When the crew arrived, my professor told them to be sure to stick around for the sheep sacrifice. Both members of the crew nodded with wide-eyed acceptance, obviously not realizing that my professor had been joking! Although this incident was funny, it is not at all humorous that for centuries, many people believed (and some still do) that matzah, the unleavened bread that Jews eat on Passover, was made with the blood of Christian children.

Yet another reason that some people hate the Jews is that, as discussed in Chapter 5, Christian Identity adherents believe that Jews are not the chosen people and are not the children of Adam. In fact, some even believe that Jews are literally Satan's offspring and that good Christians are locked with them in an apocalyptic battle of good and evil. For members of the Christian Identity church, hatred of Jews is not only acceptable—it is actually required.

A final factor that contributes to modern-day anti-Semitism is Zionism. Many people equate Jews with Israel. When Israel takes actions with which non-Jews disagree, such as Israel's treatment of Palestinians, the non-Jews blame all Jews. If reports by the ADL (2002a) are accurate, anti-Semitic beliefs and action in the United States (and worldwide) began to increase at the same time as the crisis in the Middle East began to escalate. At a May 2002 rally at San Francisco State University, Palestinian supporters reportedly told Israel supporters, "Hitler did not finish the job!" (Goldsmith, 2002).[8] White supremacist Web sites frequently highlight anti-Israel and anti-Jewish material. David Duke, for example, wrote,

> They [the Jews] don't want us to know the real reasons why Americans are so hated—because it is the Jewish bosses of American foreign policy who are the one's [sic] responsible for this growing hatred of America. . . . Let me repeat that one more time. The primary reason we are suffering from terrorism in the United States today is because our government policy is completely subordinated to a foreign power: Israel and the efforts of world-wide Jewish Supremacism. (Duke, 2001)

Although there is ample literature on anti-Semitism, few studies have focused specifically on anti-Jewish hate crimes. Those that have are primarily the work of advocacy organizations such as the ADL rather than independent scholars. Therefore, once again we do not know how common hate crime victimization is among Jews or how often it gets reported. According to the FBI, in 2000, approximately 70% of hate crimes against African Americans were crimes against persons (as opposed to crimes against property), whereas only about 36% of anti-Jewish hate crimes were crimes against persons (FBI, 2001b). What accounts for this difference, and what are the repercussions? Do events like September 11 and the Middle-East crisis really increase anti-Semitism, or do they merely serve as a vehicle for rationalizing anti-Semitism that already exists? These are only a few of the questions that remain unanswered.

HATE CRIMES AGAINST GAYS[9]

Homophobia, like anti-Semitism and racism, has a long history, although perhaps one not as well documented. Although some cultures have accepted homosexuality, at many times and in many places gays have been prohibited, persecuted, and even executed. For example, during the Nazi era in Germany, homosexual activity was a criminal offense, and gay men were sent to concentration camps.

Antigay hate is unique in several respects:

- Unlike hatred based on race or religion, it is often not covered under hate crime legislation (see Chapter 3 for a discussion of this issue).
- Unlike virtually any other targets of hatred, gays are usually a minority within their own family. Thus, in addition to animosity from strangers and other members of the community at large, gay individuals often face rejection by their own relatives. (See Box 6.5 for a discussion of attempts to "cure" homosexuality.)

NARRATIVE PORTRAIT

EIGHT BULLETS

On May 13, 1988, Claudia Brenner and Rebecca Wight were on a backpacking trip in Pennsylvania. Brenner was 31 years old and finishing a graduate education in architecture. Wight was 28 years old and a graduate student in business at the University of Virginia. Here is Claudia Brenner's account:

The first bullet: When the first bullet hit me, my arm exploded. My brain could not make the connections fast enough to realize I had been shot. I saw a lot of blood on the green tarp on which we lay and thought for a split second about earthquakes and volcanoes. But they don't make you bleed. Rebecca knew. She asked me where I had been shot. We had encountered a stranger earlier that day who had a gun. We both knew who was shooting us. Perhaps a second passed.

The second bullet: When the second bullet hit my neck I started to scream with all my strength. Somehow the second bullet was even more unbelievable than the first.

The third bullet: The third bullet came and I now know hit the other side of my neck. But I had lost track of what was happening or where we were except that I was in great danger and it was not stopping.

The fourth bullet: I now know a fourth bullet hit me in the face. Rebecca told me to get down, close to the ground.

The fifth bullet: The fifth bullet hit the top of my head. I believe that Rebecca saw that even lying flat I was vulnerable and told me to run behind a tree.

The sixth bullet: The sixth bullet hit Rebecca in the back of her head as she rose to run for the tree.

The seventh bullet: The seventh bullet hit Rebecca's back as she ran. It exploded her liver and caused her to die.

The eighth bullet missed.

It is not surprising that Stephen Roy Carr believed us both dead. He shot to kill. The neck. The head. The back. A single bolt action rifle that he loaded, shot, and unloaded eight times. Surely he believed us both dead or he would have used more of the 27 rounds of ammunition he left in his haste to get away.

He shot us from where he was hidden in the woods 85 feet away, after he stalked us, hunted us, spied on us. Later his lawyer tried to assert that our sexuality provoked him.

He shot us because he identified us as lesbians. He was a stranger with whom we had no connection.

He shot us and left us for dead. . . .

I remember distinctly as I walked alone on the trail after the shooting, how intensely silent were the normal forest sounds after the explosions of the gunshots. I wondered if the birds were communicating about the horror they had witnessed that afternoon. Less than a week later, thunderstorms covered our campsite with water, washing away the blood, cleaning the area of violence. Although the gunshots have for the most part quieted in my mind; though my wounds have healed; though I now speak widely of the homophobia that stole the life of a lover, a sister, a daughter, a friend; I will always walk with an awareness of the tragedy I knew on that silent trail.

SOURCE: Brenner (1992).

- Unlike other forms of bigotry, antigay bias is still socially acceptable. Certainly homosexuality is more socially acceptable than it was only a decade or two ago, but today even prominent public figures such as politicians do not hesitate to openly express homophobic views. Discrimination based on sexual orientation is still legal and, in several states, homosexual sexual activity remains a criminal act.
- Probably because of the potential repercussions of being openly gay, many gays choose to remain closeted to various extents. A closeted victim of a hate crime who chooses to report the crime faces the possibility of having his or her sexual orientation exposed.
- Like religion, a person's sexual orientation is usually not obvious unless he or she chooses to make it so. Targeting of victims, therefore, may take place via a different process than occurs with more apparent characteristics such as race.
- Opposition to homosexuality remains the official doctrine of many mainstream religions. Homophobes need not belong to a fringe group like Christian Identity to rationalize their views.
- Homosexuality, and especially male homosexuality, remains associated in many people's minds with AIDS, a deadly and communicable disease. Homophobia existed long before the emergence of AIDS, but the disease becomes another way to rationalize antigay bigotry.

How Common Are Antigay Hate Crimes?

As already mentioned, relatively few hate crimes ever get reported to the police. This is probably especially true for antigay hate crimes because gay victims are likely to fear further persecution by the police and, in some cases, exposure of their sexual orientation. Measuring the rate of antigay hate crimes is difficult not only because of the problems associated with police data but also because the precise size of the gay population is unknown (Green, Strolovitch, Wong, & Bailey, 2001).

Box 6.5   "Curing" Homosexuality

The *DSM*, or *Diagnostic and Statistical Manual of Mental Disorders* (2000), is the Bible of the mental health profession. Until 1973, it listed homosexuality as a mental disorder. In the current version, DSM-IV-TR, homosexuality is not considered a mental illness.

In recent years, however, a number of organizations and individuals have declared the success of "reparative therapy" or "conversion therapy" in which gays are changed into heterosexuals. Little or no scientific evidence supports these claims, however, and both the American Psychological Association and the American Psychiatric Association have issued resolutions against these therapies. Even if it were possible to change sexual orientation, it would be ethically questionable to do so.

Contrary to the claims of some, sexual orientation is probably not a choice. A growing body of recent research suggests that sexual orientation may be, at least in part, biologically determined. For example, differences have been found in the brain structures of gay and straight men, and fairly high concordance has been found in the sexual orientation of identical twins.

SOURCES: American Psychiatric Association (1998); American Psychological Association (1997); Bailey & Pillard (1991); Herek (1999); LeVay (1991).

Despite the fact that relatively few antigay hate crimes get reported to police, the official data imply that gays are one of the primary victims of hate crimes. In 2000, more than 15% of the hate crimes reported to the FBI occurred because of sexual orientation (FBI, 2001b), and the proportion is higher in cities with large gay populations (Gerstenfeld, 1998). A better source of information about hate crimes than official data may be self-reports of victims and offenders. A growing body of literature exists in this area, and it all strongly suggests that antigay hate crimes are distressingly common. In an early review of the literature, Berrill (1992) concluded that the problem was severe.

Some of the most thorough work in this field has been conducted by Gregory Herek and his colleagues. In one study (Herek, Gillis, Cogan, & Glunt, 1997), 20% of the respondents had experienced a hate crime in the previous year, and an additional 5.5% reported an attempted hate crime. More than half of the respondents had been the victim of an actual or attempted hate crime at some time in their lives. These data were consistent with the results of earlier studies (Berrill, 1992; Hershberger & D'Augelli, 1995; Pilkington & D'Augelli, 1995). In a later study that used a larger sample of 2,259 subjects, Herek, Gillis, and Cogan (1999) found that about one in four gay men and one in five lesbians had been the victim of an antigay hate crime at least once in their adult lives.

Other researchers have also found high rates of victimization among gays. Berrill (1992) reported that in the research he reviewed,

> The median proportion of respondents who were verbally harassed was 80%; 44% were threatened with violence; 33% had been chased or followed; 25% were pelted with objects; 19% experienced vandalism; 17% were physically assaulted; 13% were spat upon; and 9% experienced an assault with an object or weapon. (p. 20)

Dean, Wu, and Martin (1992) found that between 1984 and 1990, approximately 10% to 17% of the gay male population of New York City experienced a hate crime each year. In a survey of lesbian and bisexual women in San Francisco (von Schulthess, 1992), 57% of respondents had experienced threats or actual violence due to their sexual orientation. Hunter (1992) examined the data from 500 youths who were involved with a New York City agency that provides services to gay teenagers. Forty percent of the youths had experienced violent physical attacks, and 46% of those attacks were related to their sexual orientation. These high rates of victimization are not confined to the United States. One study of 527 gays in Stockholm, Sweden, found that 27% of the respondents had been victimized because of their sexual orientation (Tiby, 2001).

Whereas most research on antigay hate crimes has focused on victims, Franklin (2000) examined perpetrators. She surveyed 489 community college students in Northern California. Ten percent of the sample reported having threatened or physically assaulted someone they believed was homosexual. Men reported much higher levels of offending: 79% of the threats and 92% of the physical assaults.

Taken together, these studies provide strong evidence that antigay violence is common and widespread. Berrill's (1992) assertion that this is a severe problem appears well-founded.

The research also gives some indication of the patterns of victimization. Gay men are victimized more frequently than lesbians (Berrill, 1992; FBI, 2001c; Herek et al., 1999; Tiby, 2001). Lesbians and gay men of color experience more hate crimes than whites (Comstock, 1989; von Schulthess, 1992). Threats and physical violence are more common than property crime (FBI, 2001c; Herek et al., 1997). As with other kinds of hate crimes, most perpetrators are young males (Berrill, 1992; Franklin, 2000; Herek et al., 1997; Herek, Cogan, & Gillis, 2002). Although most hate crimes against gays are committed by lone assailants, they are more likely than ordinary crimes to involve multiple perpetrators

(Herek, Cogan, et al., 2002). Crimes against gay men tend to happen in public places, and those against lesbians tend to happen in private locations (Herek et al., 1997). The proportion of incidents that were reported to the police was low, ranging from 15% to 46% (Herek, Cogan, et al., 2002; Herek et al., 1999; Tiby, 2001; von Schulthess, 1992).

Like the official data, self-report studies have their flaws (see Herek & Berrill, 1992a). Probably the most serious of these is that it is extremely difficult to obtain a representative sample. Researchers have obtained participants from gay clubs and bars, community events, community centers and organizations, fliers, and snowball sampling (i.e., recruiting friends and acquaintances of existing participants). These methods tend to result in respondents who live in urban areas, who are open about their sexual orientation, who are active in the gay community, and who are of higher socioeconomic levels. Conversely, those who are more closeted or live in more rural areas or are poorer or less educated will not be included in the research. This is an extremely difficult problem to overcome, although Herek and his colleagues have attempted to do so by using multiple recruitment methods (Herek et al., 1999).

There are other potential problems with such surveys as well. Questions may be poorly worded and may fail to distinguish between hate crimes, noncriminal harassment such as name-calling, and discrimination. Respondents may misunderstand the questions, may be confused about the legal definitions of various acts, may have faulty memories, or may purposely over- or underreport. Of course, many of these difficulties can be mitigated through careful construction of survey instruments.

Despite their potential shortcomings, self-report studies provide an extremely valuable supplement to official hate crime data. Moreover, the similarity of the results from research conducted by a number of different scholars employing a variety of methodologies suggests that those results can be accepted with reasonable confidence.

Why Homophobic Hate Crimes?

The high prevalence of hate crime victimizations among gays raises an obvious question: Why? What causes people to commit acts of violence against others simply because of their sexual orientation? As always, there is no simple answer to this question; instead, there are a number of potential explanations, and each individual offender's behavior is probably best understood as some combination of these.

It is likely that one of the foremost causes of homophobic hate crimes is the frequency and social acceptability of antigay bias in the United States. In some respects, attitudes toward gays have improved considerably in recent years, and positive images of gays are much more common in the media. Several studies have found that increasing numbers of Americans support civil rights for gays, such as recognition of same-sex unions, prohibition of employment discrimination, and inclusion of sexual orientation within hate crime legislation (Fredella, Carroll, Chamberlain, & Melendez, 2002; Herek, 2002a, 2002b; Yang, 1998). Increasing tolerance toward gays has been reflected in several ways: A majority of states now include sexual orientation in their hate crime statutes, governments and businesses have granted increasing recognition to same-sex unions, and several states have repealed antisodomy laws or declared them unconstitutional.

All is not well, however. Although many people might support gays' civil rights in the abstract, a large proportion of the population continues to feel personal discomfort about homosexuality or to express religious or moral condemnation of it (Herek, 2000b). Olivero and Murataya (2001) report that, in their study of criminal justice students, levels of homophobia were particularly high among students planning on going into careers in law enforcement. Moreover, large numbers of people still report engaging in (noncriminal) antigay behaviors, such as verbal harassment (Franklin, 2000; Rey & Gibson, 1997; Schope & Eliason, 2000).

Antigay ideology remains institutionalized throughout America as well. Many religious organizations continue to condemn homosexuality, as do many prominent public figures. In the vast majority of jurisdictions, gays may be discriminated against in housing and employment, are prohibited from marrying their partners, and, even when committed to a long-term, monogamous relationship, are denied the benefits that spouses receive (such as tax breaks, health benefits, and inheritance rights) (Herek, 1992). Sodomy is still illegal in 18 states, five of which specifically prohibit only homosexual sodomy (Harrison, 2000). Some states prohibit gays from adopting children, and courts have denied biological parents custody of their children because those parents were gay. And under the current "don't ask, don't tell" policy, people who are openly gay are forbidden from serving in the United States military.

Thus, despite some improvements, American society remains heterosexist. As discussed in Chapter 4, cultural attitudes and images have an important impact on hate crime behavior. Those who are already homophobic can justify their behaviors as socially acceptable. To those who are not especially biased but who are seeking thrills and excitement, as appears to be the case with the majority of hate crime offenders (see Chapter 4), may believe that gays are acceptable targets. In Franklin's (2000) survey of college students, those who reported having committed antigay hate crimes were significantly more likely than other respondents to believe that their friends were opposed to homosexuality.

A second likely source of antigay hate crimes is religion. As already mentioned, many religions denounce homosexuality. Of course, religions generally denounce violence as well, but this latter point seems to be lost on some people. Consider, for example, this statement from a 17-year-old who participated in an attack on some gay men:

> I guess gay people do what they want to do and I do what I want to do. I don't think what they do is right. I can't see two guys making out. It goes against the Bible. All my life everyone told me it was wrong, and I guess that just sticks with you. But they have the same rights as I do and shouldn't be bothered. (Weissman, 1992, p. 171)

This particular individual now expresses some remorse for his actions and yet continues to view homosexuality as immoral.

Religious views color individuals' perceptions and attitudes of gays, and therefore their behavior toward them. In addition, some conservative religious organizations have pursued a specifically antigay agenda. As a result of their efforts, in 1996 Congress passed the Defense of Marriage Act, which permits states to refuse to recognize same-sex marriages from other states and, in 2000, California voters passed a constitutional amendment stating that only marriages between a man and a woman are valid (Broadus, 2000). Colorado enacted a constitutional amendment in 1992 (which was later struck down by the U.S. Supreme Court) prohibiting the inclusion of sexual orientation within antidiscrimination regulations. Again, these kinds of official actions put an implied seal of approval on antigay bigotry.

Another reason that some people commit antigay hate crimes is that they view this as a way to prove their own masculinity. By attacking a gay man, an offender simultaneously demonstrates to his peers that he rejects homosexuality and that he is "man enough" to behave violently (Perry, 2001). After all, in our culture, as in many others, maleness is equated with aggression.

There are some empirical data to support this hypothesis. Franklin (2000) found that college students who had physically attacked gays scored higher on a measure of masculine ideology than did students who had not engaged in such behaviors. Moreover, a significant number of attacks on lesbians involve a sexual assault (Herek et al., 1997). Sexual dominance of a woman (especially one who does not choose men as sexual partners) is another way for men to establish their masculinity.

Related to this factor is the evidence that men who are homophobic may also be unconsciously sexually attracted to other men (see Chapter 4). As Herek (2000b) states, sexual orientation is more accurately seen to exist on a continuum rather than as an either-or dichotomy:

> Some individuals become concerned or uncertain about their sexual identity. Because of the stigmatized status of homosexuality, such individuals may experience anxiety at the prospect of being labeled gay or lesbian, which they may externalize in hostility or overt aggression toward gay people. (pp. 253-254)

Herek (2000b) argues that sexual prejudice is incited not as much by attitudes toward gays as by attitudes toward the self and one's personal identity. Gay-bashing becomes a way of symbolically expressing those attitudes.

To some extent, antigay hate crimes, especially those committed against men, may be incited by the association many people have between homosexuality and AIDS. Again, Gregory Herek has been a major contributor to the literature in this area. Herek and Capitanio (1999) found that most Americans do associate AIDS with male homosexuality and that people tend to blame gay men with AIDS for bringing the disease on themselves. One quarter of respondents in 1999 said that they believed people who contracted HIV through sex or drug use got what they deserve. A later study (Herek, Capitanio, & Widamen, 2002) had similar results.

When invective against gays appears in extremist literature, it is frequently supported by three kinds of arguments. First, it is claimed, homosexuality is morally and biblically wrong. Second, extremists assert that gays have gone too far in their push for "special rights." Finally, hate groups argue that gays are responsible for the spread of AIDS and that this endangers innocent people. The following excerpt from the official platform of the Knights of the Ku Klux Klan is typical:

> We support the placing of all persons HIV positive into national hospitals. While the AIDS virus is almost inclusive [sic] to homosexuals and those not of European ancestry, many innocent people have contracted the virus. Despite the moral character of a person, the virus is still highly contagious with new and deadlier forms coming out constantly. *Everyone who gets it dies!* Aids [sic] carriers should receive proper medical care while a cure is being researched. this [sic] is the only way to stop the spread of the disease. They should be kept from coming into contact with uninfected people. (Knights Party, 1991)

The same platform also favors a law against the practice of homosexuality because "this is a Christian nation and the Bible condemns homosexual activity and the perversion of our society which it encourages" (Knights Party, 1991). Some extremists combine the three arguments by stating that AIDS is a punishment by God upon gays for their immoral and socially unacceptable behavior.

A question arises concerning AIDS: Are gay men stigmatized because they are associated with AIDS, or is AIDS stigmatized because it is associated with gay men? Herek and

Capitanio (1999) argue that it is the latter that is more accurate, citing as evidence the fact that overall, attitudes toward homosexuality did not change after the public became aware of the AIDS epidemic. Rather than producing homophobia, AIDS became a way of rationalizing it.

In many cases, antigay violence is probably also engendered by offenders' perceptions that gays have violated gender roles. Like most cultures, ours has established constructs of what it means to be female and, especially, male.[10] If the integrity of these constructs is to be maintained, their borders must be patrolled and trespassers sanctioned. Gays, particularly men, do not fit neatly into these constructs; gay men are not perceived to be "real men." As Perry (2001, p. 116) puts it, offenders believe that gays must be punished because they refuse to "do gender appropriately."

How does male homosexuality transgress gender roles? Arguably, gay men have voluntarily relinquished the privilege of male domination over women (Kite & Whitley, 1998). This defection from the ranks poses a threat to the established hegemony. Moreover, to allow oneself to be sexually penetrated is viewed as permitting subjugation. Extremist depictions of gays (and members of other groups) often include images of forced sodomy and penetration (Daniels, 1997), and it is not unusual for gay victims (both male and female) of hate crimes to be raped. Thus, antigay hate violence, like other kinds of hate crimes, can be seen as primarily motivated by the desire to retain power.

Evidence for this argument, although circumstantial, is strong. Heterosexual men's attitudes toward gay men are much more negative than those toward lesbians. In fact, Herek (2002a) reports that heterosexual men's attitudes towards lesbians were generally at least as favorable as those toward heterosexual women. Lesbians are seen as less threatening to masculinity and male gender roles. Moreover, heterosexual women generally have more positive attitudes toward gays (both male and female) than do heterosexual men. Heterosexual women's attitudes also do not differ between gay men and lesbians (Herek, 2000a, 2000b). Women engage in antigay behaviors much less frequently than do men (Franklin, 2000). Women, who are traditionally already in a subordinate position to men, do not perceive homosexuality as a threat to their power.

Stereotypes about gays, especially gay men, also likely fuel some hate crimes. Gay men are perceived to be weak. In fact, any man who is perceived as cowardly or feeble is apt to be labeled a "pansy" or a "faggot." Young men out for thrills may then consider a gay person to be an easy target, one who will not fight back. They may also believe (for good reason) that gays are unlikely to report an attack to the police or that the police will fail to respond meaningfully, therefore decreasing the attackers' chances of being punished for their actions.

Finally, antigay violence may be fueled in part by the gay rights movement. The gay rights movement, like the movements for African American and women's rights, took shape in the mid-20th century; it gained particular momentum in 1969 after the Stonewall Rebellion in New York City.[11] Since that time, gays in America have made significant advances in the legal and political arenas and in social acceptability. Whenever a subordinate group is perceived to be gaining influence, some members of the dominant group act to quash these advances. Earlier in this chapter, we saw how this was the case with African Americans immediately after the Civil War and again during the civil rights era. Similarly, Susan Faludi (1992) wrote of the backlash against feminism. It is quite probable that some antigay violence is spurred by the same motivations. Again, this is an issue of power.

A fair amount of research has been conducted specifically on antigay hate crimes and, consequently, we know more about these offenses than about other kinds of hate crimes.

One major question that remains, however, is how antigay hate crimes can be prevented. I discuss hate crime prevention programs in Chapter 7. However promising some of these programs may seem in the abstract, however, they are unlikely to be very successful as long as prejudice due to sexual orientation remains prevalent and socially acceptable.

OTHER VICTIMS

As I write this section, I hesitate to call it "Other Victims" because that seems to imply that these people are marginalized even as victims and that their suffering is not important. I do not intend either implication. The problem is that hate crimes against other groups are reported with less frequency, and the research tends to be scarce or nonexistent. As little as we know about hate crime victims in general, we know even less about those who are chosen for reasons other than being black, Jewish, or gay.

Ethnicity

We do know that some people have crimes committed against them because of their ethnicity, frequently because they are Americans of Latino or Asian or Middle Eastern ancestry. Hatred against these people also has a long history in America. During the same time that blacks were being persecuted (and sometimes lynched) in the South, Asian Americans and Mexican Americans were receiving similar treatment in the West and Southwest, much of it government sanctioned. For example, federal laws restricted Chinese immigration in the 1880s and 1890s, and several states passed laws in the early 1900s that prohibited noncitizens from owning land. Because federal law also barred non-whites from becoming naturalized citizens,[12] these state laws effectively kept Asian immigrants from possessing real property (Chen, 2000).

Official discrimination based on ethnic group membership continued throughout the 20th century and continues into the 21st. During World War II, Americans of Japanese ancestry (but not those of German or Italian ancestry) had their property seized and were placed in internment camps, even as their sons fought overseas under the United States flag. Very few people in my town (Turlock, California) are aware that the same grounds where they go to see prize Holsteins or a tractor pull were used during the war as an "assembly center" for Japanese Americans. The Supreme Court has permitted border patrols to stop and question those who look as if they might be Mexican, as far as 100 miles north of the Mexican border. Even more recently, laws have been passed restricting immigrants' access to social services, eliminating bilingual education, and mandating "English only."

Hate crimes on the basis of ethnicity are far from rare. You may have heard of Vincent Chin, a Chinese American man who was murdered in 1982 in Detroit by two white auto workers who apparently mistook him for Japanese. But he is only one victim; there are many more. In 1996, approximately 60 Asian American students and staff members at the University of California, Irvine, received a series of e-mails threatening them with death (Lee, 2001; see Box 6.6). I have already discussed the huge increase in crimes against Arab Americans (and those mistaken for Arab Americans) that occurred after September 11, 2001; another increase, although not quite as dramatic, had occurred during the Gulf War.

One thing that distinguishes victims chosen because of their ethnicity from those we have discussed in more detail is that the former are viewed as perpetual foreigners. A person of Asian, Latino, or Arab background, no matter how long his or her family has lived in

Box 6.6   The UCI Hate E-Mail Case

On September 20, 1996, the following e-mail was sent to Asian American students and staff at the University of California, Irvine:

Hey Stupid fucker,

As you can see in the name, I hate Asians, including you. If it weren't for asias [sic] at UCI, it would be a much more popular campus. You are responsible for ALL the crimes that occur on campus. You are responsible for the campus being all dirt. YOU ARE RESPONSIBLE. That's why I want you and your stupid ass comrades to get the fuck out of UCI. IF you don't I will hunt all of you down and Kill your stupid asses. Do you hear me? I personally will make it my life career to find and kill everyone of you personally. OK?????? That's how determined I am.

Get the fuck out,

Mother Fucker (Asian Hater).

The anonymous composer of this message, a 19-year-old former student who had been expelled, was soon identified. After his first trial ended in a mistrial, Richard Machado was found guilty, sentenced to a year in prison, a $1,000 fine, and psychiatric counseling.

SOURCE: Lee (2001).

the United States, will never be viewed as truly "American" (Chen, 2000; Gerstenfeld, 2002; Petrosino, 1999). I may be Jewish, but, by virtue of my white skin, I can assimilate and be accepted. Despite virulent racism, very few people deny that African Americans are, in fact, rightfully *American*. But when American-born Michelle Kwan participated in the 1998 Olympics, one MSNBC headline announced, "American beats out Michelle Kwan" (Chen, 2000). Hwang (2001) tells of Korean American Sylvia Kim, who was brutally attacked in San Francisco because her assailant believed she was Chinese. As she lay on the ground with a shattered hip, two white tourists approached her, and the first thing they asked her was whether she spoke English.

This perpetual otherness means that hate crimes against members of these groups often have a particularly symbolic nature. Chen (2000) argues that "hate crimes against Asian Americans take on the unique dimension of operating as a form of border control and protection of the nation against the foreign 'alien'" (p. 72). It is not only a desire to maintain power that spurs these acts but also xenophobia and the perceived need to maintain national sovereignty. People who attack these "others" may actually perceive themselves as fulfilling a patriotic duty.

Disability

In June of 1989, in the upper-middle-class suburb of Glen Ridge, New Jersey, 13 members of the high school football team lured a 17-year-old mentally retarded girl into a basement by promising her a date with one of them. While their friends cheered them on, four members of the group sexually assaulted the girl with a baseball bat, a broomstick, and a stick. The next day, a group of 30 boys attempted, unsuccessfully, to entice her into the basement again (Lefkowitz, 1998). Was this a hate crime?

Currently, about half of the states with hate crime laws include disability as one of the protected categories (ADL, 2001). Nevertheless, there have been virtually no scholarly examinations of this topic. In 2000, the FBI recorded only 36 hate crimes based on physical or mental disability (FBI, 2001b). It is likely that the true number is much higher because some victims, especially those with mental handicaps, are unable to report the crimes, and police officers may be unlikely to categorize them as bias motivated. However, there is no empirical evidence at all on the actual prevalence of hate crimes against the disabled.

The lack of attention to these crimes might seem understandable, given their small proportion of the reported total. On the other hand, there are two reasons why hate crimes against the disabled merit scrutiny. First, compared with all the other potential victims of hate crimes, these victims are truly in a vulnerable position. They are more likely to rely on other people (who might take advantage of them) for daily necessities, and they may be physically or mentally unable to protect themselves from predation. The disabled are often invisible and unheard in our society.

Second, as Grattet and Jenness (2001b) point out, people with disabilities make up one of the country's largest minority groups; about one in five Americans has some kind of disability, and about one in 10 has a severe disability. Moreover, this is a group that can potentially claim anyone at any time. My race, ethnicity, gender, sexual orientation, and religion are all more or less immutable, but I could get maimed in a car accident this very afternoon. Even if I am fortunate enough to remain relatively healthy and whole, as I age there may come a time when my physical or mental faculties fail. I ought to be interested in hate crimes against the disabled out of self-interest, if nothing else.

Some people might argue that crimes committed on the basis of disability should not be considered hate crimes because these offenders do not truly *hate* the victims; they are merely choosing them because they are vulnerable. As we have already seen, however, there is good evidence that *most* hate crimes are not really committed out of hate. People commit these crimes for excitement, or to gain social approval, or to confirm their gender or sexual identity, or to maintain power hierarchies. The rape described at the beginning of this section probably served the same symbolic function for the perpetrators as would gay-bashing or other more paradigmatic forms of hate violence.

Grattet and Jenness (2001b) argue that, in fact, disability has many of the same characteristics of other groups that are included within hate crimes legislation, including a history of discrimination and the presence of a group identity. Moreover, those who victimize the disabled, like those who victimize members of other groups, frequently seek to control their targets.

Gender

As discussed in Chapter 3, the inclusion of gender within hate crime laws has been controversial. As of 2001, 19 states included gender or sex as a protected category (ADL, 2001). The FBI hate crime data do not include gender-based hate crimes at all. In California in 2001, only 15 hate crimes (less than 1% of the total recorded) that were reported were based on gender (California Attorney General, 2002).

Of course, if one were to take an expansive view of gender-based hate crimes, these numbers would be considerably greater. In 2001, there were 9,882 forcible rapes reported

NARRATIVE PORTRAIT

SWASTIKAS IN SAN FRANCISCO

The author of the following excerpt, Victor M. Hwang, is an attorney. He directs the Hate Violence/Race Relations project of the Asian Law Caucus in San Francisco. He is also coauthor of the National Asian Pacific American Legal Consortium's annual audit on anti-Asian violence.

The Sunset District of San Francisco is an affordable residential and small-business community located in the western section of the city, running along Golden Gate Park along Irving and Judah streets. It is a culturally diverse and middle-class neighborhood with a long-established Russian-Jewish community and a rapidly growing Asian American immigrant population. The Asian American population of the Sunset District has doubled in recent years, and many now refer to the area as the New Chinatown. The area has historically prided itself on its "mom and pop" stores and has been highly resistant to the influx of chain stores and fast-food franchises.

In 1996, a Chinese American business owner opened a Burger King franchise in the area, which was immediately met with community resistance, both reasoned and racist. While some residents protested the change in the neighborhood character, others posted flyers calling for "Chinks and Burger King Out of the Sunset." The Burger King was subject to a barrage of vandalism, graffiti, and protests throughout the coming months, continuing to the present day.

In February of 1997, an individual or a group of individuals known as the SWB or "Sunset White Boys" carved swastikas into the glass storefronts of nearly two dozen Asian American businesses, mostly along Irving Street. The placement and selection of the swastikas were particularly ominous, in that primarily Asian-owned businesses were targeted and non-Asian businesses were passed over, with the exception of a Caucasian-run karate studio with Asian lettering on the storefront. The clinical precision exercised in the choice of targets not only indicated a familiarity with the community, leading people to suspect this was an "inside" job, but carried the biblical overtones of genocide and divine retribution.

The vandalism ranged from small red spray-painted swastikas accompanied by the initials SWB to three-foot-high swastikas carved with great force into the glass using some sharp instrument. A great deal of attention and energy were focused in particular upon the Bank of the Orient, with the swastika carved prominently next to the word Orient.

Surprisingly, many of the store owners were immigrants from China and Vietnam who confessed ignorance at the significance of the swastikas. All they knew was that they had been vandalized once again, and due to the indifferent or hostile treatment that they had received at the hands of police in previous cases of vandalism, most failed to even report the occurrence. Many did not realize that other Asian businesses along the street had suffered similar etchings, and more than a week passed without any action being taken; the swastikas remained prominently displayed to the public. . . .

Even after I spoke with them, some of the owners indicated that they did not intend to replace the glass panes defaced with swastikas since vandalism was rampant and they would just be hit again after spending the money. It was chilling to see customers and families coming to the area to shop or do business as usual in broad daylight with each of the storefronts marked by the violent emblems.

SOURCE: Hwang (2001).

in California (California Department of Justice, 2002). More than 90,000 rapes were reported to the FBI in 2000 (FBI, 2001a). This number is approximately 10 times greater than the total number of hate crimes reported.

From a theoretical standpoint, there is good reason to consider most rapes, as well as many cases of domestic abuse, hate crimes. Like the other forms of hate crime we have explored, these offenses are often committed out of the desire to subjugate the victim and express the perpetrator's masculinity and power. The ideologies of masculine superiority are, as we have seen, important not only to individual offenders but also to many organized hate groups as well.

On the other hand, as discussed in Chapter 3, there are potential dangers in treating gender-based crimes as hate crimes. One possibility is that, given the sheer number of rapes and instances of domestic assaults, gender-based hate crimes could overwhelm the realm of hate crimes. Other forms of bias-motivated crime might then not get the attention they deserve. Conversely, rape and domestic abuse, which surely merit consideration in their own right, could conceivably be subsumed and forgotten under the broader rubric of hate crimes. In any case, it is very clear that, whatever the laws may say, in practice it is extremely rare for a crime to be classified as a gender-based hate crime.

CONCLUSION

Surprisingly little is known about the experiences of hate crime victims. It is clear that the victimization of certain groups has a very long and cruel history. It is also clear that the motives behind this victimization, although varied, do share a common theme of the desire to maintain power hierarchies and control others.

What are the subjective experiences of hate crime victims? In what ways do hate crimes or the fear of hate crimes alter people's behavior? How many gay couples avoid holding hands in public? How many Jews and Muslims hesitate before displaying symbols of their religion? How many victims of hate crimes remain uncounted and unnoticed? Currently, we cannot even begin to answer these questions.

DISCUSSION QUESTIONS

1. What are the methods that are used to measure the number of victims of hate crimes? What are the advantages and disadvantages of these methods? If you were a researcher who wanted to study hate crime victimization, how would you go about conducting a study?

2. What patterns of hate crime victimization seem to be evident from the data? How do you think these likely compare to the *real* patterns that exist in hate crimes, including the unreported crimes? Which victims do you think are most underrepresented in the data, and why?

3. Why do you think that so little research has focused on hate crimes against African Americans? What would you propose as a solution to this problem?

4. What do you think caused the rapid increase in church burnings during the mid-1990s? What effects were these burnings likely to have on communities? Why do you think the number of burnings rapidly decreased after a few years?

5. What stereotypes about Jews have you heard or seen? What were the sources of these stereotypes? Why do you think certain stereotypes about Jews have persisted for many centuries?

6. What do you think explains the relatively high rates of anti-Semitism among people of color? Describe a policy or program you would design to help address this problem.

7. Two fairly controversial recent issues have been reparations for victims of the Holocaust and for African Americans whose forebears were slaves. What are the problems associated with these issues? If you believe that reparations are a good idea in either case, who should pay them, to whom, and how should the amounts be calculated?

8. Olivero and Murataya (2001) report that levels of homophobia were high among the law enforcement students they studied. Why do you suppose this is? What are the repercussions? What strategies or policies would you propose for dealing with this?

9. Research suggests that, although Americans today increasingly support civil rights for gays, homophobic attitudes and behaviors are still extremely common. What might explain this apparent contradiction? What methods could be used to improve those attitudes and behaviors?

10. This chapter proposes that many Asian, Latino, and Arab Americans are targeted for hate crimes because they are viewed as permanent foreigners. What evidence is there to support or refute this claim? Is there any realistic way to have a society in which citizenship is not defined by race or ethnicity?

11. Some targets of hate crimes are simply classified, both within this chapter and elsewhere, as "other" (when they are considered at all). What are the consequences of being marginalized even within a realm that seeks to battle bigotry? Is this another danger of the "identity politics" of which Jacobs and Potter (1998) speak?

INTERNET EXERCISES

1. For a good bibliography on lynching, visit www.ac.wwu.edu/~jimi/cjbib/lynching.htm or www.uky.edu/~dolph/HIS316/projects/brianbib.pdf. You can read some essays about lynching at www.crimelibrary.com/classics2/carnival/. A site with some interesting quotes about lynching (and some horrible photos) is African American Holocaust at www.maafa.org. The following site contains general information, as well as some excerpts from historical documents: www.spartacus.schoolnet.co.uk/USAlynching.htm. James Allen has compiled a photographic history of lynching; a Web version of his museum exhibit is viewable at www.journale.com/withoutsanctuary/.

2. There are several good sources of information on the Internet about the Holocaust, including The Holocaust History Project (www.holocaust-history.org/), A Teacher's Guide to the Holocaust (http://fcit.coedu.usf.edu/Holocaust/), Remember.org (www.remember.org), and the Nizkor Project (www.nizkor.org). The Museum of Tolerance at the Simon Wiesenthal Center also has extensive information available online at http://motlc.wiesenthal.com/. The following site has information on non-Jewish victims of the Holocaust: www.holocaustforgotten.com. An annotated bibliography of sources related to the Nazi persecution of homosexuals can be found at http://members.aol.com/dalembert/lgbt_history/nazi_biblio.html.

3. One major problem concerning gays is the prevalence of harassment and violence on college campuses. GLSEN (Gay, Lesbian, and Straight Education Network) is one organization dedicated to improving this situation. It has created a Safe Schools Action Network, which you can visit at http://glsen.policy.net. You can also read GLSEN's annual National School Climate Survey (www.glsen.org/binary-data/GLSEN_ARTICLES/pdf_file/1307.pdf), which documents the extreme frequency at which gay students face persecution.

4. Most people know very little about the forced internment of Japanese Americans during World War II. Here are a few Web sites that provide information on this topic: www.geocities.com/Athens/8420/main.html, www.pbs.org/childofcamp/, www.csuohio.edu/art_photos/, www.nps.gov/manz/, and www.fatherryan.org/hcompsci/.

NOTES

1. The FBI recently released national data for 2001, which can be viewed online at www.fbi.gov/ucr/01hate.pdf. The overall number of victims—12,020—was considerably higher in that year than in previous years. The distribution of victims was also somewhat different, with "other ethnicity" constituting the second largest victim group (blacks were still the largest). These numbers were very likely skewed by the large spike in crimes against Muslims and Middle Easterners after September 11.

2. The FBI data do not quite cover the entire country because some local police agencies do not report hate crime data to the FBI.

3. Even Hasidic Jews, who wear distinctive clothing, may not be properly recognized because few Americans have much knowledge of them. A few years ago, a student of mine from rural California moved to Los Angeles. When he returned for a visit, he told me about the large number of Amish people in his neighborhood. I was surprised because I hadn't realized Los Angeles had much of an Amish community. When my student described how these Amish didn't drive or turn on their lights on Saturdays, I realized that they were actually Hasidic Jews, dressed in their traditional long black coats and hats.

4. This last incident was the genesis of the U.S. Supreme Court case *R. A. V. v. St. Paul.*

5. It should be noted here that Jews were not the only victims of the Holocaust, although they were the greatest in number. Other groups that suffered great losses included the Roma (Gypsies), gays and lesbians, the disabled, and others.

6. Anti-Semitism also continues to be a serious problem in other countries. This subject will be addressed in Chapter 8.

7. A seder is a ritual meal made in observance of the Passover holiday, which celebrates the freedom of the Hebrews from bondage in Egypt. Certain foods are eaten to commemorate parts of the Passover story, and the story itself is recited by the celebrants.

8. It should be noted that some of the comments made to the Palestinian supporters were equally ugly.

9. For the sake of brevity, and unless otherwise stated, the term *gay* in this section is meant to include gay men, lesbians, bisexuals, and transgendered individuals.

10. It is intriguing to speculate why gender roles are more rigid for men than for women in our society, but such speculation is beyond the scope of our present discussion.

11. The Stonewall Rebellion occurred when police attempted to shut down the Stonewall Inn, a gay bar in Greenwich Village. Patrons of the bar resisted, and there were confrontations between police and gay people for the next few nights. This incident is considered to have helped kindle the gay rights movement.

12. This law, originally passed in 1790, was not nullified until 1952.

Chapter 7

FIGHTING HATE

Several weeks before a recent election, I received a copy of the voter's pamphlet. Included within it were brief statements of each party's platform, and I was surprised and somewhat amused to see that one party's statement of purpose included eradicating hate crimes. Like the other position statements—those supporting quality public education, good health care, and a strong economy—this one makes good campaign fodder. After all, how many voters are going to oppose it? What seems absent, in this pamphlet and in general, however, is any realistic plan for achieving this goal. As you have probably concluded by now, hate crime legislation alone has not and will not eliminate hate crimes. In fact, human nature being what it is, it is extremely unlikely that hate crimes can be eradicated.[1]

All is not gloom, however. Prejudice- and bias-motivated violence can be reduced. There is scant evidence that such reduction can be attained through hate crime laws, but those laws are not the only method of fighting hate. In this chapter, we explore other approaches to combating hate crimes. I begin with a brief summary of empirical research on decreasing prejudice. I then examine a variety of organizations and programs that have been created to fight hate crimes, including government initiatives, private antihate groups, and offender-directed programs. Finally, in light of what we have previously discussed about who commits hate crimes and why, I consider the likely efficacy of these efforts.

THE PSYCHOLOGY OF REDUCING PREJUDICE

As I mentioned in Chapter 4, no baby is born with prejudices against other people. Each of us learned our biases from family, friends, teachers, and society at large. An optimistic corollary to this fact is that if prejudice can be taught, so can tolerance. Several decades of psychological research have investigated the ways in which tolerance can be encouraged and prejudice reduced.

Theories and Models

The Contact Hypothesis, the Robber's
Cave Experiment, and the Jigsaw Classroom

One of the earliest and most influential theories on reducing prejudice was Allport's (1954) Contact Hypothesis. Allport was writing as racial segregation was both legal and

Box 7.1   The Jigsaw Classroom

Here are Elliot Aronson's 10 easy steps for implementing his Jigsaw Classroom technique:

1. Divide students into 5- or 6-person jigsaw groups. The groups should be diverse in terms of gender, ethnicity, race, and ability.

2. Appoint one student from each group as the leader. Initially, this person should be the most mature student in the group.

3. Divide the day's lesson into 5 or 6 segments. For example, if you want history students to learn about Eleanor Roosevelt, you might divide a short biography of her into stand-alone segments on (1) her childhood, (2) her family life with Franklin and their children, (3) her life after Franklin contracted polio, (4) her work in the White House as First Lady, and (5) her life and work after Franklin's death.

4. Assign each student to learn one segment, making sure students have direct access only to their own segment.

the norm in the United States.[2] He theorized that intergroup hostility could be decreased if groups were brought into contact with one another. Simply throwing them together, however, would not be sufficient; in fact, there was some evidence to indicate that prejudice between blacks and whites was *greater* when the two groups lived closer to one another.

You may remember the Robber's Cave experiment, which was described in Chapter 4. In this experiment, Sherif, Harvey, White, Hood, and Sherif (1961) created intense intergroup antagonism among previously friendly boys who were attending a summer camp. Bringing the boys together for enjoyable, noncompetitive activities such as a fireworks display not only did not decrease tensions—it actually increased rivalry.

Allport (1954) held that it is not mere contact, but the conditions surrounding that contact, that reduce prejudice. Specifically, for contact to have a desirable effect, the individuals involved must be of equal status and must cooperate to achieve a common goal. Furthermore, the contact must be sanctioned by institutional supports such as law or custom, and the contact must be of sufficient closeness and duration for meaningful relationships to form.

In the Robber's Cave experiment, the experimenters arranged for two emergencies to occur (a stuck bus and a broken water supply). The camp counselors then encouraged the rival groups of boys to work together to overcome these emergencies. The boys did so, and hostilities between the groups were greatly decreased (Sherif et al., 1961).

Brown (1995) gives a real-life example of the Contact Hypothesis at work. In 1993, a huge earthquake occurred in central India. Tens of thousands of people died, and many more people needed to be rescued. In the face of this enormous disaster, Hindus and Muslims overcame their mutual animosity and labored together to save their neighbors.

Aronson, Blaney, Stephan, Sikes, and Snapp (1978) also made use of the Contact Hypothesis in an experiment they called the Jigsaw Classroom (see Box 7.1). Children in a recently desegregated school in Texas were placed into small, racially integrated groups.

5. Give students time to read over their own segment at least twice and become familiar with it. There is no need for them to memorize it.

6. Form temporary "expert groups" by having one student from each jigsaw group join the other students assigned to the same segment. Give students in these expert groups time to discuss the main points of their segment and to rehearse the presentations they will make to their jigsaw group.

7. Bring the students back into their jigsaw groups.

8. Ask each student to present her or his segment to the group. Encourage others in the group to ask questions for clarification.

9. Float from group to group, observing the process. If any group is having trouble (e.g., a member is dominating or disruptive), make an appropriate intervention. Eventually, it's best for the group leader to handle this task. Leaders can be trained by whispering an instruction on how to intervene, until the leader gets the hang of it.

10. At the end of the session, give a quiz on the material so that students quickly come to realize that these sessions are not just fun and games but really count.

SOURCE: Aronson (2000).

Each member of the group was given a unique set of information (akin to a single piece of a puzzle), and the students were told they would all be tested on all of the information at the end. Therefore, for students to do well on the test, they had to rely on their classmates. The researchers found that children in the experiment gradually learned to work cooperatively. They also grew to like one another more and had greater self-esteem. In addition, minority group children's grades improved.

Common Ingroup Identity Model

Although many scholars still subscribe to the Contact Hypothesis, more recent research has focused on more cognitive approaches, which seek to determine why certain types of group interaction lead to reductions in bias. One of the major modern approaches is the Common Ingroup Identity Model (Gaertner, Dovidio, Anastasio, Bachman, & Rust, 1993). This model relies on the concept of social categorization.[3] According to the model, prejudice can be reduced when people decategorize others—that is, when they see them not as members of another group, but rather as differentiated individuals with their own personal characteristics (see Gaertner, Dovidio, Nier, Ward, & Banker, 1999). A problem with decategorization, however, is that attitude change toward the individuals involved does not necessarily translate to attitude change toward their entire group. An example of this phenomenon can be found in Ezekiel's (1995) study of Detroit neo-Nazis. One member of the group Ezekiel studied, Francis, had an adopted sister who was half black and half white. Francis freely announced his love for his sister and also his perception that she was white because she lived with a white family. At the same time, however, he was an active member of a white supremacist group.

As an alternative to social decategorization, the Common Ingroup Identity Model proposes that social *recategorization* can occur. When it does, existing group boundaries are redrawn and a superordinate identity is formed—that is, members of disparate groups join together to create a new, unified group (Gaertner et al., 1999).

Exactly one month after September 11, 2001, I attended a conference on diversity. One of the speakers there, who was African American, talked about a sudden change in the race relations climate. She said that in the previous month, she had noticed that white strangers who might previously have been hostile or at least indifferent to her because of her race were suddenly quite friendly. One man on an airplane gave an indication of one possible reason for this when he made a comment to her about the need for "we Americans" to fight terrorism. The borders between black and white were (momentarily, at least) no longer important but had been replaced by the larger categories of American and non-American.[4]

Recategorization does seem to lead to reductions in conflict; however, as is the case with decategorization, it is unclear the extent to which these attitude changes generalize (Brown, 1995). Furthermore, are these changes permanent? In countries such as Yugoslavia, long-standing ethnic tensions were eased considerably under the common burden of Soviet occupation but reemerged violently very soon after the Iron Curtain fell. It is especially discouraging that this occurred even though the apparent peace had lasted for several decades.

Another potential problem with all of the prejudice reduction models we have discussed so far is that they require some cooperation from both the perpetrators and the victims of bias. This requirement is problematic because perpetrators may be so entrenched in their beliefs as to refuse to participate, and victims may either distrust their oppressors or believe that only those responsible for unfairness should have to participate (Gaertner et al., 1999).

Despite these shortcomings, this body of research represents at least the possibility of reducing prejudice in a meaningful way. Real-life anecdotes also support this possibility. Recall the story of C. P. Ellis, the former Klan leader discussed in Chapter 4. Ellis was forced to work cooperatively with a person who embodied everything he loathed: a female, African American civil rights activist. After initial reluctance, Ellis began to see commonalities between his own experiences and hers. Despite a family history of racism, as well as extensive personal involvement in white supremacism, Ellis was eventually able to forge a close friendship with the woman, and he ended up renouncing his bigoted views (Davidson, 1996).

Changing Legal, Social, or Cultural Messages

Another broad area of research on prejudice reduction has concerned the effects of changing the legal, social, or cultural messages that are transmitted about people. Several times in this book I have discussed the ways in which our attitudes toward others may be adversely affected by our social environment; does it not make sense that our attitudes could be positively affected as well?

Changing the Laws

One way to try to change attitudes is to change the laws that govern behavior. This idea is controversial. When the Supreme Court declared school segregation unconstitutional in *Brown v. Board of Education* (1954), critics argued that you cannot legislate morality and that desegregation should wait until prevailing racial attitudes changed (Aronson, 1999). Perhaps the Supreme Court was responding to these concerns when, rather than requiring immediate complete action, it only required that schools act with "all deliberate speed."[5] At the same time, though, Allport (1954) was arguing that laws *can* change behavior. He stated, "We can be entirely sure that discriminative laws *increase* prejudice—why then should not legislation of the reverse order *diminish* prejudice?" (Allport, 1954, p. 469).

Allport (1954) argued that laws can change overt signs of prejudice and, in time, attitudes are likely to follow. Later, cognitive dissonance theorists agreed. As I mentioned in Chapter 4, cognitive dissonance theory states that there is an innate drive to avoid dissonant cognitions, and, contrary to popular belief, the best way to change attitudes is to change behaviors first. In the years since Allport and *Brown v. Board of Education,* for example, some research has indicated that the ending of legalized segregation, as well as the implementation of antidiscrimination laws, have resulted in reductions in prejudice (Aronson, 1999).

Hate crime legislation, then, could conceivably reduce biased beliefs by first reducing biased actions. The problem with this proposition is that there is no evidence that these laws actually do reduce biased actions, and therefore they are unlikely to affect attitudes. As Allport (1954, p. 470) pointed out, there is a big difference between a law on the books and a law in action.

Changing Perceived Cultural Norms

Another way to try to change attitudes is through changing perceived cultural norms. This can be done using mass media (see Box 7.2). As we have already seen, the largest historical surge in Klan membership was spurred in part by the media, specifically the film *The Birth of a Nation* (Griffith, 1915). But it is now nearly a century later, and we are surrounded by persuasive messages to a much greater extent than before. Amid the daily barrage of propaganda about everything from war to toothpaste, how much attention will we pay to messages promoting tolerance? Even 50 years ago, Allport (1954) was at least skeptical about the efficacy of the media in controlling bias. Nevertheless, mass media is an appealing technique because of its relative ease of implementation and potential for broad reach.

Box 7.2   Antihate Media Campaigns

In recent years, a number of organizations have undertaken media campaigns to combat hate crimes. Here are some examples:

- The Anti-Defamation League and Barnes & Noble Booksellers have organized an annual joint venture called "Close the Book on Hate." For one month each year, the bookstores sponsor bias-related educational events, distribute free copies of a brochure on fighting prejudice, and highlight an ADL book called *Hate Hurts: How Children Learn and Unlearn Prejudice* (Stern-Larosa, 2000) (see www.adl.org/presrele/education%5F01/3681%5F01.asp).
- The USA Network sponsors an "Erase the Hate" campaign. Components include bias-related programming, a Web page, and support for various local initiatives (www.usanetwork.com/functions/nohate/erasehate.html).
- In 2001, MTV had a year-long antidiscrimination campaign. It featured a program about Matthew Shepard's murder followed by 17 hours of a continuous listing of the names of hate crime victims.

(Continued)

Box 7.2 (Continued)

- Artists for a Hate Free America sells merchandise with an antihate message. Its seal appears on the albums of supporting musicians, such as Pearl Jam and Alice in Chains (see www.raspberrymultimedia.com/clients/artists/default.html).
- HBO aired a documentary about hate on the Internet titled "Hate.com: Extremists on the Internet" (DiPersio & Guttentag, 2000).
- Modesto, California, sponsors an annual Say No to Hate campaign. Activities have included a contest to design antihate posters, the distribution of those posters by the local newspaper, and marches and rallies.

Winkel (1997) briefly reviewed the literature on reducing prejudice through propaganda campaigns and concluded that sometimes these campaigns are successful, sometimes they are not, and sometimes they are even counterproductive. He described one study that did produce desirable results. Pedestrians at a mall in Amsterdam were recruited and divided into three groups, each of which was shown different versions of two videos. One group was shown videos that explicitly primed stereotypes about Muslims and people from Surinam. The second group was shown the same visuals but was also shown information that challenged the stereotypes. The third group saw the same visuals, saw the stereotype-challenging information, and heard a voiceover telling them that if they failed to look any further than stereotypes, they were like Hitler. Participants in all three groups were then given two questionnaires to measure prejudice. Winkel (1997) found that participants in the second and third groups scored significantly lower on prejudice than did those in the first group, which suggests that prompting people to ignore biases may have a positive effect.

The problem with Winkel's (1997) experiment and others like it is that the generalizability of the results is unknown. Did the prompts truly affect people's attitudes, or did they simply answer the questions in a way that they perceived to be more socially desirable? Were any effects on attitudes long-lasting? Would the same attitude changes occur if the videos were viewed under more ordinary conditions, such as a television commercial, rather than during an experiment? Did the videos affect the participants' behaviors in any meaningful way?

Clearly, the evidence linking propaganda and prejudice change is limited at best. As we will see, however, mass media appeals are one frequent method employed by those who wish to fight hate crimes. Another common, and related, method is educational efforts. Although it makes intuitive sense that bias can be reduced through diversity training, through providing information about prejudice and hate crimes, and similar programs, there is no empirical confirmation that these tactics really work.

ANTIHATE GROUPS[6]

It is somewhat comforting to know that just as there are organized hate groups, there are also organized groups dedicated to fighting hate. There are many antihate groups of various sizes, constituencies, missions, and methods. In this section, I discuss four organizations that have been most prominent in antihate activities. An interesting thing about these groups is that although they have played a major role in the hate crime legal arena and although their reports and findings are frequently cited by scholars, the groups themselves have received virtually no academic scrutiny. Worth mentioning also is the Center for Democratic Renewal; see Box 7.3.

Box 7.3   The Center for Democratic Renewal (CDR)

The CDR was founded in 1979 in Atlanta to combat hate crimes. It engages in a number of activities, including the following:

- It coordinates the Southern Coalition Against Bigotry and Racism.
- It maintains a Web site at www.thecdr.org.
- It publishes a number of pamphlets and reports, including reports on the state climate for hate crimes in Georgia, Alabama, Florida, Ohio, and North Carolina.
- In 2000 it sponsored a national youth conference.
- It has sponsored several national hate crime summits.
- It sells a publication titled *When Hate Groups Come to Town,* which is meant to help grassroots organizers.

The Southern Poverty Law Center (SPLC)

The Southern Poverty Law Center (SPLC) is headquartered in Montgomery, Alabama. The SPLC was founded in 1971 as a civil rights law firm by two attorneys: Morris Dees was the son of a farmer and the grandson of a one-time Klan member, and Joe Levin was a Jewish native of Montgomery. In 1979 the SPLC brought a civil lawsuit against a Klan group that had attacked civil rights marchers in Alabama, and in 1981 the Center founded a project called Klanwatch, which was meant to monitor organized hate.

The SPLC has sponsored a number of antihate activities. It has provided legal counsel in a number of prominent cases against white supremacists, earning multimillion-dollar judgments against Tom and John Metzger for the murder of Mulugeta Seraw, against members of a Klan group for the lynching of Michael Donald, against another Klan group for burning a black church, against yet another Klan group for attacking civil rights marchers in Georgia, against a group called the Church of the Creator for the murder of a black sailor, and against the Aryan Nations for attacking two people (Dees & Fiffer, 1993; Stanton, 1992).

The SPLC continues to monitor hate group activity today under a venture called the Intelligence Project. The project keeps track of not only traditional white supremacist groups like the Klan but also other extremists such as militias, neoconfederates, black separatists, and the Westboro Baptist Church. It uses mainstream media to compile a list of hate incidents, it publishes a quarterly magazine called the *Intelligence Report* (which is free for law enforcement officials and scholars), and it offers an online training course for police officers on hate crimes.

The SPLC also is active in educational efforts. For the past decade, it has published a semiannual magazine called *Teaching Tolerance,* which is aimed at K-12 teachers. This magazine, which is provided free to teachers, highlights antibias activities and resources.

The SPLC also provides free videos and other materials with antibias topics to educators, and it gives grants for teachers to implement projects in their classrooms. Outside its headquarters, the SPLC built a civil rights memorial that was designed by Maya Lin (who created the Vietnam Veterans Memorial in Washington, D.C.); the memorial chronicles the history of the civil rights movement and remembers 40 people who lost their lives because of the movement.

Like the other major organizations discussed in this chapter, the SPLC maintains a Web site (at www.splcenter.org). The site is particularly useful to researchers who wish to examine current hate group activity, to educators who wish to get ideas for diversity and antibias curricula, and for people who wish to get involved in antihate activities. There is also a page aimed at parents, and a portion of the site has activities for children and teens. In addition, the SPLC has a few printed antihate crime resources that it provides free for the asking.

The Anti-Defamation League (ADL)

A second major antihate organization is the Anti-Defamation League (ADL), a Jewish group founded in 1913 by a Chicago attorney named Sigmund Livingston. The ADL's earliest goals were to eliminate anti-Jewish stereotypes in the media and to combat other forms of anti-Semitism, including those perpetrated by the Klan. When fascism and anti-Semitism began their reign in the 1930s, the ADL began tracking extremists both in the United States and abroad. After World War II, the organization entered the legal arena, first in a series of lawsuits advocating the separation of church and state, and later in lobbying efforts for antihate legislation. During the 1960s, the ADL widened its scope to include civil rights activism, although its primary interest remained fighting anti-Jewish activities. Later activities included promoting Israel and sponsoring Holocaust education.

The ADL was perhaps one of the most influential organizations in lobbying for hate crime legislation. As discussed in Chapter 2, its model law was adopted and adapted by many states, and the ADL has been an ardent supporter of federal laws as well. Since 1978, the ADL has tracked anti-Semitic incidents across the United States, and it publishes an annual audit of those incidents. Like the SPLC, it tracks domestic extremist groups; it also monitors terrorist groups in the United States and internationally. During the height of the militia movement in the 1990s, the ADL actively observed those groups as well.

Today, the ADL continues its lobbying efforts. It is headquartered in New York but has local offices throughout the United States. It is very active in education; through its A World of Difference Institute, it offers diversity and antibias training to schools, universities, community organizations, and workplaces. It also offers services for law enforcement, including specialized training and various information resources. In addition, the ADL publishes many books and other materials on anti-Semitism, extremism, the Holocaust, and multicultural issues. The ADL has available for free download a piece of software called Hate Filter (www.adl.org/hatefilter/default.asp), which blocks access to hate group Web sites. And, like the SPLC, the ADL's Web site is a particularly rich source of information for researchers.

The Simon Wiesenthal Center (SWC)

The Simon Wiesenthal Center (SWC) is another organization dedicated to combating hatred. It was founded in Los Angeles in 1977 by Rabbi Martin Hier. Simon Wiesenthal himself is a Holocaust survivor who became famous after the war for his efforts to track down former Nazis. The SWC is concerned with issues of anti-Semitism and bigotry in general, and with the Holocaust especially.

The SWC focuses most of its efforts on education. In Los Angeles it operates the Museum of Tolerance, which has interactive multimedia exhibits on the Holocaust and on prejudice in general. The museum has a large library and archive that are open to researchers and students. The SWC also offers training to educators, police officers, and others on diversity and hate crimes issues. Its Task Force Against Hate engages in a variety of antihate activities, including sponsoring talks by TJ Leyden, a former racist skinhead.

One of the SWC's most useful activities from the standpoint of researchers and educators is its publishing effort; it publishes a quarterly magazine and periodic reports. The SWC also produces documentary films, and for the past few years it has issued an annual CD-ROM called "Digital Hate" (see, e.g., Simon Wiesenthal Center, 2002a) which tracks extremist sites on the Internet. In addition to its own materials, the SWC sells books and videos on issues related to anti-Semitism, prejudice, and multiculturalism.

National Gay and Lesbian Task Force (NGLTF)

The fourth major antihate organization is the National Gay and Lesbian Task Force (NGLTF). Founded in 1973 in Washington, D.C., its primary mission, as suggested by its name, is to promote the civil rights of gay, lesbian, bisexual, and transgendered people. It takes part in lobbying, advocacy, and educational activities concerning a broad range of issues important to its constituents, including hate crimes.

One of the NGLTF's major contributions to the field of hate crimes has been its efforts to track and document antigay violence. The reports of its findings have been influential in the creation of federal and state legislation. The NGLTF's Policy Institute acts as a think tank for the gay rights movement and publishes several documents, most of which are available online. In addition, the NGLTF sponsors an annual conference.

The NGLTF's Web site (www.ngltf.org) offers a wealth of useful information, including state-by-state data on legal issues relevant to gays and lesbians. For example, you can determine which states have hate crime laws that include sexual orientation, which have sodomy laws, and which have specifically prohibited same-sex marriages. A number of manuals are available to assist those who wish to engage in organizing activities.

Other Antihate Groups

There are a number of other groups that have been influential in the antihate movement as well. Jenness and Broad (1997), for example, in addition to the four groups described here, also mention the National Institute Against Prejudice and Violence, the Center for Democratic Renewal, the National Coalition Against Domestic Violence, and the National Victim Center. There are numerous smaller and local groups as well. The ADL, the SWC, and the NGLTF are not the only organizations that target hate against particular groups. A few other examples can be seen in Table 7.1.

Common Features of Antihate Groups

Combating Bigotry and Violence

Although their precise methods and goals differ, all antihate groups share certain important features. One of these, of course, is that they seek to combat bigotry and violence.

Table 7.1   Other Advocacy Groups

Organization	Web Site
American-Arab Anti-Discrimination Committee (ADC)	www.adc.org
Asian American Legal Defense and Education Fund (AALDEF)	www.aaldef.org/index.html
Asian Pacific American Legal Center (APALC)	www.apalc.org
Council on American-Islamic Relations	www.cair-net.org
LAMBDA (GLBT Community Services)	www.lambda.org
Mexican American Legal Defense and Educational Fund	www.maldef.org
National Asian Pacific American Legal Consortium (NAPALC)	www.napalc.org
National Congress of American Indians (NCAI)	www.ncai.org

Many of them tend to focus on bigotry against a particular group, and yet they appear to realize that prejudice of one particular kind—anti-Semitisim, for example, or homophobia—is best addressed not in isolation but rather as part of a larger picture of intolerance. Thus, the NGLTF, for instance, recently chose as the theme of its annual conference "Building an Anti-Racist Movement," and the SWC awarded its 2002 children's book award to a book about the African American civil rights movement.

Jenness and Broad (1997) point out that the antihate movement itself is not isolated but is a part of a convergence of many social movements. These include the civil rights movement, the women's movement, the gay rights movement, and the crime victims movement. It is no accident that the SWC, SPLC, and NGLTF were all founded in the 1970s. Thus, organizations with somewhat disparate goals and constituencies have formed coalitions and have joined in lobbying for common objectives. The ADL, the NGLTF, and the SPLC have all urged support for a federal hate crime law, for example.

Despite the common goals, it is important to remember that many of these groups do have a specific agenda, and this sometimes results in a clash over whether their own particular members are receiving proper attention as victims of hate crimes. Jacobs and Potter (1998) refer to this as "identity politics." Another repercussion is that some constituencies may have a louder voice than others. For instance, whereas the ADL and the SWC are both very concerned about anti-Semitic hate crimes, and the NGLTF concentrates on antigay crimes, no major organization has a similar focus on hate crimes against African Americans. This is somewhat surprising, given that African Americans are the most common victims of hate crimes and that powerful African American advocacy organizations, such as the National Association for the Advancement of Colored People (NAACP), do exist.

Concentrating on Education and Lobbying

Another feature that antihate groups share is that they tend to concentrate most of their efforts on two areas: education and lobbying. All such groups have significant educational efforts. These are often aimed at schoolchildren but may also focus on police officers, employers, or adults in general. The groups provide curricula, training, publications, and advertising. Many of the groups also engage in significant lobbying efforts in support of their cause. As already discussed, these efforts are probably responsible for the enactment of state and federal hate crime laws as well as other pertinent legislation. Many of the groups also make use of the legal system in general to help achieve their goals: The ADL frequently submits amicus curiae (friend of the court) briefs in legal cases (see Box 7.4), and the SPLC has provided legal assistance to the victims of hate crimes in several cases.

Box 7.4   Friend of the Court

Advocacy organizations often get involved in prominent legal disputes by filing amicus curiae briefs. *Amicus curiae* literally means "friend of the court." Filing amicus curiae briefs is a procedure by which a person or organization that is not actually a party directly involved in a particular case is allowed to file with the court a legal brief (a written document containing facts or arguments). The purpose of this brief is to provide the court with additional information pertinent to the case and to allow the participation of parties that, although not actually involved in a particular case, do have a stake in the outcome of the case.

The ADL has been particularly active in filing amicus briefs. In addition to cases involving such issues as discrimination and the separation of church and state, the ADL has contributed amicus briefs in hate crime cases, including *Wisconsin v. Mitchell*. These briefs were often prepared by a coalition of organizations, including the NGLTF, the Arab-American Anti-Discrimination Committee, and the National Asian Pacific American Legal Consortium. You can access some briefs to Supreme Court cases at http://supreme.lp.findlaw.com/supreme_court/briefs/.

Monitoring and Reporting on Hate Groups

A third feature shared by many (although not all) of the organizations is the attention they pay to organized hate groups. Organizations such as the SWC, ADL, and SPLC all spend significant amounts of effort monitoring extremist groups in various ways, and reporting the results of that monitoring. Some organizations, such as the ADL and the SWC, maintain electronic databases containing this information; others, such as the SPLC, publish articles about extremists and extremist groups in their magazines. For many of these organizations, the focus on extremist groups accounts for a very significant proportion of their antihate efforts.

Being Private and Nonprofit

Finally, all of these organizations are private, nonprofit groups. This may give them a greater degree of freedom than groups that are largely or entirely public. A government-based organization is, arguably, more affected by the pressures of politics and may also have more stake in maintaining current power hierarchies and the general status quo. On the other hand, private groups are forced to rely on donations and sales for their operating expenses, so they probably have to face the additional issue of how to market themselves effectively. Smaller groups, especially, may also be limited in their activities because of limited funding.

GOVERNMENT INITIATIVES FOR FIGHTING HATE

The simplest way for the government to address the problem of hate crimes is also the most common: A hate crime law can be enacted. As discussed earlier in this book, however, there is no evidence that such laws actually reduce hate crimes, and there is good reason to doubt that they do. Even the laws' most ardent supporters would never argue that the laws will be so effective that they would eliminate bias-motivated offenses altogether. In addition to creating laws, then, government entities at the federal, state, and local level have created various programs and initiatives that are meant to complement hate crime laws and add further ammunition to the battle against bigotry. In this section, I review several of these programs. The Web pages of the government programs mentioned in this section can be found in Table 7.2.

One federal initiative related to hate crimes was already mentioned in Chapter 6. The National Church Arson Task Force involves many agencies, including components from the Department of Justice, the Treasury Department, and state and local law enforcement. Created in 1996, it was charged with reacting to the sudden increase in attacks on churches, especially those with African American congregations. Its primary activities are to investigate and prosecute these crimes, although it also assists victims in rebuilding efforts. According to its 2000 report, the Task Force has an arrest rate of

Table 7.2   Government Antihate Initiatives

California Attorney General Civil Rights Commission on Hate Crimes	http://caag.state.ca.us/publications/civilrights/reportingHC.pdf
Chicago Commission on Human Relations	www.ci.chi.il.us/HumanRelations/HumanRelations.html
Healing the Hate	www.edc.org/HHD/hatecrime/HTH.pdf
Lieutenant Governor's Commission for One California	www.ltg.ca.gov/programs/1ca/index.asp
Massachusetts Governor's Task Force on Hate Crime	www.state.ma.us/StopHate/
Montgomery County, Maryland, Committee on Hate/Violence	www.montgomerycountymd.gov/mcgtmpl.asp?url=/content/HumanRights/hateviolence.asp&v=false
National Center for Hate Crime Prevention	www2.edc.org/themes/view.asp?id=379&web=&pub=True
National Church Arson Task Force	www.atf.treas.gov/contact/churcharson/index.htm
Preventing Youth Hate Crime	www.ed.gov/pubs/HateCrime/start.html
Protecting Students From Harassment and Hate Crime: A Guide for Schools	www.ed.gov/pubs/Harassment/
Responding to Hate Crime: A Multidisciplinary Curriculum for Law Enforcement and Victim Assistance Professionals	www.ojp.gov/ovc/publications/infores/responding/
Stop the Hate	www.stopthehate.org
U.S. Department of Justice Community Relations Service	www.usdoj.gov/crs/pubs/htecrm.htm

more than 36%, which is twice the national average for arson (National Church Arson Task Force, 2000).

The Department of Justice implements broader antihate efforts as well. For example, the Community Relations Service helps local government and school officials deal with specific instances of racial or ethnic conflict. Its services include mediation, training, and public education. The Office for Victims of Crime has developed an extensive training manual on hate crimes titled *Responding to Hate Crime: A Multidisciplinary Curriculum for Law Enforcement and Victim Assistance Professionals,* and the Department of Justice sponsors a National Hate Crime Training Initiative for police. The National Center for Hate Crime Prevention aims to prevent youth involvement in hate crimes by providing training and publications.

Although some federal efforts have focused on law enforcement, others have centered on education. Through its Safe and Drug-Free Schools program, the Department of Education publishes guides for preventing hate crimes in schools. Guides for school administrators include *Preventing Youth Hate Crime* and *Protecting Students From Harassment and Hate Crime: A Guide for Schools.* For teachers, there is *Healing the Hate,* a hate crime curriculum for middle-school students sponsored by the Office of Juvenile Justice and Delinquency Prevention.

In addition to its own efforts, the federal government has provided grants and funding to support the antihate crime activities of state and local agencies and private organizations. For example, the Simon Wiesenthal Center's Tools for Tolerance Program received financial support from the Bureau of Justice Assistance, which also funded antihate programs in Omaha, Los Angeles, San Diego, Massachusetts, and Maine.

Several states have their own antihate programs. For example, the Massachusetts Governor's Office sponsors a Task Force on Hate Crime. Founded in 1991, this group engages in a wide variety of activities. Some are aimed at police, such as efforts to ensure police reporting of hate crimes. Others are meant to increase victim reporting of hate crimes; such efforts include community outreach and media campaigns. Other activities focus on youth. The Task Force has engaged in public awareness efforts in schools. Its Stop the Hate Web site, which was created by high school students working with the Task Force's Student Civil Rights Team program, contains information about hate crimes for students, educators, police officers, and the general public. A particularly interesting Task Force project is aimed at hate crime offenders. Massachusetts law requires all people convicted of hate crimes to participate in a diversity awareness program. The Task Force created a weeklong curriculum for this.

Massachusetts's program is perhaps the most ambitious, but it is not the only statewide program in existence. For example, the California Lieutenant Governor's Commission for One California, which seeks to promote unity and diversity and fight hate, holds periodic meetings that are open to the public. In addition, California has the Attorney General's Civil Rights Commission on Hate Crimes, founded in 1999. The purpose of the Commission is to advise the Attorney General on issues related to hate crimes and tolerance. As might be expected, its primary focus has been law enforcement issues. As its initial project (spanning 1999 through 2001), the Commission held a series of forums across the state on the topic of reporting hate crimes and then published its findings (see California Attorney General, 2001). The report documented and described the underreporting of hate crimes in California and offered several recommendations to address the problem.

A great many places in the United States also have hate crime programs at the county or city level. As discussed in Chapter 3, there are bias crime units in perhaps several hundred local law enforcement agencies. Other programs exist as well, often in conjunction

NARRATIVE PORTRAIT

A MOTHER FIGHTS HATE

The following piece was written by Gabi Clayton, cofounder of the organization Families Against Hate. In it, she discusses what led her to establish this group and the impact her work has had.

Our youngest son, Bill, came out as bisexual when he was 14 years old. At 17 he was assaulted in a hate crime based on his sexual orientation. Bill committed suicide a month later, on May 8, 1995, believing that death was a better option than being hated and never safe.

How can a parent survive something like this? I had to do something with my pain. I found that working to end the hate that killed Bill and supporting other people who are dealing with hate violence, and with the death and destruction of lives that comes from it, is the only thing that made sense for me. One thing I did was to publish Bill's story on a Web site that I created in his honor (www.youth-guard.org/gabi/). There are now responses there from people all over the world—most have been positive, supportive heartwarming letters, with an occasional hateful one, which I leave there to educate people. The author of one of those hate letters wrote me years later and he asked me to take it down. He wrote:

I must say how disgusted I was at my ignorance then. . . . an ignorant kid who thought he knew it all. . . . First, I want to ask for your forgiveness, for any hurt or pain I caused you for my heartless comments. I now understand that this type of behavior breeds nothing but hate and separation. I truly am sorry.

We can never know how many lives this one man might have hurt with hatred or how many lives he can have a positive impact on now. Knowing that I may have been a part of helping this man to change by being public about our son means a lot to me.

In 1998 I was filmed talking to four self-identified homophobic youth for a documentary called "Teen Files: The Truth About Hate" (Shapiro, 1999). These young people met Bill through my stories and photographs, and I saw them start to connect with him as a person. Then I told them about the hate crime and Bill's suicide. In the end they said things like "I never thought of a gay person having a family—having a mother who loved them—before." I talked to someone later who stayed in touch with those youths; that person told me that the experience of talking to me and the other things they did as part of that film had a lasting impact on them.

Supporting victims, families, and communities that have been impacted by bias-motivated violence has become my way of living with the ongoing grief of Bill's death. This has lead to my being a cofounder of Families United Against Hate (see www.fuah.org), a grassroots organization that offers support, guidance, and assistance to families (biological, extended, and chosen) and individuals dealing with incidents based on bias; and to the people, organizations and agencies who serve and support them.

with local human relations commissions and frequently as a coalition of several local governmental units and agencies. An example of a typical program is the Chicago Commission on Human Relations, which tracks and reports hate crimes in the city. In conjunction with the city police and the State Attorney's office, it also provides victim assistance, such as legal assistance and referrals and mediation. The Commission also works with local community-service agencies to provide victim services.

Another example of a local antihate program is the Montgomery County, Maryland, Committee on Hate/Violence. It is composed of residents of the county as well as representatives from local educational, political, and law enforcement bodies. Its primary responsibilities are to advise the county on hate crime issues and to develop and distribute information on hate crimes and diversity.

OTHER ANTIHATE APPROACHES

Not all antihate initiatives are conducted on such a grand scale as those of the ADL, SPLC, and similar organizations, nor are all of them sponsored by government agencies. Some exist as the result of efforts of local, grassroots organizers. (See Box 7.5 for ideas on how to start an antihate campaign if you want to get involved.)

Box 7.5   How to Start Your Own Antihate Campaign

Although some of the efforts to fight hate crimes have been conducted on a national or international scale, others have started with only a single person acting at the local level. If you want to get involved, here are a few ideas for getting started:

- Check to see whether any local initiatives already exist. Some places to check include campus organizations, county human rights commissions, and police bias crime units. You might also see if any of the national organizations such as the NGLTF, the NAACP, or the ADL have an office near you, or look for local civil rights groups.
- Consider modeling your own program on an existing one, such as Project Lemonade (www.artsandmusicpa.com/world_cultures/teachtolerance.htm) or Not in Our Town (www.pbs.org/niot/). You can check out these projects' Web sites for contacts and ideas.
- Several organizations have information on activism. Check out One People's Project, the SPLC's Tolerance.org, and the Center for Democratic Renewal for some ideas.
- The ADL and Barnes & Noble published a pamphlet on 101 ways to combat prejudice. You can pick up a free copy at a Barnes & Noble Bookstore, or download it from the ADL's Web site.

Local Initiatives

Tang, Ho, Thompson, Kim, and Ganz (2001) spoke at a symposium about the need for organizing social justice movements, including those intended to combat hate crime. Not only can these movements be particularly sensitive to local needs, but they can also effectively challenge existing power hierarchies, and they can provide a voice for those who are not adequately represented by other groups. They can also address the broader issues of intolerance and prejudice, including intolerance sanctioned by the government, rather than focusing exclusively on the problem of hate crimes, and they can build coalitions among diverse groups.

One example of a local initiative that later became national is "Not in Our Town." This movement began in Billings, Montana, in 1993, when there was a rash of hate incidents. Swastikas were painted on the home of a Native American family, a Jewish cemetery was vandalized, skinheads showed up during services at a black church, and someone threw a brick through the window of the home of a young boy who was displaying a Hanukkah menorah. Among other actions, several community organizations passed antihate resolutions, people participated in pro-tolerance marches and vigils, and local residents donated time and paint to clean up racist graffiti. The newspaper printed a picture of a menorah, which 10,000 residents and businesses placed in their windows. A PBS documentary (O'Neill & Miller, 1995) was made about these events and, eventually, a national organization arose. The organization goes to communities in need and offers educational and organizational assistance for dealing with racism; it also sponsors a Web page (www.pbs.org/niot/).

Another campaign that started locally is "Project Lemonade." In 1994, a Jewish couple in Springfield, Illinois, collected pledges from local residents to donate money for every minute that a local Klan rally lasted. The longer the hate event, the more money would be raised. They raised more than $10,000, which went to antihate groups and was used to buy library books on diversity issues. The couple created a kit to help other communities conduct similar activities. According to the SPLC, pledges in other areas have been used to pay for library books on black history, to build a Holocaust memorial (complete with a plaque crediting the Aryan Nations for making the memorial possible!), for victim compensation, and for police sensitivity training (SPLC, 2002a). In November 2001, Ann Coulter came to speak at Earlham College in Indiana. Coulter had recently advocated deporting all Arabs and Muslims, and had defamed the Muslim faith in general. A student group collected pledges for every minute she spoke, and the money went to help international students at the college ("Ann Coulter Angers Students," 2001). When the Klan planned a rally in Manhattan in 1999, a U.S. Senator asked residents to pledge money ($1 for each minute a KKK member spoke) to the ADL, the NAACP, a Catholic charity, or the Puerto Rican Legal Defense and Education Fund ("In Protest of KKK Rally," 1999).

Wassmuth and Bryant (2001) discuss another community's efforts to fight hate. In the 1980s, the area around Coeur d'Alene, Idaho, became a sanctuary for several hate groups, and the community at large experienced increased racism and anti-Semitism. A task force was created, and it engaged in numerous activities meant to combat hatred, build coalitions, and improve respect for diversity.

The Internet

Just as the Internet has been used as a tool for hate groups, it has also made it possible for a wide variety of antihate groups to spread their message. There are at least dozens of these groups; because no index of them exists, it is impossible to get an accurate count. Some, like Families United Against Hate (www.fuah.org/resources.html), aim to assist victims. Others, like the USA Network's "Erase the Hate" campaign (www.usanetwork.com/functions/nohate/erasehate.html), are primarily focused on education or public information efforts. Some, like the Hate Crimes Research Network (www.hatecrime.net) and Political Research Associates (www.publiceye.org), are academic in nature, intended to facilitate research on hate crimes. Some, such as One People's Project (www.onepeoplesproject.com/about.htm), encourage activism. And some, such as Community United Against Violence (www.cuav.org)—which concentrates on crimes against lesbians, gays, bisexuals, and transgendered people—tend to focus on a specific type of hate crime.[7]

Chapter 3 discussed the fact that the extent to which hate sites influence the public at large to accept or adopt bigoted views is unknown. Similarly, it is unclear what effect antihate sites have on increasing tolerance and decreasing hate crimes. There is the possibility that many of them may simply be preaching to the choir. In fact, antihate sites are probably less attractive to casual browsers than hate sites. The provocative nature of hate sites' content may inspire people to visit them out of curiosity, whereas there is nothing particularly intriguing about antihate sites except to those who are already interested in promoting tolerance. On the other hand, antihate sites may be effective because they can serve as a resource or inspiration for those who wish to take personal actions to combat hate. To date, there has been no research on the impact of antihate sites (and very little on the impact of hate sites), so this issue remains open.

DEALING WITH HATE CRIME OFFENDERS

An approach to combating hate crimes that is entirely different from antihate groups and local initiatives is one that has not been used as frequently—namely, to deal directly with the offenders. Of course, they can be "dealt with" simply by sentencing them to prison, but few people believe that this will encourage them to behave with more tolerance in the future. In fact, prisons are usually full of racial and ethnic divisions, and they are also breeding grounds for hate group recruitment and activity. Therefore, some people advocate that, in addition to receiving punishment for their deeds, hate crime offenders ought to be exposed to some sort of rehabilitative efforts.

Ways to Promote Rehabilitation

Restorative Justice Model

One way to promote rehabilitation is through the restorative justice model and the use of victim-offender mediation. *Restorative justice,* which has achieved a fair degree of popularity in recent years, seeks to heal the harms of crime by actively involving offenders, victims, and the community (Johnstone, 2002; Van Ness & Strong, 2001). One component of restorative justice is *victim-offender mediation,* which has been used in hundreds of jurisdictions across the United States and worldwide (Umbreit, Coates, & Vos, 2001). During mediation, the victim and offender are brought together, and the victim has the opportunity to describe the harm he or she has experienced, and to ask questions. The offender can offer apologies or explanations. The goal is for the parties to reach a mutually acceptable agreement for reconciliation (Shenk, 2001).

Shenk (2001) argues that victim-offender mediation is an ideal way to handle hate crimes because it is likely to have three primary benefits. First, by humanizing the justice system, it will enable the offender to realize the harm he has caused and to see the victim as an individual, perhaps breaking down stereotypes. Second, it will provide emotional release for both parties. And third, it may fill the gaps of existing hate crime legislation by encouraging reporting by victims and by reducing recidivism.

Victim-offender mediation is an approach that would likely receive support from those involved in combating hate. As already mentioned, it is popular right now in the criminal justice community in general. The antihate movement is allied with the victim's rights movement, so many private organizations would probably welcome this process because it involves victims directly (Jenness & Broad, 1997). Victim-offender mediation also does not necessarily preclude additional, more punitive approaches, which may appeal to supporters of hate crime laws.

Although some research indicates that victim-offender mediation has proved fruitful (Umbreit, Coates, & Vos, 2001), other scholars argue that its efficacy remains unproven (Lemley, 2001). Furthermore, its success in hate crime cases specifically is completely unknown. It is possible that many offenders' preexisting levels of bias make this solution unworkable. Even if this method does work in some cases, it requires the cooperation of both the victim and the offender, and such cooperation may sometimes be very difficult to obtain.

Counseling or Education Programs

A second approach to rehabilitating offenders is to provide them with some sort of counseling or educational program. As mentioned previously, at least one state—Massachusetts—requires such a program for all convicted hate crime offenders (see Box 7.6). Although this program is not required in other jurisdictions, some judges have imposed it as a condition of sentencing.

Box 7.6   Requiring Diversity Training

Massachusetts requires that all people convicted of a hate crime undergo diversity training. Here is the relevant law:

> There shall be a surcharge of one hundred dollars on a fine assessed against a defendant convicted of a violation of this section; provided, however, that moneys from such surcharge shall be delivered forthwith to the treasurer of the commonwealth and deposited in the Diversity Awareness Education Trust Fund established under the provisions of section thirty-nine Q of chapter ten. In the case of convictions for multiple offenses, said surcharge shall be assessed for each such conviction.
>
> A person convicted under the provisions of this section shall complete a diversity awareness program designed by the secretary of the executive office of public safety in consultation with the Massachusetts commission against discrimination and approved by the chief justice for administration and management of the trial court. A person so convicted shall complete such program prior to release from incarceration or prior to completion of the terms of probation, whichever is applicable.

SOURCE: Massachusetts General Laws, Chapter 265, § 39(b).

Many counseling programs focus on juvenile offenders. This emphasis makes sense for several reasons: Most hate crime offenders are young, young people may be less set in their ways and so more amenable to rehabilitation, and those who are subject to juvenile court jurisdiction and who commit relatively minor crimes are likely to receive only minimal punishments anyway.

One such program, known as Juvenile Offenders Learning Tolerance (JOLT), is run by the Los Angeles County District Attorney's office. In this program, young offenders and their parents receive counseling on topics such as anger management and parenting skills, are taught a curriculum on civil rights and human relations, and are required to give victims apologies and restitution. Offenders who successfully complete the program can avoid being prosecuted for the crime or receiving other sanctions.

The ADL also operates diversion and intervention programs in several areas. Generally, participants in these programs learn about anger management, listen to speakers from different ethnic and social groups, listen to victims of hate crime, and learn about the law. They also participate in community service, usually related to the community that they have harmed.

Less Formal Rehabilitation Efforts

In other cases, rehabilitation efforts are less formal. An offender might be ordered to tour a Holocaust museum, for example, or listen to a Holocaust survivor or hate crime victim speak. In some cases, offenders are matched with mentors who are members of the victim's group. The goal is for the offender and mentor to develop a friendship, which is what ex-Klansman Larry Trapp experienced (without court intervention) (Aho, 1994), as was discussed in Chapter 5.

Civil Remedies

A completely different tactic for dealing with hate crime offenders is to pursue civil remedies against them. Many states have legislation that specifically permits hate crime victims to sue offenders, and federal law also allows victims to sue under certain circumstances. In addition, victims may bring traditional tort claims against their attackers, such as battery, negligence, or trespassing (B. Levin, 2001b).

Sometimes, these lawsuits are brought against the particular individuals who committed the crime (or their parents, if the offenders are minors). More often, however, the suits target hate organizations. Not only are the organizations likely to have deeper pockets, but these lawsuits may prove more effective in battling hate because they give organizations incentive to control the actions of their many members. In previous chapters, several examples of successful cases against extremist groups were discussed, including multi-million-dollar judgments against the Klan, Aryan Nations, and White Aryan Resistance. Not only have these cases met with success in the courtrooms, but they have also withstood legal scrutiny on First Amendment and other grounds (Leavitt, 2001).

But what if the perpetrator does not belong to an organized hate group? In many cases, victims have pursued legal action against other types of organizations (see Box 7.7). One approach has been to sue offenders' employers. For example, a Palestinian American man was badly beaten by a United Parcel Service (UPS) employee on September 11, 2001, probably because of his ethnicity. He sued UPS for $270,000 (Cambanis, 2002). In San Leandro, California, seven employees of a dairy plant sued the plant for failing to take action to stop a series of racist incidents, including the placement of a noose near the workspace of an African American employee (Yi, 2001). In at least two separate incidents, men who had been beaten by members of the Chicago Police Department sued the department; the victims were targeted because they were thought to be gay (Warmbir, 2001a, 2001b).

> **Box 7.7   Vicarious Liability for Hate Crimes**
>
> When one person or organization is held legally responsible for the actions of another, it is known as *vicarious liability*. Although people rarely face criminal charges based on a claim of vicarious liability, suits on the basis of civil liability are not at all uncommon. In hate crimes cases, vicarious liability has mainly been used in four types of situations:
>
> - Parents have had to pay damages for the actions of their minor children.
> - Employers have had to pay for the actions of their employees.
> - Organizations, such as the Klan, have had to pay for the actions of their members.
> - Government agencies, such as police departments and schools, have had to pay for failing to adequately stop the actions of others.
>
> There are a number of legal rules and limitations for vicarious liability, all of which are too complicated to discuss here.[8] However, there are a number of issues that you might want to think about. Under what circumstances do you think a person or group should have to pay for the actions of another? What are the potential advantages of using vicarious liability, and what are the potential dangers? How can people or groups protect themselves from vicarious liability? How should appropriate damages be computed? Is criminal liability ever appropriate in these cases?

In addition to suing offenders, victims have sued government agencies for failing to respond appropriately to hate incidents. Several of these lawsuits have been directed at schools. In Santa Fe, Texas, for instance, Jewish parents sued the school district after their son experienced several years of anti-Semitic harassment and attacks (Moran, 2002). Others have targeted other agencies. After Brandon Teena, a transgendered individual, was raped and later murdered in Nebraska in 1993, Teena's mother sued the local sheriff for not arresting the rapists (who were the same men who later killed Teena) (Burbach, 2001).[9] In a recent case that is quite interesting from a legal perspective, the United States Department of Justice sued the San Francisco Housing Authority. The lawsuit claims that the Housing Authority did not take appropriate actions to end assaults and harassment of an interracial couple who were tenants in a public housing complex (Gordon, 2002).[10]

ASSESSING EFFORTS TO FIGHT HATE

On the one hand, prejudice makes sense. From a sociobiological view, it is sensible that humans should fear strangers and favor people who are most like themselves. From a cognitive standpoint, prejudices stem from shortcuts that conserve mental efforts. On the other hand, however, human beings are deeply social animals. We are also quite diverse in many ways. It makes sense, then, that we should be capable of coexisting in a peaceful manner. And, if prejudice is learned, it is logical that it can be unlearned.

The programs and organizations described in this chapter, and dozens (perhaps hundreds) more like them, strive to encourage peaceful coexistence and to teach tolerance. The big question is whether they are successful in these efforts. Unfortunately, because no research has been done to evaluate the effectiveness of antihate efforts, this question remains unanswered. By examining what we know about hate crime offenders and how best to reduce prejudice, however, we can hypothesize about the probable results of such research.

Effectiveness of Major Approaches to Fighting Hate

To summarize the preceding sections, antihate efforts have taken the following major approaches:

- Education of youth or the public in general
- Lobbying for antihate laws
- Monitoring organized hate groups
- Strengthening law enforcement efforts
- Public relations campaigns
- Creative approaches, such as Project Lemonade
- Victim-offender mediation
- Offender counseling and education
- Lawsuits against offenders and others

How effective are these remedies likely to be? In Chapter 4, I presented evidence that most hate crime offenders are not hard-core bigots, nor do most of them belong to organized hate groups. Instead, most appear to be young people with more or less ordinary levels of prejudice who are seeking excitement and a way to impress their peers. Hate crime laws themselves do not seem to be an effective deterrent to these offenders, and so enhancing law enforcement efforts will probably have relatively little effect. Activities aimed at organized hate groups, such as monitoring their activities and raising money through programs like Project Lemonade, will also not directly affect these offenders. Educational efforts and public relations campaigns aimed at reducing prejudice, encouraging tolerance, and respecting diversity may have little impact because bigotry does not seem to be the primary motivating factor for these perpetrators. It is also unlikely that lawsuits will be a deterrent because few of these young offenders have many assets to lose and because the prospect of being sued probably seems even more remote than the prospect of being criminally prosecuted.

The two remaining approaches—victim-offender mediation and offender counseling and education—appear to hold more promise for changing the attitudes of hate crime offenders. Both solutions require the offenders' cooperation, and, because these particular perpetrators do not appear to be particularly entrenched in hatred, they might be willing to participate. Both approaches might help the offenders realize the effect of their actions on their victims, which the perpetrators very likely had not thought much about before. And both strategies should encourage the offenders to carefully examine their own thoughts and behaviors; introspective self-analysis is also probably uncommon among this group. The significant drawback to these approaches is that they require that the offenders actually be apprehended first. As has already been made clear, a small percentage of those who commit hate crimes are ever caught, so the opportunity to use these approaches rarely arises.

What about the less common type of offender, who is involved with organized hate and whose actions are motivated primarily by bias? Cooperation with mediation or counseling is less likely. As you might recall from the discussion in Chapter 5, the research suggests that most people initially join hate groups for social rather than ideological reasons. Therefore, people might be discouraged from joining if alternative social outlets were available to them, but none of the common approaches to fighting hate offer this option to any great extent.

You might also remember that studies indicate that people leave organized hate groups when faced with social pushes and pulls against hate; circumstances make it difficult for them to continue their hate group activities, and some force leads them in another direction. Threatening offenders with prison time if they do not participate in mediation or counseling may constitute a sufficient push and pull against hate. This is especially true if their family members or friends can be enlisted to help discourage their activities. Again, though, this requires that the perpetrators be apprehended for their actions. And it does not stop people from joining the groups in the first place.

The Contact Hypothesis

As discussed at the beginning of this chapter, social psychological research suggests several ways that prejudice can be reduced. According to the Contact Hypothesis, one way to achieve this goal is to put members of disparate groups together on equal footing for sufficient periods of time and to provide support for their cooperation in achieving a common goal. Although this approach has achieved success with Aronson's Jigsaw Classroom model, few of the antihate initiatives make use of it.

A few antihate programs have employed the Contact Hypothesis to a limited extent. Recently, for example, the SPLC's Teaching Tolerance project sponsored a campaign encouraging students to sit at lunch one day with a different group. In addition, a few local organizations have encouraged residents to invite families of a different ethnicity to their homes for dinner. It is unclear, however, whether either of these types of programs produces contact of a sufficiently long duration to have the desired effect. Nor does either program involve cooperation toward a goal, and this cooperation appears to be essential for the diminishment of prejudice.

Victim-offender mediation may include all the necessary components for contact to be effective. The victims and offenders meet repeatedly; they are required not only to communicate with each other but also to reach a mutually satisfactory resolution; and their efforts are supported by mediators and official court sanctions. Those offender counseling and education programs that pair offenders with mentors from a different group may also have these characteristics. Again, then, for those who have already been apprehended for their crime, and who are cooperative, these tactics are promising.

Social Recategorization

Recent research indicates that another way to successfully reduce bias is to encourage social recategorization, to get people to redraw existing group boundaries and form a new, broader identity. Some antihate literature does this implicitly by stressing that we are all one people. A few programs, such as Not in Our Town, also emphasize unity. Most of the approaches to fighting hate crimes, however, do not make any use of this model, perhaps because social recategorization is difficult to achieve in practice.

Changing the Laws

A third tactic suggested by the research is to change the laws, in the hope that social mores will follow. As Aronson (1999) put it, stateways *can* change folkways. Many of the antihate efforts discussed in this chapter make excellent use of this method. Lobbying is a major activity of many organized antihate groups, and government activities often focus on passing and enforcing laws, including not only hate crime laws, but also other legislation that promotes tolerance, such as antidiscrimination laws.

The advantages of this approach are that it is relatively easy to implement, and it can have an effect on an entire state or country. The disadvantages are that its effects are likely to be slow and indirect,[11] and the legislative process is heavily influenced by politics. Furthermore, simply passing laws does not challenge the existing power structures that sanction bias and can even strengthen those power structures. Laws ostensibly intended to fight hate and promote equality may actually result in restrictions of civil liberties, and in the government quashing dissent and exercising a greater degree of totalitarianism.[12]

Changing Cultural Norms

The final method of reducing prejudice that we discussed earlier was to change cultural norms to make bias less common and less socially acceptable. Of course, this is much easier said than done. Mass media campaigns, like those used by several antihate groups, are of questionable value. Educational approaches, such as curricula that teach tolerance and respect for diversity, may help. However, to be effective, they must be concerted and long term. Moreover, even if youths receive antihate messages at school, the effect of these messages will be greatly diluted if youths continue to be exposed to prejudice at home and in the media. Some of the antihate approaches discussed here, such as Not in Our Town, are a good first step in removing bias from the culture in general.

CONCLUSION

Based on the discussion in this chapter, several recommendations can be made concerning future antihate efforts. First and foremost, much research needs to be done to evaluate the effectiveness of existing programs and policies. There is no shortage of individuals and organizations that wish to combat hate crimes, but there is almost a complete lack of assessment of their efforts. If we knew which endeavors work and which do not, these people could channel their energies and finances in much more useful ways. It will not necessarily be easy to determine what works, but, at this point, any addition to existing knowledge would be of great benefit.

A second recommendation is that those who engage in antihate activities use the growing literature on hate crimes and on prejudice reduction to shape their policies. Those policies that seem immediately most obvious, such as attempting to control hate groups, may often not be the most effective policies to reduce the occurrence of hate crimes. Furthermore, because a variety of factors influence hate crime activities, and because each method of dealing with the problem has distinct advantages and disadvantages, a variety of approaches ought to be considered, depending on the particular issues that a group or individual wants to address.

Third, antihate activists should pursue further coalition building. Of course, many such coalitions exist already. But in other cases, activists remain uninformed about the successes of others, replicate one another's efforts unnecessarily, or even work at cross-purposes to one another, which wastes precious resources and weakens the movement as a whole. Furthermore, increased coalition building might help to give a louder voice to those groups that are largely unrepresented in the current movement.

Finally, some antihate approaches appear, at least on their face, to be more promising than others. These include comprehensive and sustained educational efforts and offender counseling and mediation. Until more research is done to indicate what really does work, antihate groups might want to consider concentrating their resources in these areas.

NARRATIVE PORTRAIT

THE PROBLEM WITH "HATE CRIMES"

Michael Novick is a longtime antiracist activist and author of the 1995 book *White Lies, White Power: The Fight Against White Supremacy and Reactionary Violence*. He is also editor of *Turning the Tide: Journal of Anti-Racist Action, Research & Education,* now entering its 16th year of publication, available from Anti-Racist Action/People Against Racist Terror, P.O. Box 1055, Culver City, CA 90232. Novick is a public school teacher in Los Angeles and represents adult education teachers in United Teachers Los Angeles, a joint affiliate of the American Federation of Teachers (AFT) and the National Education Association (NEA). In this piece, he describes some of his experiences as an activist.

I have been involved in opposing racism and reactionary violence, as a grassroots activist, since the 1960s. For almost 7 years, I participated in the Hate Crimes Network of the Los Angeles County Human Relations Coalition. In my work, and especially through that experience, I have become increasingly frustrated and uncomfortable with the very concept of "hate crimes" and the way this concept is used and enforced in our society.

Bigoted violence is a very real problem in the United States and in other countries. But the model that hate crime legislation and enforcement offers for dealing with that problem will in fact never solve it; it may even serve to worsen it. How? By diverting and confusing the efforts of people who are appalled by bigoted violence from dealing with the roots of the problem within the institutions and relationships of our society as a whole. The hate crimes approach stigmatizes and criminalizes the perpetrators of "hate" and proposes increased incarceration as the method for dealing with them.

There are three major flaws with this hate crimes model. First of all, this mainstream, police-focused approach ignores the fact that some of the worst bigoted violence we see in the United States is actually carried out by law enforcement. Furthermore, it tends to equate the actions of oppressed people who may hate their oppressors and the oppression they suffer with the violence of the privileged seeking to maintain and enforce their ability to oppress others. Finally, by defining "enhanced penalties" as the solution to bigoted violence, the hate crimes approach falls in with the trend toward mass incarceration, longer sentences, and prisons within prisons that has resulted in the placement of 6.6 million adults in the United States, predominantly people of color, into prison, probation, or parole.

When I participated in the Hate Crimes Network in Los Angeles, I discovered that almost every other participant was a government employee or nonprofit agency staff member on paid time. Many members of the network were active-duty law enforcement personnel. There were no other grassroots activists or independent community members at any of the meetings. There were two predominant activities that the network as a whole and its constituent organizations were engaged in. One was educating "victims" and potential victims in how to file police complaints of hate crimes, thereby generating and accumulating quantifiable data that could demonstrate, in various annual reports, the scope and seriousness of the problem. The other was "sensitizing" law enforcement agencies to recognize, deal with, and collect and report valid data on "hate crime" incidents.

Little or no thought was given to prevention of such acts, beyond the unspoken shared assumption that enforcement of hate crime statutes would act as a deterrent. Beyond the celebration of "diversity," there was no discussion of reducing, combating, or ameliorating racism in our society, either in ways it is institutionalized or in ways it is internalized by individual members. Any discussion I tried to raise of racism, sexism, or homophobia within law enforcement was greeted with discomfort at best. Nor were members interested in probing the economic and sociopolitical roots of individual racists or racist acts. Members of the network generally shared the view that they, police included, represented the good guys and the high ideals of a society that was, in the main, on the right track. Hate crime perpetrators were seen as aberrant, antisocial individuals motivated by outmoded personal prejudices to carry out acts of senseless violence. Network members were, generally speaking, enjoined by their agencies and organizations from any form of political activism.

My work prior to and throughout this period followed a different analysis and approach, and it continues to do. I see random acts of racist, sexist, and homophobic violence by individuals as a manifestation of deeper social ills. The consciousness that produces such acts is rooted in material realities and social relationships of oppression, exploitation, and privilege that are institutionalized in American life and government. Law enforcement and correctional institutions are particularly culpable agencies of government because they are shot through with racism and sexism, so they are highly unlikely to be effective actors in solving the problem.

What's more, as a grassroots antiracist activist committed to uprooting racism, I am interested in educating and reaching out to the very young people, often alienated whites from middle-class and working-class backgrounds who are the perpetrators of hate crimes. I want to counterorganize against the conscious white supremacist groups that seek to foment such racist violence. Law enforcement is content to round them up, or occasionally to infiltrate their ranks. But the perpetrators of far more damaging forms of racism within the banks, the real estate, and corporate interests, and the schools, the government, and agencies that reinforce redlining, racial divisions, and colonial relations in our society are invisible within the hate crimes model. These perpetuators of racism are above reproach and beyond reach or accountability in this model. For example, the Los Angeles County Human Relations Commission in the early '90s worked, through the Network, with the Los Angeles County Office of Education to produce a report detailing the incidence of hate crimes and hate incidents in elementary and secondary schools in the county. But the far greater racism involved in the economic disparities facing black and Latino students went unremarked. The racism in disproportionate school expulsions and other punishments, in tracking, and in test scores was beyond their purview.

The final straw for me came in 1992, on the heels of the civil unrest in Los Angeles in the wake of the acquittal (in Simi Valley, a nearly all-white suburb heavily populated by police) of several of the officers involved in the videotaped beating of Rodney King. The Nationalist Movement, the Ku Klux Klan, and other white supremacist forces descended on the Ventura County community where the trial had taken place, seeking to take advantage of the trial's notoriety as well as to organize young white kids they felt would be sympathetic to their racist message. I got involved in a coalition that included some Los Angeles residents such as myself but that was based primarily in the Ventura County suburbs (which includes Simi Valley itself as well as somewhat more multiethnic areas such as Oxnard). This group, called Neighbors Against Nazis, opposed police brutality, condemned the acquittals, and mobilized against the proposed neo-Nazi rally in Simi Valley. It brought together, mostly from Ventura, blacks, Jews, gays, Mexicanos, Native Americans, Asians, women's organizations, labor unionists, immigrants, and other concerned people.

While engaged in this work, which involved frequent drives out to Ventura County, I missed a meeting of the Hate Crime Network. I returned the following month to inform the members of the activities and plans of Neighbors Against Nazis and to ask them to notify any contacts or affiliates they had in Ventura County of our planned counterdemonstration against the racist Nationalist Movement and our follow-up "unity and solidarity" program. Scheduled speakers at the two events included local rabbis, priests, and other clergy in Ventura County, farm workers, campus gay organizations, the National Organization for Women (NOW) chapter, Chicano activists against police abuse, the NAACP, local Chumash Indian elders, and others. To my chagrin, I discovered that during the previous month's meeting, when I had been absent, the Los Angeles County Hate Crimes Network took it upon itself to oppose the Ventura County counterdemonstration and issue a call for people to stay away!

This was not even a case where there were two rival "anti-Klan" events (one of which would be held away from the site of the Klan rally), which later came to be a common occurrence. Our coalition, Neighbors Against Nazis, had preempted the field. We were the only people organizing around the issue in Ventura County and had built a broad coalition that united more radical and more mainstream forces, that connected opposition to organized white supremacist hatemongers with opposition to police brutality and injustice within the criminal courts system, and that brought everyone together in a joint show of solidarity and unity *and* direct face-to-face opposition to the racist groups. We knew that many residents would inevitably show up, if only out of curiosity, at the white supremacist rally, and we wanted to be there in large numbers to welcome them with a message of antiracist solidarity as well as to convey to the racists that they were not welcome to exploit Simi Valley or promote further racist violence.

The Hate Crimes Network, which generally speaking never took "political" stands, suddenly had found the ability to speak out—IN OPPOSITION to this rally! They claimed that counterdemonstrating only gave the white supremacists the "attention and notoriety" they craved. In fact, this argument was strictly a cover for the real political stance of the Hate Crimes Network and its constituent organizations: Don't rock the boat. The Hate Crimes Network sees hate crimes as an embarrassment that destabilizes a status quo that they uphold and find desirable. They see antiracist activists as part of the problem, perhaps even a bigger problem than the perpetrators of hate crimes or the organizers of neo-Nazi groups. A network, half of whose members came from law enforcement agencies, was not interested in a rally that opposed police brutality and injustice in the courts, and that identified those problems as being rooted in systemic racism fully as much as the Ku Klux Klan is. Calling for the community to turn its back on the Nazis was a convenient way to cover the real basis of its disagreement with the Neighbors Against Nazis coalition and rally.

Really, the contrast in the composition and politics of the two groups is the clearest evidence of the difference between the two approaches. The members of the Ventura County-based Neighbors Against Nazis Coalition were predominantly concerned individuals and families in Simi Valley and surrounding towns. The groups that got involved were grassroots activist organizations with a history in protest politics and labor and community organizing. The community spokespeople we reached out to and involved as speakers at our rallies recognized the threat that organized white supremacists posed to their communities and also came to recognize that forms of institutional racism in Ventura County, as an area of "white flight," including racism in the courts and law enforcement, were also part of the same problem. The Hate Crimes Network, in contrast, was composed of social service agencies dependent on government funds, active duty law enforcement officers, and paid "human relations" staffers. Not content with seeking to circumscribe and define the problems of racism and sexism as individual criminal behavior, they felt compelled to actively oppose and sabotage an organizing effort having nothing to do with hate crimes as they defined them, and taking place in another jurisdiction!

In the wake of these developments, I ended my involvement with the Los Angeles County Hate Crimes Network. I have continued to take an alternate approach to dealing with the issues the Network raises, as well as many other related issues they would not touch with a 10-foot pole. Anti-Racist Action staffs tables at many youth cultural events and concerts, organizes militantly against white power music shows and other public organizing, and is involved in several ongoing coalitions against police abuse and other systemic social ills. Because the members of this group have seen racism promoted by corporate economic hegemony, and by militarism, we participate in coalitions against "globalization" and war. We also know that bigoted violence is a form of racist terror. There is a standard paradigm, and it reflects the reality that the overwhelming number of such cases involve the violence of the privileged against the oppressed, carried out in circumstances conducive to bullying by larger, stronger, numerically superior racists—whites attacking isolated black people, men abusing women, straights brutalizing gays, citizens and native born stigmatizing immigrants and language minority members. When was the last time you saw a group of lesbians or gays target and beat a straight man? Or a bunch of Mexicano farm workers gang up on a few survivalist campers playing war games? Yet the hate crime laws are predicated on the phony "evenhandedness" or "colorblindness" that criminalizes blacks or Chicanos to a disproportionate degree.

Based on my up-close experience, there is a need for a drastic revision and rethinking of the entire hate crimes model. One final illustration of this need is the recent release from prison of a number of the old Peni Skins and Nazi wannabe Low Riders. Locked up for drug dealing and an assortment of hate crimes (or bigoted violence, as I usually refer to it), they have more recently been coming out of the prisons as battle-hardened Nazi veterans. With the old-line Aryan Brotherhood prison fascists locked up in "security housing" (the SHU), the younger generation quickly became hard-core white power boneheads and drug dealers. Now they are coming back on the streets seeking to toughen up a new generation of "style-nazis" into a white power fighting force that could carry out a racial and ethnic cleansing of the United States. This demonstrates the ultimate self-defeating futility of the entire "hate crimes/lock 'em up" approach.

DISCUSSION QUESTIONS

1. In the early part of the 20th century, sociologist William Gordon Sumner said, "Stateways cannot change folkways" (see Allport, 1954). Many scholars have disagreed, however, asserting that legislation can reduce prejudice. With which argument do you most agree, and why? Should the laws push social change upon a society that is not ready for it? Or is doing so the only realistic way to effect social change? Although this discussion was particularly pertinent to issues related to desegregation, can you think of some more modern issues to which it might apply?

2. Describe a media campaign you would create to reduce prejudice. In what ways would it be similar to and different from the campaigns you have already seen? How would you assess its effectiveness?

3. What commonalities and differences do you see among the means and goals of the advocacy groups described in this chapter? In what ways do you believe these groups can work together, and where do you think conflicts are most likely to arise?

4. Certain organizations concern themselves specifically with hate crimes against specific minority groups, such as Jews, gays and lesbians, Arab Americans, and Asian Americans. On the other hand, other groups such as African Americans and Latinos are less well represented. What do you think is the reason for this? What are the repercussions?

5. At least one state (Massachusetts) requires all people convicted of hate crimes to take part in a diversity awareness program. Do you believe that such a program will reduce recidivism, as it is intended to do? What information and activities should such a program include? How long should the program be, and where should it take place? Who should provide the training, and what should their qualifications be? What should be the consequences for offenders who refuse to participate? Would you support a similar law in your jurisdiction?

6. Suppose you have just been elected governor of your state. Would you create some kind of hate crime task force? If so, what individuals or agencies would it include? What would be its primary mission? What are some of the projects or programs it would take part in? Do you feel that this is an effective way to combat hate crimes?

7. What hate crime agencies and programs exist in your area? How do they compare with those in similar jurisdictions across the United States? Do you believe they are sufficient to meet the needs in your area? If not, what would you change or add?

8. Imagine that you have been asked to develop a program to counsel or educate hate crime offenders. At what sorts of offenders would your program be aimed? What would your program entail? How would you assess its effectiveness? Are there ethical problems associated with forcing people to take part in these types of programs? In other words, don't people have the right to hate?

9. What conclusions can you draw about the probable effectiveness of existing antihate efforts, given the research on hate crimes and prejudice? Why do you think the research plays relatively little part in shaping programs and policies? If you headed an organization dedicated to combating hate crimes, what would your plan be?

10. This chapter ends with four general recommendations. Do you agree with them? Are there any that you would change, delete, or add?

INTERNET EXERCISES

1. Elliot Aronson has a Web site about his cooperative learning technique, which he calls the Jigsaw Classroom: www.jigsaw.org. This technique may even be useful at the college level; an example of a professor who uses it for one of her psychology courses can be found here: siop.org/tip/backissues/TIPApr02/pdf/394_109to112.pdf.

2. Look at the Web pages from several antihate organizations. A good place to start looking to find such groups is the National Criminal Justice Reference Service's directory at www.ncjrs.org/hate_crimes/additional.html. Political Research Associates is another good place to look. Go to www.publiceye.org, scroll down to "Links," and click on "Challenging the Extreme Right & Hate Groups." What commonalities and differences do you find among the sites you visited?

NOTES

1. The same could be said for most other forms of crime as well: Although we can often reduce crime levels significantly, we can probably never completely deter any crime.

2. In fact, his landmark book, *The Nature of Prejudice,* was published in 1954, the same year that the United States Supreme Court held in *Brown v. Board of Education* that segregated public schools were unconstitutional.

3. Social categorization was discussed in Chapter 4. Briefly, it describes the tendency of humans to divide the world into us (ingroups) and them (outgroups).

4. Of course, this is not to say prejudice had miraculously disappeared after 9/11. This woman's experiences would likely have been radically different if she had appeared to be Muslim or Arab.

5. Changes turned out to be much more deliberate than speedy. Half a century later, courts are still dealing with de facto school segregation.

6. For the contact information and Web addresses of the groups discussed in this section, see Appendix B.

7. One of my personal favorite antihate sites is a parody of the Westboro Baptist Church. As you may remember from Chapter 5, the Church, which is fervently homophobic, maintains a Web site at www.godhatesfags.com. The parody site, which appears to be the work of an individual, is at www.godhatesfigs.com.

8. For more information on vicarious liability, see Keeton, Dobbs, Keeton, & Owen (1984).

9. Teena was a biological female who preferred to present herself as a man. After two men discovered that she was actually female, they beat and raped her. She reported the rape to the police, but the men were not arrested. The men, John Lotter and Marvin Thomas Nissen, later killed her and two witnesses to her murder. The film *Boys Don't Cry* (Pierce, 1999) was a depiction of these events.

10. The Santa Fe case was settled out of court; after a complicated legal battle, Brandon Teena's mother was awarded $98,223 in damages; and the San Francisco case is still pending.

11. As discussed earlier, for example, many American schools remain segregated half a century after legal segregation was outlawed.

12. Consider, for example, the Patriot Act of 2001, signed into law less than two months after 9/11. This law, which supposedly is meant to combat terrorism, permits expanded government surveillance of Americans and increases governmental power to detain or deport non-U.S. citizens. It also reduces judiciary and due process constraints on federal law enforcement powers. For a brief and useful critique of the Patriot Act, see Herman (2001).

Chapter 8

GLOBAL HATE
International Problems and Solutions

Fifty million people. Take the entire population of the 15 largest cities in the United States—that is, New York, Los Angeles, Chicago, Houston, Philadelphia, San Diego, Phoenix, San Antonio, Dallas, Detroit, San Jose, San Francisco, Indianapolis, Jacksonville, and Columbus. Now *double* the total. You are still a little short of 50 million. But according to one estimate (Cohan, 2002), more than 50 million people died during the 20th century due to genocidal campaigns.[1]

This mind-numbing statistic aside, a glance at any day's newspaper will tell you that violence based on race, ethnicity, religion, gender, and so on is hardly rare, nor is it a uniquely American phenomenon (see Box 8.1). A single page of my local paper this morning contained articles about such violence in six countries on three continents.

DISTINGUISHING HATE CRIMES FROM OTHER EVENTS

In this chapter, I explore the topic of hate crimes in countries other than the United States. In doing so, we are faced with a dilemma: How can hate crimes be distinguished from terrorism, political crimes, war, and similar events? On the one hand, such a distinction is necessary if the chapter is to remain reasonable in scope—after all, massive volumes have been written on each of these subjects. On the other hand, any distinction is going to be somewhat arbitrary because all of these events are, at the very least, extremely close cousins.

MacGinty (2001) recently tackled this issue. Comparing hate crimes in the United States to events in areas such as Northern Ireland, Israel/Palestine, and South Africa, he concluded,

> The major difference between the United States and the other conflict areas is that the former is host to what may be described as an undeclared war, or an identity-driven conflict that does not possess a cloak of political legitimacy. The other conflict areas are host to more formal and politicized conflicts. (MacGinty, 2001, p. 639)

Whereas hate crimes in countries like the United States are apt to be viewed on the personal level, those in these other areas are larger in scale, more political, and more volatile.

Box 8.1   A Daily Snapshot of Hate

How widespread is hatred and bias? Here are a few events featured in recent world news headlines:

- Australia has raided several homes in hopes of finding the radical Muslims resposible for a deadly bombing in Bali.
- Canada has warned Canadian citizens who were born in the Middle East to use caution when traveling to the United States in case they are persecuted under new U.S. antiterrorism laws.
- Explosions erupted in Soweto, South Africa, apparently caused by a right-wing group attempting to destabilize the government.
- Two Islamic militants were sentenced to life in prison in France for their role in 1995 subway bombings.
- A leader of the Chechen rebellion was arrested in Denmark on charges he helped plan a siege of a theater in Moscow in which 119 hostages died.
- The Irish Republican Army broke off negotiations with an international body charged with attempting disarmament in Northern Ireland.
- An Israeli coalition government collapsed, endangering the possibility of peace talks with the Palestinians.
- White farmers who were evicted from their farms in Zimbabwe have been forbidden to take their farm equipment with them as they attempt to relocate to Zambia.
- Debate continues in Nigeria over whether a woman convicted of having sex outside of marriage will be stoned to death.

MacGinty himself, however, notes that the distinction he makes may be "a little artificial" (MacGinty, 2001, p. 650) in that all hate crime is, arguably, political. Nevertheless, he argues that the disparities do imply differences in the way hate crimes should be handled from a policy perspective.

For the purposes of this book, I constrain our discussion to those incidents which most closely resemble the American concept of hate crimes in that (a) the acts are committed by individuals or small groups (not the government itself); (b) the victims are chosen because of their race, ethnicity, religion, or other group; (c) the acts are a violation of criminal laws; and (d) the acts are not part of a civil war.[2] Although other countries do not necessarily use the term *hate crimes* to describe these incidents, many nations experience problems that are very similar to those described in the previous chapters of this book. In this chapter, we examine some of those problems and the approaches taken by other countries in trying to deal with hate crimes. We will also investigate some of the issues related to the globalization of hate, including the growing links between domestic and international extremists, and the difficulties inherent in policing transborder hate crimes.

THE NARROW STUDY OF GLOBAL HATE

Almost all of the scholarship on the kinds of acts classified in this chapter as hate crimes has concerned European or English-speaking countries. For example, in one edited collection on international hate crime (Hamm, 1994), every chapter focuses on North American or Western European countries; in another (Kelly & Maghan, 1998), five of the eight chapters concern the United States or Western European countries.[3] Recent issues of several journals have been devoted to hate crimes as well[4] but, again, almost all of the articles concern European countries or the United States.

The lack of attention to hate crimes in other areas may reflect Americocentrism and Eurocentrism within academia itself. In many cases, too (although by no means all), hate crime activity in African, Asian, and South American countries is closely linked to large-scale terrorism or political instability, so it is difficult or impossible to study as an isolated phenomenon. In Rwanda, for example, more than 750,000 Tutsis and moderate Hutus were murdered by the Hutu-led armed forces and militia (Berman, 2001). In India, tension between Hindus, Muslims, and Sikhs has led to secession movements and to acts of violence and destruction, some of which have been government-sponsored (Rahman, 1998). Another possible explanation for the almost complete focus on hate crimes in European and English-speaking countries is that many of the offenders—skinheads, neo-Nazis, and the like—are common to all these countries, so scholars feel a familiarity with them; in contrast, an English-speaking researcher might know relatively little about the main offender groups in, say, Peru.

Whatever the reason for the lack of literature on hate crimes in Asia, Africa, and South America, it is a serious deficiency. Not only does it mean that most people are largely unaware of the hate crime situations in most of the world, but it also may limit our understanding of hate crimes and prejudice in general. Although the precise targets of hate crimes may differ from place to place, there are likely to be commonalities in bias regardless of geographic location. There will likely be interesting discrepancies as well. As Jackson, Brown, and Kirby (1998) conclude, "understanding the nature of prejudice and racism, and the psychological and social consequences in any given country, can be accomplished only be studying similarities and differences in the relationships between dominant and subordinate groups across countries" (p. 101).

Unfortunately, because of this limitation in the literature, the discussion in this chapter must be confined primarily to Europe, North America, and Australia. I will especially focus on Germany, the United Kingdom, Australia, and Canada. See Boxes 8.2 and 8.3 for information about hate crime in Hungary and Colombia.

Box 8.2   Hungarian Minorities Abroad

Hungary is a country facing many challenges. It continues to suffer physical wounds from both World Wars and economic, political, and social difficulties brought about by decades of Soviet domination. Unemployment in much of the country is high, wages are extremely low by European standards, the reigning political party has changed with every election since the Iron Curtain fell, there are various pressures from some of the country's even less prosperous neighbors, and Hungary is struggling with the requirements for acceptance into the European Union. Despite all these very serious problems, however, one of the Hungarian government's primary priorities is protecting the rights of Hungarian minorities abroad,[5] who are entitled to receive certain benefits from the Hungarian government, including monetary stipends if they teach their children the Hungarian language and subsidized annual travel to Hungary.

(Continued)

Box 8.2 (Continued)

When I first learned about this policy, I was somewhat mystified. Why would a country facing so many challenges choose to spend a large amount of their resources in this way? To my American eyes, the distinctions between Hungarians and their ethnic neighbors seemed pretty trivial; apart from the language they spoke, I would find it impossible to tell the difference between a Hungarian and, say, a Romanian.[6] Upon further reflection, however, I realized that this was only a single example of the importance that race, religion, ethnicity, and similar distinctions continue to play for human beings worldwide. Unfortunately, although Hungary manages to deal with the issue of ethnicity through diplomacy, all too often ethnic differences lead to violence.[7]

THE RISE OF NATIONALISM AND THE RADICAL RIGHT

In the 1990s, at approximately the same time as the United States was experiencing an increase in hate group membership and political conservatism in general, a parallel movement was occurring in Europe. In both areas, nationalism increased dramatically, and the radical right gained new footholds.

On the one hand, it seems as if the early 1990s should have been a time of peace and optimism. The Iron Curtain had fallen, relieving some countries of a powerful antagonist and giving others new independence. The ongoing growth of the European Union promised political stability and economic power for its members. And yet, on the other hand, sectarianism grew considerably, frequently with devastating results.

Box 8.3   Colombian Gamíns

Wilson and Greider-Durango (1998) argue that the murder of street children, or *gamíns,* in Colombia is a form of social cleansing that constitutes a hate crime. Hundreds, perhaps thousands, of children live on the streets of Bogotá, supporting themselves however they can. Their lives are miserable and full of violence.

Certain Colombians find the street children undesirable, both because they often steal to support themselves and because they interfere with the city's desired image of being modern and clean. Consequently, police and vigilantes have beaten and murdered many of these children. There is even a paramilitary group called "Death to Gamíns."

Violence against the street children has largely gone undocumented and unmeasured. Wilson and Greider-Durango (1998) do report, however, that between February 1 and May 14, 1991, the bodies of 187 young people were found in Bogotá.

Table 8.1 Immigration in European Countries in 2001

Country	Asylum Seekers	Illegal Immigrants
Austria	30,135	48,659
Belgium	24,549	90,000
Denmark	12,403	N/A
England	70,135	1 million
France	47,263	400,000
Germany	88,363	500,000 to 1.5 million
Greece	5,499	1 million plus
Ireland	10,324	10,000
Italy	10,000	300,000
Netherlands	32,579	N/A
Norway	14,383	N/A
Portugal	192	N/A
Spain	9,219	N/A
Sweden	23,513	N/A

SOURCE: Ford (2002).

Weinberg (1998) reviews some of the factors that converged to foster right-wing extremism in the last decade. Both the United States and Western Europe were relatively wealthy areas surrounded by poorer countries. This led to immigration, which, in turn, often resulted in xenophobia and resentment. In addition, family structures were under increased strain, changes in the economy led to production jobs being replaced by low-paying service jobs, and confidence in public institutions (especially the government) decreased.

There were other triggers as well. The birth rate among whites continued to decline, leading to negative population growth in many areas and an increase in the proportion of nonwhite residents. The disintegration of the Soviet Union uncovered conflicts that were buried but not dead (Hockenos, 1993). Furthermore, as that common enemy disappeared, recategorization occurred, and new boundaries between ingroups and outgroups formed. Many people in both Europe and the United States feared increasing globalization. They felt that a New World Order was at hand, under which their personal interests would be overridden, and their cultures would be lost. Finally, even though many extremists were alarmed about globalization, they were able to take advantage of some of its components (most notably the improved communication made possible by the Internet) to spread their messages to a much greater extent than ever before. In addition, the number of asylum seekers and illegal immigrants in Western European countries increased during the 1990s and beyond (see Table 8.1 for data for 2001).

Special situations existed in Central and Eastern European countries that had formerly been part of the communist bloc. Many of these countries had been suffering under oppression and economic stress for decades.[8] Although their citizens may have initially felt great optimism when the Iron Curtain fell, this optimism flagged considerably when they saw they would not immediately obtain the relative prosperity of their neighbors to the west. The Soviet Union's rather sudden absence led to potential political, economic, and social chaos, and to a vacuum that was easily filled with the familiarity of nationalism (Hockenos, 1993). The far right enjoyed popularity in part because of its vehemently anticommunist doctrines. Ironically, too, the freedom that came with the collapse of police states permitted people to speak and write openly about their beliefs, so extremism found a newly public voice.

As a result of all these disparate forces, nationalism did increase throughout Europe,[9] and with it, so did intolerance. In fact, it has been argued that nationalism necessarily means intolerance (see, e.g., Caplan & Feffer, 1996; Ramet, 1999). Of this phenomenon, Ramet (1999) wrote,

> The most powerful arguments are often those that do not make any sense at all. They appeal not to reason, which labors only with effort, but to prejudice and emotion, which require no emotion at all. It is the specific "genius" of the far right, if one may use that lofty word to describe the far right's studied flight from reason and moderation, that it understands this principle. . . . The Other lies at the heart of radical right politics, and for the radical right, which understands the world in terms of struggle, in terms of "us" versus "them," the Other is translated into "the Enemy." (pp. 3-4)

Clearly, in the 1990s, these arguments resonated with many of Europe's inhabitants. One result of these circumstances was that adherents to far-right doctrines gained political power in several countries. In Italy in 1994, a coalition of far-right parties gained a majority of the seats in Parliament. In Russia, the party of neo-fascist, nationalist Vladimir Zhirinovsky gained 23% of the parliamentary vote in 1993, and an additional 20% of voters chose other extreme right-wing candidates (Hockenos, 1993). In Belgium, the blatantly xenophobic Vlaams Blok Party received 25% of the vote in Antwerp in 1991, making it the largest political party in the city (Lee, 1997). In Austria, the far right Freedom Party received nearly 28% of the vote for the European parliament (Lee, 1997). Even where the extreme right did not actually gain power, it influenced policymaking. In France, a law granting citizenship to anyone born in France was revoked (Lee, 1997), and, in Germany and Austria, more restrictive immigration laws were passed (Fijalkowski, 1996; Lee, 1997).[10]

Although some members of the far right attempted to further their agenda through politically legitimate means, others worked outside the laws. Skinhead and neo-Nazi youth gangs became active not only in Britain, Germany, and the United States, but in other countries as well. Among the countries experiencing organized hate group activity in the 1990s were Australia, Austria, Canada, the Czech Republic, Denmark, France, Hungary, Italy, the Netherlands, Norway, Russia, Slovakia, and Sweden (Fangen, 1998; Kürti, 1998; Morton, 2001; White & Perrone, 2001). According to one study, in 1994 there were approximately 191 extraparliamentary right-wing groups (i.e., groups that operated outside legal political boundaries) in Western Europe alone, with an estimated 20,000 members (Weinberg, 1998).

European countries in the 1990s experienced what has been called a wave or explosion of hate crimes (see Box 8.4). In the first 10 months of 1995, for example, in the Czech Republic, police recorded more than 150 hate crimes (Kürti, 1998). There were official reports of 1,483 violent hate crimes in Germany in 1991 (Lee, 1997).[11] The precise targets differed somewhat from place to place. Whereas Asians were being attacked in England, Turks were victims in Germany, and Romany (Gypsies) were the target of preference across Eastern Europe. In general, however, as in the United States, the victims of hate crimes across Europe were immigrants, ethnic and religious minorities, people of color, and gays and lesbians.

Box 8.4   Examples of Hate Crimes in Europe in the 1990s

- Kielce, Poland, 1990: Smoke bombs are thrown into a concert of a Ukrainian Jewish group and the meeting of a Jewish Solidarity group.
- Hoyerswerda, Germany, 1991: A large group of skinheads attacks immigrants from Mozambique and Vietnam, including throwing Molotov cocktails at an asylum hotel.
- Mölln, Germany, 1992: Skinheads throw Molotov cocktails at the home of a Turkish family, killing two women and a child.
- Rostock, Germany, 1992: Skinheads set fire to a hostel inhabited by asylum seekers.
- Hadareni, Romania, 1993: Seventeen Romany houses are burned to the ground, killing four people.
- Solingen, Germany, 1993: The home of a Turkish family is burned; two women and three children are killed.
- Czech Republic, 1994: Group of skinheads kills a 42-year-old Romany man.
- Lubeck, Germany, 1994: Synagogue is firebombed by neo-Nazis.
- Lubeck, Germany, 1996: Fire set at a home for immigrants kills 10 and injures 35.
- Riga, Latvia, 1998: A synagogue is firebombed.
- London, England, 1999: A bomb in a black neighborhood injures 40 people; two weeks later, a bomb in a racially mixed neighborhood kills three and injures 65.
- Moscow, Russia, 1999: One synagogue is bombed and another is the scene of an attempted arson.

It is unclear whether most of the hate crimes in Europe in the 1990s were committed by members of organized hate groups. If so, that would be a different pattern from the one found in the United States, in which most offenders do not belong to organized groups. Certainly, a great many of the European crimes were blamed on skinheads and neo-Nazis. It is possible, however, that authorities simply assumed that the perpetrators were members of these infamous groups. Furthermore, even in some of the cases in which the crimes were almost certainly committed by skinheads, such as the attacks in Hoyerswerda, local residents cheered on the skinheads and jeered police efforts to stop them. Bias in these countries, as in the United States, was certainly not the exclusive territory of extremist groups.

Although hate crimes themselves may have peaked in Europe in the mid-1990s, the far right continues to have a strong influence. In France in 2002, Jean-Marie Le Pen received 17% of the presidential vote, placing him second behind the incumbent Chirac and qualifying him for a runoff election (which he ultimately lost). Le Pen, who founded the far-right National Front Party, has been outspokenly xenophobic and anti-Semitic. In Austria in 2000, the far-right Freedom Party was able to gain seats in government as part of a coalition government. In the Netherlands in 2002, the far-right LPF (Lijst Pim Fortuyn) Party won second place in parliamentary elections and formed a coalition government with two other parties (the government collapsed less than three months later as a result of infighting within the LPF). In Italy, the National Alliance is part of the ruling coalition. Again, even when far-right and nationalist parties have not actually gained governmental control, they have influenced governmental policies, leading to anti-immigrant legislation in several European countries (Ford, 2002).

Much of the current popularity of the extreme right in Europe appears to stem from its anti-immigrant stance. Immigration has swelled in many European countries, and politicians

---

Box 8.5   Hate Crimes Against the Roma

People in America probably do not think very often about the *Roma* (or *Romany*) people. In fact, most Americans probably do not even know who the Roma are; they are more familiar with the term *Gypsy,* which is avoided today by the Roma people themselves because of its negative connotations. In Europe, however, especially the Central and Eastern portions, the Roma constitute an important ethnic group. In fact, numbering between 7 and 9 million, they are the largest ethnic minority in Europe (Brearly, 2001).

The Roma emigrated to Europe from India perhaps 1,000 years ago (Crowe, 1996). Despite their long sojourn on the European continent, however, they maintained a separate ethnic identity and never fully assimilated into the larger communities. This separation was due partially to their different language and cultural practices, and partially because they looked different, tending to be of darker complexion than their neighbors.

---

have emphasized supposed links between immigration, crime, and economic woes. The events of September 11 in the United States only added to the xenophobic fire in Europe (as in they did in the United States as well).

Anti-Semitism appears to be increasing dramatically in Europe at this time, probably fueled by the crisis in Israel. According to the Anti-Defamation League (ADL) and the Simon Wiesenthal Center (SWC), anti-Semitism is at its highest levels worldwide since World War II (ADL, 2002e; SWC, 2002c). In France alone, according to the SWC, there were 300 anti-Semitic incidents during the first two weeks of April 2002 (SWC, 2002b), and the ADL reported a spate of anti-Semitism in Russia in 2002 (ADL, 2002b). A recent poll showed an increase in anti-Semitism in Germany as well, with more than a quarter of respondents saying they believe Jewish influence is too great ("Poll: Anti-Semitism Rising in Germany," 2002). In 2002, there was also an outbreak of violence directed at synagogues, Jewish cemeteries, and Holocaust memorials. Bombings occurred in locations as diverse as Canberra, Australia; Paris, France; Kiev, Ukraine; London, England; Tunis, Tunisia; Quebec City, Canada; and Belower Woods, Germany.

It is clear that hatred is not a phenomenon that is unique to any one country, nor are its manifestations necessarily different in various locations. The rise of the right in Europe, as in the United States, has been stimulated by a variety of factors. In both regions it may be associated with increases in bias and hate crimes. Many of the targets of these crimes remain the same from country to country (see Box 8.5 for a discussion of hate crimes against the Roma in Europe). Despite these similarities, each country does have a distinct political and social situation, and each has taken a different approach to combating hate crimes. In the next section, I examine several of these individual approaches in more detail.

FOUR COUNTRIES' APPROACHES TO HATE CRIMES

Germany

Perhaps more than any other country, Germany is forced to deal with a legacy of hate. It was, after all, a German government that led the planned extermination of more than

The history of Roma in Europe is a history of discrimination and persecution. At various times and in various countries, they were imprisoned, expelled, or murdered. Their children were stolen from them (although, ironically, the stereotype is that Gypsies steal other people's children). During the Holocaust, the "Gypsy Nuisance" was scheduled for genocide, and 200,000 to 500,000 Roma were killed during the war (Brearly, 2001).

During the communist era, governments in Central and Eastern Europe attempted to forcibly assimilate the Roma population. After the Soviet Union disintegrated, conditions became even worse as Roma became handy scapegoats for social and, especially, economic woes. Today, anti-Roma prejudice is largely publicly acceptable across Europe in a way that few other biases are. The Roma people remain economically, educationally, medically, and socially disadvantaged, and they are the primary victims of hate crimes in many European countries (Brearly, 2001; Crowe, 1996; Lee, 1997).

12 million human beings. To a great extent, Germans have acknowledged this legacy. There are Holocaust memorials at many locations within the country and, during a recent trip to Germany, I saw plaques dedicated to the memory of German Jews and a statue in honor of a Nazi resistance fighter. The German government, in partnership with several corporations, is currently paying several billion dollars in restitution to surviving victims of the Holocaust. Nevertheless, the specter of the Holocaust has not erased hatred from all hearts and minds in Germany, and bias continues to be a serious problem.

At a conference I recently attended, another participant commented that she did not understand why Germany had a problem with hate crimes because almost everyone in that country was white. This misconception, which she likely shares with many Americans, seriously underestimates Germany's ethnic diversity. Although, by American standards, Germany has a small number of residents of African or Asian ancestry, that does not mean the country is ethnically homogeneous. Roughly 9% of German residents today are immigrants (German Ministry of the Interior, 2001). This is quite comparable to the United States, in which the 2000 census estimated that 10.4% of the population was foreign-born (U.S. Census, 2001). And, also as in the United States, the number of immigrants in Germany has increased dramatically in the last 2 decades, rising by nearly two thirds since 1980.

Immigration

Germany's large immigrant population is primarily attributable to three sources. First, there was the gastarbeiter program, begun in Germany in 1956. Under this program, people from other, poorer countries were encouraged to come to Germany as guest workers, primarily to solve Germany's labor shortage. Although they came from all over, by far the biggest number came from Turkey; today, roughly 2 million German residents have Turkish citizenship. At the Stuttgart Airport, as in other places in Germany, the signs are in German, English, and Turkish.

The idea was that the guest workers would work for a few years and then return to their home countries. What happened, in fact, is that they stayed and raised families, many of which are now in their third generation as German residents. The children and

NARRATIVE PORTRAIT

ETHNIC CLEANSING IN BOSNIA-HERZEGOVINA

A particularly poignant recent example of ethnic conflict occurred in the former Yugoslavia. When communism collapsed, nationalists won control of the government, and the result was "ethnic cleansing." Stevan M. Weine is a clinical psychiatrist who works with Bosnian refugees in the United States. The following excerpts come from his interviews with those refugees.

Survivors recall that [before ethnic cleansing] in both the cities and countryside of Bosnia-Herzegovina, Muslims, Serbs, and Croats shared their lives with one another.

We worked together. We slept together. Lived together. Ate together.

Our neighbors—we were like a family. Our flats were like one home. They had two kids. We lived together. We vacationed together. Spent holidays together. Overnight, they changed. . . .

H. says his troubles began on June 20, 1992, when Serb forces came into the factory where he had worked for years and fired all the Muslims. The next morning they went to Muslim homes, rounded up the men, and brought them to a concentration camp on the grounds of an old manor called Omarska.

I was one of the first people brought to Omarska. After 10 days there were 20,000 prisoners. They put us in a big yard. . . . The guards shot anyone who tried to leave and sometimes shot for no apparent reason.

After 3 days the "investigations" began. I was in them 25 times and only one time did I leave on my own legs. In the investigations they always beat us. Guilty or not guilty. . . .

For the first 10 days they didn't give us any food. Then they fed us once a day at 6:00. They gave us 3 minutes to get from our building to the kitchen. Some of us were more than 50 meters away, and we had no chance to reach the kitchen. Those who came had 3 minutes to eat. Those who did not had no food. The guards formed a line that we had to run through to get to the kitchen. As we ran they beat us with guns, wheels, and tools. . . .

Sometimes they put us in a 4 × 4 meter room—700 people. They told us to lie down and they closed the windows and the doors. It was summer. We lay like sardines in a can. Those on top were in the best position. Every morning some on the bottom were dead. Every morning a guard came with a list and called people's names. Those they brought out never came back.

One day they came at 3 A.M. and they brought 174 people. I was with them. They lined us up behind a building and they started to shoot us. Only three of us survived.

The worst event was when I watched a young man as they castrated him. Right now I can hear his cry and his prayers to be killed. And every night it wakes me. He was a nice young man. His executioner was his friend from school. He cut his body and he licked his blood. He asked him just to kill and to stop all that suffering. All day and all night we heard his prayers and his crying until he died. This is something that I cannot forget. It gives me nightmares and makes sleep almost impossible. I can't remember the people who were the executioners. For me all of them in those uniforms were the same. I can't remember who was who.

SOURCE: Weine (1999, pp. 15-35).

grandchildren of these guest workers speak German and are generally as assimilated into German culture as I, a third-generation American, am into American culture. In the eyes of many Germans, however, these people remain foreigners, and they become easy scapegoats when the country faces difficult economic times.[12]

A second source of immigration to Germany is asylum seekers. In 2001, more than 88,000 people applied for asylum in Germany, seeking to escape conditions in their home countries; perhaps as many as 1.5 million were there illegally (Ford, 2002). A great many of these people have immigrated from Eastern European countries embroiled in conflict, such as Serbia-Montenegro, Macedonia, and Bosnia and Herzegovina. In response to pressures from the right, the German government passed a law in 1993 that restricted the right to asylum, but Germany's policy is still among the most generous in Europe (Bergmann, 1997). Although the German government attempted to force or encourage repatriation of asylum seekers through a number of means, including offering financial incentives to those who would leave and reducing benefits to those who stayed, the majority of these immigrants still find conditions in Germany preferable to those in the countries of their birth.

The third major source of immigrants to Germany is ethnic Germans from abroad—that is, people born in other countries (primarily Eastern Europe) who can claim German ancestry (Aronowitz, 1994). Although these people may have legitimate blood ties to Germany, their families may long since have adopted the language and cultural practices of other countries. In some cases, their families may have emigrated from Germany centuries earlier (Harnishmacher & Kelly, 1998). When they arrive in Germany, therefore, many Germans consider them foreigners.

Unification

Aside from the large numbers of immigrants, another factor that has had a large effect on hate crimes in Germany is the country's unification. In 1989, the Berlin Wall fell, and East and West Germany were formally rejoined as a single country in 1990. Amid some trepidation in other countries regarding a single Germany's potential for relapse into pre–World Word II efforts at domination, a wave of nationalism swept through Germany. Furthermore, unlike West Germany, East Germany previously had few immigrants. Many citizens of the former East Germany were now confronted for the first time with those who were not ethnically German (Bergmann, 1997). They were also faced with the clear disparities between their own standard of living and that of their Western compatriots. Former West Germans, on the other hand, feared tax increases and other economic and social hardships that they might encounter in the effort to equalize the two halves of the country.

As a consequence of these conditions, xenophobia flourished in Germany, and there was a precipitous rise in hate crimes in the years following unification (Bergmann, 1997; Kühnel, 1998; Watts, 2001). Some of these attacks were mentioned in Box 8.4. Moreover, although the overall number of hate crimes peaked in 1995 and has decreased slightly in more recent years, the German government concluded in 2001 that "xenophobic and extreme right-wing violence have retained the same stable base since 1995 and temporarily reached a new high in the second half of 2000" (German Ministry of the Interior, 2001, p. 21).

Interestingly, anti-Semitism continues to play an important part in German extremist rhetoric, even though the Jewish population of that country is miniscule. There are an estimated 55,000 Jews in Germany today, one tenth the number that existed before World War II (S. Stern, 2001). There are fewer Jews in Germany than there are residents in the town I live in (Turlock, California), a town insignificant enough by California standards that most people have never heard of it.

Comparisons With the United States

In many ways, hate violence in Germany resembles that in the United States—that is, it is committed primarily by young men. Although much of the media attention has focused on skinheads and neo-Nazis, existing data suggest that most hate crimes are not committed by members of organized groups (Aronowitz, 1994; Watts, 2001). This does not mean, however, that these organized groups are unimportant; as Watts (2001) suggests, they likely contribute to an overall atmosphere of intolerance that fosters acts of violence even among nonmembers. Wetzel (1997) estimates that in 1995 in Germany, there were about 2,500 members of neo-Nazi organizations, 6,200 militant right-wing extremists, and 35,000 members of far-right political parties.

Another similarity between Germany and the United States is that it is difficult to interpret the data. In some respects, German data are more complete, in that reporting is mandatory (it is not in most U.S. states). German reports also tend to be much more systematic in recording the motivation or hate group membership of offenders (Watts, 2001). Furthermore, Germany has a uniform federal criminal law system, unlike the United States, where laws can (and do) differ considerably from state to state. On the other hand, in both countries, underreporting (both by victims and by police) may be a serious problem, especially where bias is present among governmental units themselves. The German government acknowledges that xenophobia probably exists among members of its police and military forces, but it has little empirical data to test this (German Ministry of the Interior, 2001).

Despite the similarities between hate crimes in the United States and those in Germany, it is difficult to make any direct comparisons of the two countries because the legal approach in Germany is very different from that of the United States. In fact, the term *hate crime* itself was not used in Germany until 2001; instead, offenses were classified as politically motivated, anti-Semitic, or xenophobic.

Germany has no equivalent to the penalty-enhancement type of hate crime laws that exist in most of the United States. Hate motivation can, however, be considered an aggravating factor when choosing the offender's sentence (Gilmour, 1994). In addition, there are a number of discrete offenses that are related to bias (see Appendix A for the text of these laws, part of the Penal Code of the Federal Republic of Germany). Among these are incitements to racial or ethnic violence; arousal of hatred; Holocaust denial; and production, distribution, or display of Nazi symbols. Political parties that are found to be a danger to free democratic order can be banned by the German Constitutional Court. To deter hate crime, antiextremist police units were created in 1994, and Holocaust education is compulsory in German schools.

Many of the German laws would be unconstitutional in the United States, but, for better or worse, German protections on freedom of expression are less broad than those in the United States.[13] Although the German constitution protects the right to express oneself freely, that right is limited by certain provisions, including citizens' rights to personal respect. The German Constitutional Court held that Holocaust denial is not protected speech because, in this case, the importance of free speech is outweighed by the potential harm denial inflicts upon the Jewish community (German Embassy, 2001). As a result of this situation, extremist literature, music, Web sites, and other materials that are legal in the United States are illegal in Germany. Problems have arisen when these materials have crossed international borders; we will discuss these problems later in this chapter.

United Kingdom

You might expect that the British approach to hate crimes would be quite similar to our own. The legal system in the United States began as an adaptation of the British system. Two of the pillars of extremism in America today were also originally British imports: skinheads and Christian Identity. However, the trip across the Atlantic, as well as the years that have passed since these things were imported, have resulted in significant differences in their British and American manifestations.

Sources of Diversity

The citizenry of the United Kingdom is noticeably diverse. The United Kingdom itself encompasses not only England but also Scotland, Wales, and Northern Ireland. In a total landmass slightly smaller than the state of Oregon, it includes 60 million inhabitants. In 2001, about 107,000 people were granted permanent settlement in the United Kingdom. Of these, the largest portion came from Asia (about 43,000) or Africa (about 31,000) (Mallourides & Turner, 2002). The increasing ethnic mixture of the country may, perhaps, be exemplified by the fact that some official government documents are published not only in English but also in Welsh, Chinese, Hindi, Punjabi, Bengali, Gujarati, and Urdu. The single largest ethnic minority within the United Kingdom is people from India, Pakistan, Bangladesh, and the West Indies.[14] About 1.5 million Britons are Muslim, and there are about half a million people each who are Hindu or Sikh.

Prejudice, the Far Right, and Extremist Groups

Just as World War II left a lasting legacy in Germany, in the United Kingdom it signaled the beginning of the end of the British Empire; after the war, the United Kingdom was only a minor player in the power struggle between the superpowers of the United States and the Soviet Union. Diminishing world power, combined with a collapsing industrial base and increasing immigration of non-Europeans, set the stage for the youth movement that eventually coalesced into the skinheads. As discussed in Chapter 5, the British skins were primarily urban and working class (as opposed to their American counterparts, who often come from the middle class). They quickly became infamous for their violence against South Asians and gays, among others.

An interesting thing about the United Kingdom is that parties of the far right have never enjoyed the degree of popularity and success that they have achieved in other European countries. To be sure, the Conservative Party, led by Margaret Thatcher, instituted a number of restrictions on immigration in the 1980s, but that government was no more radical than its American contemporary, the Reagan-led Republicans. Kitschelt and McGann (1995) conclude that the far right has held little sway in the United Kingdom in part because it is too deeply divided and in part because the more mainstream parties have been able to co-opt the more moderate portions of the far right's platform. As a result of the radical right's rather insignificant status, the mainstream parties have had less need than in other countries to appease the far right in their policies.

As the existence of the skinheads demonstrates, however, the absence of the far right as a political power does not necessarily mean that there is a lack of bigotry among the population at large. As Lawrence (2000a) points out, bigotry is nothing new in the British isles. During the 12th and 13th centuries, Jews were massacred or expelled and, in the 16th century, British residents of color were also deported. More recently, racist riots took place during the late 1950s.

Prejudice and its consequences continue to be a part of life in the United Kingdom just as they are in the United States. For example, a study issued in 2001 by the Home Office (Weller, Feldman, & Purdam, 2001) concluded that religious minorities, especially Muslims, Sikhs, and Hindus, experienced relatively high levels of discrimination. This unfair treatment occurred at the hands of government agents, such as police officers and teachers, as well as the media and private individuals.

Measuring Racially Motivated Incidents

Official data show that the annual numbers of *racially motivated incidents* (the British term for hate crimes) are generally high, with a peak in 1995, roughly the same time ethnic violence in Germany hit a high point, as did membership in U.S. "patriot" groups (e.g., the Christian Patriots). Although it cautioned that the reliability of the data is suspect, the Home Office estimated that in 1999, there were 280,000 racially motivated incidents in the United Kingdom (Clancy, Hough, Aust, & Kershaw, 2001). The fact that this is several orders of magnitude higher than the number of hate crimes recorded in the United States that same year (9,301) is attributable to the different methodologies used in the two countries. The U.S. data are the result of voluntary reporting to the FBI by local law enforcement, whereas the U.K. data are the result of a survey in which crime victims were asked whether they felt the incidents they experienced were racially motivated. Although the U.S. data certainly undercount hate crimes, it is quite possible that the U.K. data overcount them.[15]

Several studies have attempted to measure hate crimes in the United Kingdom. Bowling (1994) surveyed residents of an ethnically diverse London neighborhood. He found that incidents of racial and ethnic harassment or violence were fairly common, that this was perceived by the respondents to be a serious problem in their neighborhood, and that only a small percentage (less than 5%) of the incidents were reported to the authorities. Although there were limits to the methodology (e.g., it's unclear whether his sample was representative and whether the survey was reliable) of this study, and although the study concerned only one neighborhood in one city, the results suggest that hate crime is a fairly common occurrence in at least some parts of the United Kingdom. And, as in the United States, it is likely that official crime statistics do not accurately depict the problem.

Sibbitt (1997) studied the perpetrators of racial violence and harassment in two London boroughs. Contrary to most American research, which has found that typical offenders are teenagers and young men, she found that the offenders were of a wide variety of ages. Possible reasons for this discrepancy include the following: (a) the Sibbitt study encompassed not only hate crimes but also noncriminal hate incidents, and (b) Sibbitt relied heavily on field interviews, and the methodology was more qualitative than quantitative. Sibbitt also concluded that offenders' attitudes toward members of other ethnic groups were reflected within the community at large.

A much more comprehensive study was conducted in Scotland (Clark & Moody, 2002). This study made use of a number of methodologies, including statistics from police and other agencies, questionnaires for minority organizations and individuals, and focus group interviews with representatives from minority organizations. There were a variety of findings, among them that racist incidents occurred fairly frequently; that rates of reporting to the authorities had increased recently but were still low; that there were difficulties in police training, reporting, and sensitivity; that prosecuting hate crime cases was a high priority but was still beset by problems; and that there were gaps in support services available to victims.

Legislation to Address Hate-Motivated Violence

The first British laws to specifically address hate-motivated violence were passed in the 1960s and 1970s. These laws criminalized incitement to racial violence or hatred, much like some of the German laws. However, the laws against inciting racial hatred have rarely been enforced, primarily because prosecutors are reluctant to try these cases (Lawrence, 2000a). An additional law, passed in 1991 as part of the Football Act, makes it illegal to engage in indecent or "racialist" chanting. This provision was clearly a response to football hooliganism, much of which has been racially charged and which has often led to violence.

On April 22, 1993, Stephen Lawrence was stabbed to death on a street corner in London. Lawrence was black and his assailants were white; the victim's race was clearly the sole motive for the crime. After a police investigation and prosecution that were bungled and tainted with racism, three people were acquitted of the murder. Nobody was ever convicted. In response to public outcry, a government inquiry into the murder was commissioned, and a report—the Macpherson Report (1999), named after the commission's chairman—was issued. The report was quite damning in its appraisal of the British criminal justice system's response to racist violence.

In 1998, as the Stephen Lawrence inquiry was being conducted, the United Kingdom's first comprehensive hate crimes laws were being passed (see Appendix A). The Crime and Disorder Act created the category *racially aggravated offences*. This law acted as a penalty enhancer, increasing sentences by as much as several years and also increasing potential fines when a crime was motivated by the victim's race or ethnicity. The purpose in enacting the law was threefold: to deter hate crimes, to promote social cohesion, and to serve as the impetus for the criminal justice system to respond to hate crimes more effectively (Iganski, 1999b).

In the last purpose, at least, the law seems to have succeeded. In response to the law itself and the Macpherson Report, the Home Secretary created an action plan and issues a progress report on the plan annually. The plan called for a number of courses of action, including improving relations between the police and minority communities and advancing better police training and practices. Perhaps the most influential part of the plan is that it calls for a very broad definition of hate crimes: "A racist incident is any incident which is perceived to be racist by the victim or any other person" (Home Secretary, 1999).

Other relevant legislation has also been considered. For example, there has been discussion of making Holocaust denial a crime, as it is in some other European countries (e.g., Germany and France). However, many critics have opposed such a law on free speech grounds (Iganski, 1999a). The penalty enhancement laws, on the other hand, have met with considerably less opposition than they have in the United States (Iganski, 1999b).

Comparisons With the United States

The United Kingdom's approach to hate crimes is considerably closer to the American approach than is Germany's. However, there are at least four important differences between the British and American systems. First, as in Germany, the British system is federal, which means it is uniform throughout the country.[16] This is very unlike the United States, where, as we have seen, there are often rather substantial discrepancies in the laws of different jurisdictions. The practical repercussions of this distinction are unclear, but there may be advantages in having a single, consistent law. It certainly makes comparisons of crimes within various geographical areas simpler.

A second thing that distinguishes the British system from the American is that the working definition of a *hate crime* in Britain is considerably broader. Nowhere in the

United States are police instructed that an act is a hate crime whenever the victim or anybody else thinks it is. Instead, American police officers are expected to exercise their own judgment to determine what the offenders' motives were. In the United Kingdom, it is the victim's (or bystander's) perception that counts; in the United States, it is the officer's. Again, the effects of this difference are unclear, other than that it undoubtedly results in higher numbers of reported crimes in the United Kingdom than in the United States.

Third, the British law includes only crimes based on the victim's race or ethnicity, although the supporting government documents note that in some cases (such as crimes against Jews or Sikhs) a religion might be categorized as an ethnic group. Thus, religion is generally not protected, nor are sexual orientation, gender, disability, or other categories frequently included within American hate crime laws. It is extremely unlikely that attacks based on victims' membership in these groups do not occur in Britain, and it is unclear why these groups were not included within the law.[17]

Finally, American hate crime laws were mostly enacted in a rather piecemeal fashion and, apart from sometimes being partnered with civil and/or data collection provisions, they generally stand alone. The British laws, however, serve as only a part of a more comprehensive plan to reduce bias. The British scheme encompasses not only the prosecution of those who commit acts of hate violence but also includes efforts to improve community relations in general, especially between police and minority communities. If it is well implemented, the U.K. approach might be expected to be more successful.

Australia

Like the United States, Australia has linguistic, cultural, and legal ties to the United Kingdom. Although Australia remains a commonwealth of the United Kingdom, it has evolved its own approach to dealing with hate crimes, as the United States has.

Sources of Diversity

Fully one quarter of Australian residents were born overseas, and 40% of Australians are migrants or the children of migrants. Traditionally, the largest proportion of these originated in the United Kingdom and other European countries. In fact, from 1901 until the 1970s, a "White Australia" policy severely restricted immigration from non-European countries. The ethnic makeup of Australia began to change in the 1980s, when Australia experienced a large influx of immigrants from Asia. There were particularly large increases in immigrants from Cambodia, Fiji, Malaysia, the Philippines, and Vietnam (Mukherjee, 1999). Another small but significant contributor (about 2% of the total population) to Australia's diversity is its population of indigenous people. Like the Native Americans, indigenous Australians have suffered a long history of institutionalized persecution (Cunneen, 1997).

There are other sources of diversity in Australia as well. About 3.5% of Australians are affiliated with a non-Christian religion; Islam and Buddhism are the largest of these. There is also an active gay and lesbian community in Australia. In 2002, the international Gay Games were held in Sydney.

Assessing the Level of Bias

It is difficult to assess the level of bias in Australia. On the one hand, the country prides itself on its multiculturalism and tolerance, and certainly it has a less ample history of bigotry than do most European countries or the United States. Commenting on the fact that Jews generally "enjoy a level of equality unsurpassed in almost any other country," Nemes (1997) notes, "The political environment [in Australia], with its emphasis on multicultur-alism and tolerance, preaches tolerance and understanding to its citizens" (p. 57). However, Australia, too, has been influenced by prejudice, nationalism, and the far right. Many of the patterns there mirror those found in other countries.

As mentioned previously, Australia's aboriginal people have suffered and continue to suffer from discrimination, harassment, and violence. As Cunneen (1997) concludes, "Aboriginal people suffer from racism every day of their lives in employment, education, cultural facilities, on public transport and at the hand of government officials, including police" (p. 147). Cunneen also concludes that racist violence against this group is an "endemic problem" (p. 137).

Australia has had its share of far-right extremist organizations, including the League of Rights, which spouts anti-Semitic, anti-immigrant, and anti-Communist dogma; an assort-ment of neo-Nazi groups (including skinheads); and a segment of the gun rights lobby that, like its American counterpart, seems obsessed with racist conspiracies (Greason, 1997). There are even branches of the Klan in New South Wales, Victoria, and Queensland (McGlade, 1997). During the 1990s, at the same time the right was experiencing an increase in power elsewhere around the globe, it was gaining some popularity in Australia as well, primarily on account of the right's anti-immigrant stance. Pauline Hanson, a particularly outspoken opponent of indigenous people and immigration who had frequently vilified the "Asianization" of Australia, was elected to Parliament in 1996 (White & Perrone, 2001). Other anti-immigration candidates also received substantial shares of the vote in some areas (Cunneen, Fraser, & Tomsen, 1997; Greason, 1997).

Some scholars argue that Australia's political climate has led to increased bigotry and hate crime:

> There has been increased reporting of racial vilification, many new stories of physical violence directed against the person and property of "outsiders" . . . and the imposition of official crackdowns on particular communities alleged to be rorting [cheating or taking advantage of] the welfare system or engaging in criminal activity. It is a climate where "difference" is being highlighted, the "Other" further entrenched with outsider status, and fear and loathing promoted as part of the mainstream of media and political debate. . . . For many, the reality of life in contemporary Australia is shaped by what can only be described as a climate of hate. (White & Perrone, 2001, pp. 162-163)

There appear to be differences of opinion, therefore, about the degree to which Australia is currently experiencing a "climate of hate."

Assessing the Frequency of Hate Crimes

Even if prejudice is a major problem in Australia today, the frequency with which hate crimes occur in Australia and their trends are largely unknown because the government does not systematically keep track of these crimes. Certainly, compared with the United States, Australia is a less violent nation on the whole. As one author put it, "Australians often take refuge in the relative comfort of this not being America, with its generally high level of violence" (Nemes, 1997, p. 45). But hate-motivated crimes do occur in Australia; the question is, how often?

In 1989, the Australian government commissioned the National Inquiry Into Racist Violence. Among its findings were that police were often involved in racist violence and harassment, that few victims of hate crimes reported the crimes to the police, that media coverage encouraged stereotyping, and that hate crimes and related incidents were particularly common against aboriginal peoples (Fewster, 1999). On the other hand, the Inquiry also found that the crimes were not nearly as common as in other countries (Cunneen & de Rome, 1993).

Cunneen and de Rome (1993) also reported on a pilot project in New South Wales for police collection of hate crime data. Victims of crime were asked whether they felt that the crime had been motivated by prejudice; the officers themselves also assessed the offenders' motives. Victim and officer ratings appeared to be fairly similar. Depending on the jurisdiction, 2% to 7% of all reported crimes were identified as motivated by prejudice, and as many as one quarter of assaults were seen to be hate crimes. If these results are accurate and generalizable, they would indicate a high rate of hate crimes.

Mouzos and Thompson (2000) examined data on bias-motivated homicides of gay men in New South Wales. They concluded that approximately four men a year were killed in that state because of their sexual orientation. The rate remained stable over a 10-year period. Compared with ordinary murders, homophobic murders were more likely to involve multiple offenders and to be characterized by higher levels of brutality, the victim and offender were more likely to be strangers, and the offenders were younger and more likely than other homicide offenders to be unemployed. In general, these findings appear consistent with those of American studies.

Several Australian researchers have commented on the difficulties in assessing hate crimes. White and Perrone (2001), for example, noted that, although conflict between youth gangs may be related to ethnicity (because such gangs are usually organized along ethnic boundaries), ethnicity is usually not the primary motivating factor: "Fights between groups of relatively powerless sections of the community is less a matter of 'hate crime' per se than that of social dislocation and marginality" (White & Perrone, 2001, p. 179). Similarly, Tomsen (2001) argued that hate crimes should not be examined in isolation but rather as part of the influence of class and power upon young men. Hate crimes, he asserts, are not so much irrational acts as they are a means of attainment of masculine identity. These authors' arguments support the assertion made in Chapter 4 that hate crime should not be considered an exceptional or deviant behavior but rather a manifestation of cultural and societal norms.

Hate Crime Legislation

Australia currently has no federal hate crimes law; although there have been attempts to enact one, they have failed, primarily due to concerns about freedom of speech. The Federal Racial Anti-Discrimination Act, which was passed in 1975 (see Appendix A), does make it unlawful to engage in a public act that is likely to offend or insult someone on the basis of race or ethnicity. There are a number of exceptions, however, such as works of art or acts with academic purpose (Fewster, 1999). In Western Australia, a law prohibits racial vilification, which involves publishing or distributing materials that are intended to promote hatred of a racial group (Jones, 1997). Fewster (1999) contends that laws such as these are not likely to be effective in reducing hate for two reasons: They are rarely enforced, and they fail to address issues of structural bias.

New South Wales and the Australian Capital Territory have identical laws that prohibit incitement of racial, ethnic, or religious hate (see the New South Wales Anti-Discrimination Act, Sections 20C and 20D, in Appendix A; the language of sections 65 though 67 of the Australian Capital Territory's Discrimination Act 1991 is the same). As is true of other countries discussed here, sexual orientation, gender, and disability are not included.[18] As in Western Australia, there are numerous exceptions to these laws. And the process for enforcing these laws is ponderous: A complaint has to be made by a victim or representative of a particular racial group to the Anti-Discrimination Board, which then determines whether the law has been violated. If such a determination is made, the Board attempts to solve the dispute through mediation; acts involving threats or actual physical harm are referred to the Attorney General, who must consent to prosecution. Not surprisingly, there have been extremely few criminal prosecutions under these laws. Jones (1997) does assert, however, that the civil provisions have been relatively successful.

Comparisons With the United States and the United Kingdom

Compared with American and British approaches to hate crimes, the Australian approach is more circumscribed. Federal and state legislation is limited and rarely enforced. No comprehensive effort at data collection is made. Although multicultural affairs in general are a priority of the Australian government, hate crime laws have not been made part of any comprehensive effort to fight bigotry. The legislative efforts appear to focus primarily on the activities of organized extremist groups with little attention paid to the types of crimes that are committed by ordinary people and that are targeted by penalty enhancement laws (such as those that exist in the United States).

More than a decade ago, the National Inquiry Into Racist Violence made a number of recommendations for addressing hate crimes. This course of action is similar to the situation in the United Kingdom, where the Macpherson Report made similar suggestions. Unlike the United Kingdom, however, there seems to have been little attempt in Australia to follow the recommendations. It is unclear whether the reasons for this are rooted in politics, indifference, or other factors.

Canada

Because of Canada's proximity to the United States as well as our common cultural, linguistic, and legal heritage, it is a bit tempting to assume that the situation regarding hate crimes is the same in Canada as it is in the United States. Such an assumption would be false, however, because there are significant differences between these neighboring countries.

Sources of Diversity

Despite its large geographical size, Canada has a population of only 32 million, slightly less than the state of California. Within this relatively small population, however, there is a fair amount of diversity, the most obvious manifestation of which is the distinction between English-speaking and French-speaking Canadians. Not only does Quebec use a different language than that used in the rest of the country, but Quebec also has a different legal system (based on a French, rather than a British, model) and feels strongly enough about its separate identity that there have been nearly successful attempts at secession.[19]

Immigration also plays a major role in Canada. Approximately 17% of Canadian residents were born in another country; the comparable figure for the United States is only about 10%. Traditionally, most migrants to Canada came from the United Kingdom, the United States, and Europe, but, in the 1990s, more than half of all immigrants came from Asia or the Middle East. Furthermore, those Europeans who did immigrate to Canada in recent years tended to come from Eastern, rather than Western, Europe. Urban areas of Canada were especially likely to attract immigrants, and, as a result, Toronto, Montreal, and Vancouver are very ethnically diverse. More than 11% of Canadians consider themselves to be *visible minorities,* or people of color, with Chinese being the largest of these groups.

In addition to the immigrants, of course, there are Native Canadians, who make up about 3% of the population (a larger proportion than Native Americans do in the United States). The majority of these aboriginal people live in the northern and western portions of the country, many in rural areas.

There are also small groups of religious minorities in Canada. About 1% of Canadians are Muslim, and a slightly higher percentage are Jewish. Buddhists, Hindus, and Sikhs each make up about .05% of the population.

Prejudice and Extremist Groups

As in the United States, the far right has had only an incidental influence on mainstream Canadian politics. However, a variety of extremist organizations, many of which parallel those found in Canada's neighbor to the south, exist: Identity adherents, skinheads, Klans, Holocaust deniers (see Box 8.6), and assorted white nationalists (Kinsella, 1994). In fact, many of these Canadian groups have had ties with their U.S. counterparts.

Box 8.6   The Trials of Ernst Zündel

Ernst Zündel was born in Germany and immigrated to Canada as a teenager. He worked as a commercial artist but is famous for his activities to promote Nazism and deny the Holocaust. Zündel's actions, which included a zealous publishing campaign, resulted in three criminal trials in Canada.

In the first trial, in 1985, Zündel was charged with violating a law that prohibited knowingly publishing false news. He was convicted, but the conviction was over-turned. During his retrial, he produced as his "expert" witness Fred Leuchter, who claimed that the gas chambers did not exist. Leuchter's exploits were chronicled in the excellent documentary *Mr. Death: The Rise and Fall of Fred A. Leuchter, Jr.* (Morris, 1999). Despite the assistance of Leuchter and other Holocaust deniers, Zündel was convicted and sentenced to 9 months in jail.

In 1996, Zündel was tried again, this time for violating Section 13 of Canada's Human Rights Act (which prohibits exposing people to hatred or contempt; see Appendix A). This time, it was the content of a Web site bearing Zündel's name that was in question. Zündel ultimately lost that trial as well, and the court ordered that the offensive sections of the Web site be removed. However, that order has proved unenforceable because the Web pages are hosted on a U.S.-based Internet server, and Zündel himself has moved to Tennessee, where he continues his work.

SOURCE: Adapted from Anti-Defamation League (2002d).

Furthermore, prejudice continues to play a part in the lives of everyday Canadians. Perhaps not surprisingly, given the large numbers of immigrants and the changing ethnic composition of those immigrants, anti-immigrant sentiment has achieved popularity in Canada. Esses, Dovidio, Jackson, and Armstrong (2001) hypothesize that this anti-immigrant sentiment is largely due to perceived economic competition: Immigrants who do well financially are seen to be stealing opportunities from natives, whereas those who are struggling are seen to be financial drains on the country. Li (2001), however, argues that much of the discourse that is ostensibly about immigration is really about race. Like Australia and the United States, for a long time Canada had immigration policies that especially restricted migrants from Asia. However, Canada's policies were changed to be less discriminatory in 1962, quite a bit earlier than in the other countries, which did not change their policies until the 1970s.

In general, research indicates that Canadians are supportive of multiculturalism and are tolerant of other groups (Berry & Kalin, 1995). Moreover, these attitudes appear to have remained stable over the last quarter century (Kalin & Berry, 1996). However, Europeans are still viewed more positively than people with origins on other continents. Moreover, as discussed in Chapter 4, positive ratings of other groups on attitude surveys does not necessarily translate into positive behaviors toward others in real life. Dion (2001), for example, found that immigrants, especially those from non-European countries, experienced discrimination in finding housing in Toronto.

Assessing the Incidence of Hate Crimes

In 1995, the Canadian Department of Justice attempted to assess the nature and incidence of hate crimes in Canada (Roberts, 1995). Because Canada did not have a data collection law akin to the U.S. Hate Crime Statistics Act, this assessment relied on voluntary reports from several law enforcement agencies and audits from Jewish and gay and lesbian groups. Although acknowledging the many problems with measuring hate crime, the resulting report concludes that the patterns of offending in Canada are very similar to those in the United States and that hate crimes themselves are severely underreported.

A victimization survey estimated that in 1999, there were nearly one quarter of a million crimes in Canada motivated by hate (Roberts & Hastings, 2001). As we discussed earlier, there are difficulties with relying on victim surveys. However, if these data are at all reliable, they would indicate that hate crimes occur at quite a high rate.

The League for Human Rights of B'Nai B'rith Canada is an organization very similar to the ADL in the United States. Like the ADL, it publishes an annual audit of anti-Semitic incidents. In recent years, the League has reported an increase in anti-Semitic incidents, with 197 such incidents having been recorded in the first 6 months of 2002 (League for Human Rights of B'Nai B'rith Canada, 2002). Whether other groups also experienced a rise in hate crimes is unknown.

Legislation for Hate Crimes

Canada has devised a variety of legal means for handling hate crimes (see Appendix A). As early as 1970, a law was enacted specifically for the purpose of curtailing hate propaganda. Among other things, this law prohibits publicly promoting genocide as well as publicly inciting hatred against an identifiable group. The law also specifically authorizes the seizure of hate propaganda. There was intense debate over this law and its subsequent amendments; some critics argued that it was too broad, whereas others argued that it did not go far enough. It was generally perceived as unworkable, and prosecutions under it are rare (Ross, 1994).

In addition, the Canadian Human Rights Act includes within its definition of discrimination telecommunications that are likely to expose others to hatred or contempt on the basis of race, religion, and so on. Again, it is unclear whether such a broad law is enforceable in any meaningful way; if it were, it would quite likely apply to extremist Web pages.

New antiterrorism legislation that was enacted in light of the events of September 11 includes provisions prohibiting mischief against places of religious worship. This legislation is similar to antidesecration laws that exist in many U.S. jurisdictions.

Finally, Canadian law also permits hate motivation to be used as a penalty enhancer. Originally, this provision was not specifically sanctioned by any particular code, but the right for judges to consider bias as an aggravating factor was recognized by case law (Gilmour, 1994). In 1996, however, a statutory provision was enacted that did expressly allow judges to use hate motivation to increase sentences. Roberts and Hastings (2001) conclude that this codification resulted in more offenders receiving harsher sentences when they committed crimes because of bias. However, these authors also caution that it is often difficult to determine the offender's motive.

Comparisons With the United States

There are some clear similarities between American and Canadian legal approaches to hate crimes. In both countries, offenders may receive enhanced penalties if their crimes were motivated by bias toward the victim's group. On the other hand, there are significant differences to the approaches as well. Canada's law operates at the federal, rather than state, level, which theoretically results in increased consistency between jurisdictions. Canada also has what Americans would consider hate *speech* laws, which would almost certainly be held unconstitutional in the United States. Finally, at this time, Canada has no law requiring the collection of hate crime data.

Commonalities and Differences

The four countries profiled in this section clearly do not represent the entire possible range of approaches to hate crimes. It is interesting and worthwhile, however, to explore some of the common themes and trends found among them, as well as their differences.

Each of these countries enacted hate crime legislation in recent years in response to similar problems. All have experienced increasing immigration and, concomitantly, increasing diversity of citizenship. The far right has also enjoyed popularity to varying degrees in each country. Together, this has appeared to result in increased bias, both within the culture at large (including social and government institutions) and in the form of hate-motivated violence. The precise targets of this bias vary a bit from place to place, but the underlying causes, as well as the manifestations of that bias, are very similar.

Not surprisingly, measuring hate crimes has turned out to be as problematic in other countries as in the United States. In every jurisdiction, hate crimes are severely underreported to police, and other methods of assessing the scope of the problem have not proved entirely satisfactory. To the extent that hate crimes have been measured, the patterns appear fairly similar in all four countries and the United States. Hate activities seem to have peaked in the early to mid-1990s, but, although the numbers of hate crimes are no longer at peak levels, they do not seem to have tapered off appreciably. The offender profile seems similar as well from country to country: young men, often working class, often acting in small groups, but not necessarily affiliated with organized hate.

Each of the four countries and the United States also has had to contend with a similar assortment of extremist groups. Skinheads appear to be ubiquitous; indeed, it would be interesting to undertake more studies to determine why this particular lifestyle seems to have such international appeal. But other factions exist as well, including nationalists and even that seemingly American-specific group, the Ku Klux Klan. The fact that similar groups exist in so many countries has made global networks of hatred a reality in the 21st century; these networks will be discussed later in this chapter.

Although the pattern of hate is quite similar in the countries discussed here, each country has adopted a different tactic for dealing with it. This diversity is a bit surprising, especially because three of the four countries, as well as the United States, share a common historical and legal heritage. Most of the countries do have some sort of law prohibiting certain racist speech, but the wording and application of these laws vary quite a bit. Some legislation also specifically forbids Holocaust denial. Of course, because of America's rather expansive right to freedom of expression, the United States does not use either of these methods to combat hate crime.

In the United States, the primary method of addressing hate crime has been the use of penalty enhancement laws. For reasons that are unclear, this type of legislation is not as popular elsewhere. One possible partial explanation is that other countries seem to be particularly concerned with the activities of organized hate groups. Or perhaps other countries are skeptical about the practical effects of penalty enhancers. The difference might also be explained by Americans' fondness for punitive measures; in general, we seem to enthusiastically embrace harsh penalties as an appropriate antidote to many social ills, and our sentences are among the most severe in the world.

Another characteristic that distinguishes American hate crime laws from those abroad is their scope. Whereas most other countries protect only race and ethnicity (and, to some extent, religion), many of our states include several other categories within their legislation as well. It is perhaps most notable that none of the nations examined in this chapter considers crimes committed because of the victims' sexual orientation to be hate crimes. As discussed in Chapter 6, these crimes are relatively common in the United States, and anecdotal evidence indicates that they are not unknown abroad, either. The reasons for the exclusion of sexual orientation from international hate crime laws is unclear. Less tolerant attitudes toward homosexuality may be an explanation, or perhaps the gay communities in these countries are less politically active. Of course, other categories, such as gender, disability, and age, are also excluded in the laws of the other countries mentioned here, which may reflect a narrower conception of what the term *hate crime* means in those countries. It is possible, also, that the identity politics of which Jacobs and Potter (1998) wrote are less prevalent in other countries than in the United States.

The fervor with which the fight against hate crimes has been waged varies from country to country as well. Most of the countries seem willing to at least study the subject, but they are not necessarily anxious to follow the recommendations of those studies. The methods that are adopted are generally rather piecemeal and isolated. The exception to this pattern is the United Kingdom, which uses the broadest definition of hate crime, has enacted one of the most comprehensive legislative approaches to hate crimes, and has made its criminal laws only one part of a larger program aimed at reducing all forms of prejudice. Because so few attempts have been made to evaluate the influence of governmental approaches to hate, it is uncertain which approach, if any, is preferable, nor do we know what variables might affect the success of each approach.

A GLOBAL HATE NETWORK

If it is 3 A.M. and I have a sudden need to know the lyrics to Bob Dylan's *Quinn the Eskimo* or to find out the major exports of Bhutan, no problem—it just takes a few clicks of the mouse. If home starts to feel a little boring, I can hop on a plane, and in a matter of hours I can be touring the temples at Angkor Wat or shopping at a souq (market) in

Marrakesh. Unfortunately, those who wish to spread hatred can travel the world with equal ease, either in person or virtually via their computers. As a result, organized hate groups have developed a web of connections across the planet. The United States, with its lack of restrictions on hate propaganda and its particularly fertile climate for extremist groups, is at the center of that web.

Links Between Hate Groups

Although it has been obvious for many decades that international links between organized hate groups exist, there has been little academic study of them. Prior to World War II, the pro-National Socialist German-American Bund (an American group that supported the Nazis) was reportedly funded in part by Hitler (Lee, 1997). After the war, George Lincoln Rockwell, head of the American Nazi Party, had supporters from abroad (Schmaltz, 1999). More recently, skinheads have appeared in numerous countries on at least three continents, and even the Ku Klux Klan has established itself in places like New Zealand and Germany.

A number of questions arise concerning hate groups' international presence: Exactly which groups have such a presence? Where? Why do some groups appear to have more global appeal? What is the nature of the groups' international ties, and how are they established and maintained? To what extent are international chapters of groups like the Klan or the skinheads truly allied? That is, are they truly affiliated with one another, or have parallel organizations with similar names and practices been established separately? Because of the lack of research in this area, most of these questions remain unanswered. What little we do know tends to come from watchdog organizations and anecdotal reports.

The fall 2001 issue of the Southern Poverty Law Center's *Intelligence Report* documented several examples of hate groups' worldwide ties. According the Southern Poverty Law Center (2001), several factors led to an increase in such links during the last decade: the rise of the philosophy of "pan-Aryanism," the international appeal of Holocaust denial, modern communications such as the Internet and shortwave radio, and the popularity of white power music. Among the recent examples of extremist groups' international ties cited by the SPLC are appearances in the United States at meetings of American hate groups by leaders of the British National Party, the German National Democratic Party (NDP), and the Danish National Socialist Movement; visits by David Duke to Russia, William Pierce to Germany and Greece, and Jared Taylor (editor of *American Renaissance*) to England; and international conferences of Holocaust deniers.

Even more recently, an article in *Intelligence Report* (Lee, 2002) concerns ties between neo-Nazi and Muslim extremist groups. Actually, such ties are not exactly new; George Lincoln Rockwell spoke at a Black Muslim convention in 1962, and Tom Metzger pledged money and support to the Nation of Islam (Lee, 2002; Schmaltz, 1999). In the last few years, however, there have been increasing links between these white nationalists and Muslim extremists. Despite the apparent differences between these two groups (and the fact that many white supremacists have long characterized Muslims as "mud people"), they are united in a common hatred of Jews. They have joined in espousing Jewish conspiracy theories, denying the existence of the Holocaust, and denouncing Israel. As mentioned in Chapter 5, the Web sites of some white supremacist groups now provide links to Muslim extremist sites.

Another indication of the spread of hatred is the legal troubles faced by those who have attempted to extend their influence beyond national boundaries. As mentioned earlier, after being found guilty of violating Canada's Human Rights Act, Holocaust denier Ernst Zündel moved to Tennessee, where he continues to publish his materials. In 1995,

American neo-Nazi Gary Lauck was arrested in Denmark and extradited to Germany to face charges of inciting racial hatred; the charges stemmed from his attempts while visiting Germany to distribute literature and to organize local youths. He was convicted and served several years in a German prison before returning to his home in Nebraska. In 1999, Hendrik Möbus, a German white power musician who had been convicted of murdering a 15-year-old boy, violated parole and fled to the United States. After spending some time with the National Alliance, he was denied political asylum and extradited to Germany.

The Internet

One of the unifying factors for organized hate has clearly been the Internet. As we have seen, racist propaganda and symbols are illegal in many countries but not in the United States. Gary Lauck does not need to travel to Germany and face arrest if he wants to spread his word; he needs only to create a Web site (which he has done). Finding a host within the United States for such a Web site is not difficult. Although some companies, such as Geocities, attempt to prohibit extremist content, enforcing such prohibitions is difficult. Moreover, other Internet servers have noncensorship policies, and a few, such as Don Black's Stormfront, actually specialize in providing Web hosting for extremists.

Once a Web site has been created, it can be viewed by anyone in the world who has Internet access. As I will discuss in the next section, it is virtually impossible to block people from viewing materials that are illegal in their own country but legal in the country in which the Web page is hosted. Extremist groups are well aware of this fact, and many of them make efforts to appeal to an international audience. In the content analysis my colleagues and I conducted of extremist Web sites, we found that more than one quarter of the sites had content in languages other than English (Gerstenfeld, Grant, & Chiang, 2003). Other sites were all in English but had links that would translate the content into other languages. We also found that more than half of the sites contained links to sites in other countries.

White Power Music

White power music has also unified white supremacists worldwide. According to the SPLC,

> In many ways, this remarkably violent music is accomplishing for the radical right what decades of racist theorizing didn't: It has given Skinheads and many other extremists around the world a common language and a unifying ideology—an ideology that replaces old-fashioned, state-based nativism with the concept of "pan-Aryanism." (SPLC, 2001)

As can be seen from the international popularity of many mainstream musicians, most music can be enjoyed equally by listeners from many countries. It is not even necessary to understand the lyrics. This very minute I am listening to a CD containing songs from Brazil, Iran, and Uganda, and I do not speak the languages of any of those countries.

White power music is produced by several labels that specialize in such music. It is sold over the Internet, often by U.S.-based companies, which means customers can buy music that it is illegal to distribute in their own countries. Companies such as Resistance Records (which is owned by the National Alliance) promise overseas buyers on their Web site (www.resistance.com) that they will "do their best to minimize the risk of orders being seized by customs." Visitors to the Web sites can frequently download digital versions of songs, which avoids customs issues altogether. Many of the Web sites also sell other racist paraphernalia, such as clothing and books.

In addition, white power concerts that often feature both local bands and those from other countries are held in the United States and abroad. According to the SPLC (2001), recently in Germany there have been about 180 white power concerts a year. Not only do concerts occur in their own right, but they may also be used as a draw to lure potential recruits to extremist group meetings. For example, following an August 2002 rally in Washington, D.C., the National Alliance promised a "Rock Against Israel" concert.

If the watchdog organizations are correct, connections between hate groups worldwide are a growing problem. Not only is there strength in numbers, but groups may be able to work together across borders to contravene individual countries' laws. As recent events have made clear, one consequence of this easy information exchange could be international acts of terrorism. Clearly, this is an issue that is potentially of great concern and therefore deserves more research.

POLICING HATE ACROSS BORDERS

Throughout this book, I have explored some of the many challenges that arise when attempts are made to prohibit hate crimes. How much more difficult the problem becomes, then, when hatred reaches across international borders.

Without delving too far into the depths of international law, suffice it to say that, in general, a country does not have the power to criminally prosecute people for acts they commit in other countries. If I print and distribute within the United States a book denying the Holocaust, for example, criminal charges cannot be brought against me in Germany, even though the book would be illegal there. Of course, if I travel to Germany and take copies of that book with me and try to sell them there, then I will be subject to German law. This is what happened to Gary Lauck.

The Internet has made this situation immensely more complicated because it allows a person to penetrate another country's borders without physically leaving his or her home country. Suppose I write a manifesto against Jews and place it on my Web page, which is hosted by an Internet service provider (ISP) in the United States. I am well aware that Germans who are surfing the Internet will be able to access my manifesto, and perhaps I even encourage them by making it available in German. Now may I be prosecuted? Or may the German government take legal action against either my ISP or ISPs in Germany that are allowing their customers to access my Web page? That is the dilemma, which is intensified by the fact that the United States (as the Constitution requires) takes a generally more lenient approach to hate speech than most other countries.[20]

In fact, there have been attempts to regulate hate speech over the Internet. Ernst Zündel was prosecuted in Canada for the content of a Web site that was located on a U.S. server. The prosecution was successful because of evidence that Zündel had engaged in activities in Canada to support the site. However, Zündel then simply left the country, and the Canadian court had no means to enforce its order that the Web site be shut down.

In 1999, Frederick Tobin, an Australian resident and prominent Holocaust denier, traveled to Germany. After he arrived, he was arrested for violating German law, both by passing out leaflets in Germany that denied the Holocaust and also for maintaining a Holocaust denial Web site in Australia that was accessible to Germans. Although he was convicted on the leaflets charge, the court threw out the Web site charge. On appeal, however, a higher court held that in fact Tobin and others could be criminally liable in Germany for materials posted on Web sites in other countries.

In 2000, a French court ordered the ISP Yahoo! to ban French users from accessing auctions of Nazi paraphernalia that Yahoo! hosted. Yahoo! is an American company, the people placing the items for sale were Americans, and it is legal to sell Nazi items in the United States but doing so is illegal in France. Yahoo! was subject to a penalty of 100,000 Euros (about $100,000) a day if it violated the order. The French court also threatened to try Yahoo!'s president for inciting racial hatred.[21] Yahoo! argued that there was no practical way to block only French users from viewing the auctions, and it filed suit in a California court to block the French court's order. The California court found that it violated the U.S. Constitution for a foreign nation to regulate speech within the United States. Because no international treaty encompasses this complicated situation, the ultimate resolution is unclear.

There have been other cases in other countries as well that involve the Internet and hate groups; these are just a few prominent examples. Attempts to resolve some of these issues through the use of international treaties have not been successful. This area is clearly one that will keep many lawyers busy for a long time and one that remains a particularly troubling problem.

One disturbing aspect of this dilemma is the potentially chilling effect that international prosecution of hate may have on expression. The students in my hate crimes class have created an informational Web site on hate crimes, which I maintain. I suspect that some of the content on that site would be illegal in other countries. Should I fear that the next time I travel abroad, I might be arrested? Even if I stay in the United States, will foreign courts start levying large fines against me? If such fears seem justified, I may be tempted to shut down the site. The situation becomes even worse when I realize that there is an extremely wide spectrum of material that is banned somewhere in the world. China, Singapore, and several Southeast Asian countries prohibit Internet content deemed detrimental, immoral, or contrary to "Asian values." Some Middle Eastern nations forbid content that conflicts with Islamic principles (Siegel, 1999). If all Internet communications were to accept the lowest common denominator and exclude all material that might be illegal *somewhere,* the Internet itself would become both hopelessly insipid and nearly useless as an information source.

The potential for a chilling effect is not merely hypothetical. Shortly after the French lawsuit, Yahoo! removed all auctions of Nazi paraphernalia. And a recent study of the massive Internet search engine Google (Zittrain & Edelman, 2002) revealed that Google's French and German versions excluded certain Web sites. A search for the term "Stormfront" on the American version of Google results in a list of sites with the white supremacist Web site Stormfront first on the list, but an identical search on the German version results in a list in which the white supremacist site is absent. Of course, there is nothing to stop Germans from avoiding this screening by simply using the American Google site, and this sort of action raises the very real possibility that Google and other search engines might at some future point entirely block any site viewed as potentially objectionable.

It does seem inevitable, however, that nations will continue to attempt to police hate material that enters their borders, whether they do so physically or electronically. If these countries do not take such actions, their own legislation against hate will become useless. It takes very little effort for a bigot in Germany or France or Canada (or anywhere else) to create a Web site containing propaganda that is illegal in his or her own country and to find an American ISP that will host that Web site. And it takes even less effort for a bigot in the United States to spread his or her views to people abroad.[22] Does

NARRATIVE PORTRAIT

WHEN HATE LEADS TO GENOCIDE

When those who hate obtain political or military power, they are able to commit hate crimes in their ultimate manifestation: genocide. One recent example occurred in Rwanda in 1994, when perhaps as many as a million Tutsis were murdered by members of another ethnic group, the Hutus. In the following chilling passage, BBC journalist Fergal Keane describes the aftermath of the massacre of 3,000 people who had sought refuge at a church in Nyarubuye.

"This is Nyarubuye," says Frank. Moses begins to slow the car down and Glenn is preparing his camera to film. As we drive closer the front porch of the church comes into view. There is a white marble statue of Christ above the door with hands outstretched. Below it is a banner proclaiming the celebration on Easter, and below there is the body of a man lying across the steps, his knees buckled underneath his body and his arms cast above his head. Moses stops the car but he stays hunched over the wheel and I notice he is looking down at his feet.

I get out and start to follow Frank across the open ground in front of the church. . . . As I walk toward the gate, I must make a detour to avoid the bodies of several people. There is a child who has been decapitated and there are three other bodies splayed on the ground. Closer to the gate Frank lifts his handkerchief to his nose because there is a smell unlike anything I have ever experienced. I stop for a moment and pull out my own piece of cloth, pressing it to my face. Inside the gate the trail continues. The dead lie on either side of the pathway. A woman on her side, an expression of surprise on her face, her mouth open and a deep gash in her head. She is wearing a red cardigan and a blue dress but the clothes have begun to rot away, revealing the decaying body underneath. I must walk on, stepping over the corpse of a tall man who lies directly across the path, and feeling the grass brush against my legs, I look down to my left and see a child who has been hacked almost in two pieces. The body is in an advanced state of decay and I cannot tell if it is a girl or a boy. I begin to pray to myself. "Our Father who art in heaven . . . " These are prayers I have not said since my childhood but I need them now. We come to an area of wildly overgrown vegetation where there are many flies in the air. The smell is unbearable here. I feel my stomach heave and my throat is completely dry. And then in front of me I see a group of corpses. They are young and old, men and women, and they are gathered in front of the door of the church offices. How many are there? I think perhaps a hundred, but it is hard to tell. The bodies seem to be melting away. Such terrible faces. Horror, fear, pain, and abandonment. I cannot think of prayers now.

SOURCE: Keane (1995, pp. 77-79).

a sovereign nation not have the right to prohibit whatever it views as detrimental to its well-being?

This is, to say the least, a difficult dilemma. On a more optimistic note, it does make for lively debate, and the legal and ideological battles will probably prove to be interesting for some time.

CONCLUSION

Although the term *hate crime* is used almost exclusively in the United States, offenses based on the victim's group occur worldwide. Different countries have taken varying approaches to this problem, and it is worth examining those approaches to determine if perhaps improvements can be made upon our own models.

It is perhaps ironic that many extremists who are most opposed to globalism and are the most adamant about maintaining national borders are taking advantage of technology and other means to share information and viewpoints with those people abroad who hold kindred beliefs. In any case, the need for an international perspective on controlling hate crimes is particularly acute today.

DISCUSSION QUESTIONS

1. At the beginning of this chapter, I struggled with the difficulty of distinguishing hate crimes from similar events, such as terrorism and war. What do you see as the differences, if any? What are their policy implications? Do you agree with MacGinty (2001) that all hate crime is political?

2. Why do you believe that so little research has examined hate crimes in non-European or non-English-speaking countries? How would you propose that this situation be remedied? What similarities and differences would you expect to find between hate crimes in Europe, North America, and Australia, and those in Asia, Africa, and South America?

3. Compare and contrast the conditions that led to the rise of the radical right in Europe and in the United States. Are there other parts of the world that you believe are currently in a situation similar to that of Europe in the 1990s and, if so, have hate crimes resulted? In your assessment, will the most recent political, economic, and social conditions in Europe (and in the United States) result in an increase or decrease in extremism?

4. During a recent trip to the Czech Republic, I visited a newly founded boarding school for Romany high schoolers. In addition to teaching a traditional curriculum, the school provided a place where Romany youths could learn about their language and culture, and be in a place somewhat separate from the discrimination and harassment they normally faced. What do you think about schools like this? What do you feel will be their effect, if any, on bigotry in general? Would you support a similar type of school for an American ethnic minority group?

5. Because freedom of expression is generally less broadly protected in places other than the United States, countries such as Germany are able to criminalize acts such as Holocaust denial and distributing racist propaganda, whereas the United States cannot. What are the benefits and disadvantages of each approach? Which do you see as preferable? Do you agree with the German court that the harms of Holocaust denial outweigh the right to free speech?

6. In what ways is the British approach to hate crimes similar to and different from the American approach? What are likely to be the advantages and disadvantages of each? If you were a policymaker, would you choose one of them, a combination, or something else entirely?

7. In the United Kingdom, an offense is classified as a hate crime whenever the victim or anybody else thinks the incident was racially motivated. Why do you think such a

broad definition was adopted? Is this scheme generally preferable to relying on the police officer's perception for classification purposes? In what ways are victims' and witnesses' perceptions of motive likely to differ from those of the police, and why? Which do you think will be more accurate?

8. Why do you feel the recommendations of Australia's National Inquiry Into Racist Violence were not followed more closely, whereas Britain did adopt most of the suggestions in its Macpherson Report? If you were to make a similar report for the United States, what would your findings and recommendations be? Do you think it is likely that your suggestions would be followed?

9. In contrast with the countries discussed in this chapter, the United States has little in the way of federal hate crime legislation; instead, most laws (and most prosecutions) are at the state level. What do you see as the advantages and disadvantages of this system?

10. What common themes do you see among the countries discussed here, including the United States? In what ways have the political and social pressures on the government to respond to bias been similar, and in what ways have they been different? How do you think the reactions to these pressures (e.g., enactment of new laws and special studies by the government) would compare to other countries, especially those that are non-European and non-English-speaking?

11. Describe the apparent link between immigration and hate crimes. Why do you believe that immigrants are so often the target of animosity, even in places like the United States, Canada, and Australia, where so many citizens' families arrived recently from other countries themselves? Can you think of ways to reduce hostility toward immigrants?

12. How would you conduct research on international links between hate groups? What do you see as the potential dangers of such links?

13. Describe the dilemma concerning the international regulation of hate. What different interests are involved? What possible resolutions do you envision, and which ones would you advocate?

INTERNET EXERCISES

1. A variety of reports on hate crimes in the United Kingdom can be viewed on the Web site of the British Home Office: www.homeoffice.gov.uk/rds/racerelate6.html. Compare the kinds of information available there to the types that are available within the United States.

2. You can learn more about Roma history and persecution online. Here are a few places to start: European Roma Rights Center (www.errc.org); Romani.org (http://romani.org); Roma Rights and Access to Justice in Europe (www.rraje.org); links on Roma and the Holocaust (http://dmoz.org/Society/Ethnicity/Romani/Holocaust/); and The Roma: Myth and Reality (www.geocities.com/Paris/5121/mythandreality.htm).

3. The ADL issued a report on the legal issues concerning restricting hate speech on the Internet. It is available at www.adl.org/Civil_Rights/newcyber.pdf.

NOTES

1. Another author has placed the estimate as high as 60 million (Waller, 2001). For excellent material on many major 20th-century genocidal campaigns, see Totten, Parsons, & Charny (1997).

2. This definition still leaves the categorization of certain acts in doubt. For example, what about cases of civil war or rebellion, in which a group declares itself to be a governmental authority, and that group attacks its enemies, but its authority is not officially recognized? Or what about situations in which independent individuals attack others and the government unofficially sanctions those acts or at least turns a blind eye toward them? This is arguably the situation in Zimbabwe today and was the case in the United States during the height of Klan violence.

I should also note that I've violated this definition in previous chapters when I referred to government-endorsed actions like slavery, Japanese internment camps, and massacres of Native Americans as hate crimes. As I said, the boundaries here are diaphanous.

3. The remaining three chapters are on Palestinians, India, and Colombia.

4. For example, volume 58, number 2, of the *Journal of Social Issues,* and volume 45, number 4, and volume 46, number 1, of *American Behavioral Scientist.*

5. The phrase "Hungarian minorities abroad" refers to people of Hungarian ethnicity who are currently living in other countries, such as Romania, Croatia, and Slovakia. Many of these people live in areas that were part of Hungary until the end of World War I, when Hungary lost nearly three quarters of its territory and two thirds of its inhabitants under the Treaty of Trianon.

6. This is a good example of outgroup homogeneity, a concept discussed in Chapter 4.

7. Ethnic conflict has led to violence in Hungary's neighbor, the former Yugoslavia.

8. Even today, in Budapest the damage from World War II artillery and the Hungarian Revolution is still clearly visible on many buildings, and in both Budapest and Prague, people still leave flowers at memorials to those killed opposing the communists.

9. There is some debate over whether more people became intolerant or whether certain subsections of the population became more radicalized. Bergmann (1997) argues that, in Germany at least, the latter was true; overall opinions about Jews and ethnic minorities remained stable among the German population at large, but young members of the ultra-right wing precipitated a wave of violence.

10. Of course, at about the same time in the United States, both the federal government and individual states were also passing anti-immigrant laws, such as California's Proposition 187 (which denied health care, education, and other publicly funded benefits to illegal aliens and required that certain public employees report illegal aliens to the government).

11. The Czech Republic has a population of approximately 10 million people, which is slightly less than the combined populations of New York City and Chicago. Germany has about 83 million inhabitants, about 2.5 times the population of California.

12. Fijalkowski (1996) points out that anti-immigrant sentiment in Germany on economic grounds is certainly misplaced because large numbers of immigrants remain necessary to make up for Germany's declining workforce.

13. It should be noted that other European countries also have laws similar to Germany's. For example, it is illegal in Belgium, France, Spain, and Switzerland to deny or justify the Holocaust (Boyle, 2001).

14. Actually, the largest minority is the Scottish, followed by Irish and Welsh, but each of these groups constitute ethnic majorities within their historical homelands.

15. A more direct comparison of the two countries would be possible if the U.S. National Crime Victimization Study asked victims the same question as does the British survey (that is, whether the crime was motivated by their race). However, at this point anyway, it does not.

16. Technically, Scotland has its own law, but it is essentially the same as the rest of the United Kingdom's.

17. Indeed, attacks based on religion have long been very common in at least one part of the United Kingdom: Northern Ireland.

18. For a discussion of the policy issues related to including gender in Australian legislations, see Gelber (2000).

19. It is interesting to note that unlike many other countries experiencing similar divisions, Canada has been marked by relatively little separatist violence. Unfortunately, speculation about the reasons for this is beyond the scope of this book.

20. It is worth noting, however, that this situation has presented itself in reverse concerning child pornography, because U.S. law in that matter is *more* restrictive than that of some other countries.

21. In contrast, a German court held that Yahoo! could not be held liable in Germany under similar circumstances.

22. The fact that some material on a Web site is banned in a surfer's own country might only make it all the more appealing. If you lived in Germany, where selling Hitler's *Mein Kampf* is illegal, wouldn't you be tempted to take a look at a copy online just to see what's so horrible about it?

Chapter 9

THE FUTURE OF HATE

On October 3, 2002, Gwen Araujo was beaten and strangled to death in the Silicon Valley town of Newark, California. Her body was later found in a shallow grave in the Sierra foothills, and four young men (three 22-year-olds and one 19-year-old) were arrested for her murder. Araujo, who was only 17, was apparently killed solely because she was transgendered. As of this writing, prosecutors are seeking a hate crime penalty enhancement against three of the offenders; the fourth has pled guilty to voluntary manslaughter.

By the time you read this, Gwen Araujo may have faded to a distant memory for all but her family and friends. It is very likely, however, that her place in the public consciousness will be taken by someone else who suffered because of his or her race or ethnicity or religion or sexual orientation.

It is always rather risky to make predictions about the future, but I can say with complete confidence that there will be more Gwen Araujos, and Matthew Shepards, and James Byrds, and thousands more victims of hate crime whose stories never make the news. There will also be more William Pierces and Bob Matthewses, who will spout essentially the same messages of hate, and there will continue to be those who listen to those messages. There will be outcries for tougher hate crime legislation. Occasionally, and with public fanfare, politicians will respond. There will also be groups and individuals who dedicate their efforts in a variety of ways toward fighting hate.

If true progress is to be made against hate crimes, however, something needs to change. In this chapter, I examine several suggested directions for the future, including new research, better data collection, law enforcement responses, legislation, and other ways to fight a hate that is increasingly global in scope.

NEW RESEARCH

Imagine that you make an appointment to see your doctor. As soon as you walk into the examining room, she takes one look at you and, without asking a single question or performing a single test, hands you a prescription. You dutifully take the pills but continue to feel unwell. When you go to see her again, she just hands you the same prescription, maybe in a stronger dosage this time.

Clearly, this would be ridiculous. You would find another doctor (and your original physician would probably face malpractice charges very soon). In a way, though, this type of method is how we have tried to "cure" society of hate crimes: Although we believe a problem exists, we have asked very few questions about it, we have opted for overly simplistic solutions, and, when those solutions have failed, we have simply tried them again, sometimes in stronger forms.[1]

Thus far, almost all of the approaches that have been made to combating hate crimes have been based on certain assumptions. Some of these assumptions—that most people who commit hate crimes belong to organized hate groups, or that hate crime laws will directly deter people, for instance—are almost certainly false. Others, such as the assumption that hate crimes are more harmful than ordinary crimes, seem perfectly reasonable, but that does not mean that they are true. The empirical evidence to support these assumptions is scanty at best, nonexistent at worst. Like doctors attempting to treat a sickness, we cannot possibly hope to treat the problem of hate crimes effectively unless we know something about their scope and nature.

In the past few years, there has been a small explosion in the number of studies about hate crimes. If you take a quick glance at the References section at the back of this book, you will see how many of the references were published in just the last couple of years. However, because so little was previously known about hate crimes, these studies have made only a small amount of progress toward understanding the problem. There are a number of areas in which more research is especially needed, including hate crime perpetrators, hate crime victims, and what works.[2]

Areas Where Research Is Needed

Hate Crime Perpetrators

To be able to shape an effective policy against hate crimes, it is essential to know who commits these crimes and why. The existing research, both in the United States and abroad, agrees that most perpetrators are male, and most are in their teens or early twenties. The same could be said, of course, for nearly all kinds of crime, so this should come as no particular surprise. What may be more enlightening is that a large majority of those who commit hate crimes are not affiliated with any organized hate groups. Although this finding might be somewhat unexpected, it is robust, having been discovered in a variety of studies in a variety of settings. Not only does this fact have significant policy implications; it also suggests that more research is needed to find out why seemingly ordinary people sometimes seem moved to attack others because of their victims' group affiliation.

McDevitt, J. Levin, and Bennett's (2002) work on typologies of hate crime offenders appears quite promising and is cited frequently in the literature. Their findings have also been confirmed, at least to some extent, by other research by Byers and Crider (2002) and Franklin (2000). However, these are only three studies, each of which was conducted among a fairly limited population (Boston in the early 1990s, eight young men who had committed crimes against the Amish, and a sample of community college students in California, respectively). Before the results can be accepted with confidence, the studies must be replicated in other jurisdictions and among other populations.

If the typology is sound, it has several implications. Recall that these researchers all concluded that most hate crimes are "thrill-seeking" crimes, committed not so much out of heightened bias as out of boredom and the desire to impress friends and gain excitement. Therefore, hate crime prevention programs that focus on reducing prejudice may not be particularly effective. Nor will initiatives that target organized extremist groups.

In fact, if there are indeed several distinct types of hate crime offenders, then no single approach to combating hate crimes will work. For example, one approach might be helpful with thrill seekers but of little use toward retaliatory, defensive, or mission crime perpetrators. Different programs should be crafted to reach different types of offenders.

Some research has focused on a related theme: determining the factors that lead individuals to become hate crime offenders. Many people have assumed that such behavior is linked to aspects of the individual personality, such as authoritarianism or parenting styles, which cause some people to be particularly prejudiced. Others have assumed that the causes lie in discrete situational factors, such as economic conflict. Existing research, however, questions both of these assumptions.

What seems more likely is that hate crime is the result of several forces. One of the most promising areas of research on this topic, but one that has so far received little attention, is the role of the group. Scholars from different disciplines might approach this issue in different ways. Social psychologists, for example, might look at aspects such as conformity and obedience. Sociologists, on the other hand, might examine factors such as differential association or the influence of the culture in general. Any of these perspectives could prove useful and enlightening.

Finally, for those offenders who are affiliated with organized hate groups, more research should be conducted on how those groups recruit new members, on what attracts some segment of the society to extremism, and on when and why members eventually leave the groups. Again, there are some promising studies in this area, especially those conducted by Aho (1988, 1990), Blazak (2001), and Hamm (1993). In particular, research should focus on what distinguishes extremist groups from nonracist gangs and noncriminal social groups. Studies should also be conducted to explore what influence, if any, extremist groups have upon the attitudes and behavior of nonmembers.

Hate Crime Victims

Concerning hate crime victims, one basic question that needs to be addressed is "Who are they?" The data that exist suggest that the most common victims in most jurisdictions are blacks, gays and lesbians, and Jews. However, there are regional and temporal variations, such as the steep increase in crimes against Muslims and Arabs (and those perceived to be Muslim or Arab) after September 11. Moreover, as we have discussed in detail, hate crimes are severely underreported, and there are other significant problems with hate crime victimization data, such as wide variations in the extent to which police officers and police agencies record these data. Therefore, the true nature of hate crime victims remains unknown, as does their number.

Other topics that remain virtually unexplored are the social, economic, and behavioral patterns of hate crime victims. Under what circumstances is someone likely to be targeted as a victim of hate crime? Are attacks truly random, as some have asserted, or are there precipitating factors? Are these victims distinguishable in any meaningful way (demographically, behaviorally, and so on) from other victims of crime?

A second area that needs more study is the effects of hate crimes upon victims. As discussed in Chapter 2, many scholars, policymakers, and commentators have asserted that hate crimes are more harmful than other crimes, making this widely accepted as a truism. The truth, however, is that there is little empirical evidence to support this claim. This is an important issue because this reasoning is one of the primary arguments given in support of enacting special hate crime legislation and giving these offenses special treatment.

Thus far, the most convincing research on the effects of hate crime victimization has been conducted by Herek and his colleagues (Herek, Cogan, & Gillis, 2002; Herek,

Gillis, & Cogan, 1999; Herek, Gillis, Cogan, & Glunt, 1997), and it does suggest that hate crimes are more harmful to victims than ordinary crimes. It is unclear how generalizable these results are, however, because Herek's studies have been conducted only among gays and lesbians in the Sacramento area, and because there are difficulties associated with obtaining a representative sample of victims.

Not only does more work need to be done on this issue, but that research must be carefully constructed. As representative a sample as possible must be obtained. Questions must be worded so that it can be determined not only whether hate crime victimization is harmful (as it most certainly can be), but whether it is *more* harmful than other types of victimization. Jacobs and Potter (1998) have hypothesized that perhaps hate crimes have greater effects on their victims (and on the community at large) than ordinary crimes when the offenses are relatively minor, but not when they are severe. This proposition is worth testing as well.

Further investigation should also focus on particular kinds of victims. Because of the work of a few researchers, as well as some advocacy groups, we know something about people who were chosen to be victims of crime because of their sexual orientation or because they were Jewish. We know virtually nothing, however, about other groups. One of the most glaring deficiencies is the lack of research on African American victims. African Americans are probably the most frequent victims of hate crimes, and their history of oppression in the United States is among the longest and most blatant. Yet only a tiny number of studies have focused on African Americans' experiences as victims of hate crimes.

As discussed in Chapter 3, one topic that has generated much debate is which categories to include in hate crime statutes. Sexual orientation is perhaps the most contentious of these categories, but gender is a close second. There has also been discussion concerning disability and age. Aside from gays and lesbians, virtually no research has examined those who are victimized because of their gender. This is unfortunate because further research might help inform the debates. For example, can hate crimes against women be distinguished in any meaningful way from "ordinary" crimes, such as domestic violence and sexual assault? Are the disabled frequently targeted as victims of crime in the same way that gays and people of color are and, if so, in what ways (if any) do their disabilities preclude them from being able to pursue remedies against the perpetrators?

Research from the United States and from other countries indicates that victims rarely report hate crimes to the police. A few researchers have explored the reasons for this, but not in any large-scale or systematic manner. Studies should be conducted concerning the rates at which different kinds of victims report hate crimes, as well as the reasons they fail to report them. Such research would not only result in more accurate statistics, but it could also help law enforcement and others craft more effective approaches to dealing with these offenses.

One rather simple way to obtain more information about hate crime victims would be to include some appropriate questions in the National Crime Victimization Survey (NCVS). As discussed in Chapter 8, a similar effort in Great Britain resulted in an estimated number of hate crimes that was 30 times higher than the number the FBI reported in the United States the same year, despite the facts that (a) the United States has a much larger population[3] and (b) the United Kingdom counted only those crimes committed because of the victims' race, whereas the United States included crimes committed on other grounds as well. Of course, NCVS data have their own limitations, including the inability to verify reports and the potential for survey participants to misunderstand or misrepresent. However, such information would be a valuable addition to existing victimization data, which are absent or deeply flawed.

What Works

The third major area in which more research about hate crimes is badly needed is program assessment. Most states now have hate crime legislation of some sort and, as discussed in Chapter 7, there are also a large variety of government and private initiatives aimed at reducing hate crimes. What we do not know is whether any of them work.

In a way, these programs are part of a vicious circle. Most of the prevention programs are not informed by empirical research, and their effectiveness is not evaluated by such research, either. Of course, some of them might be just wonderful. We simply do not know.

It is common for scholars from a number of disciplines—sociology, psychology, criminology, and political science, among others—to assess programs and policies related to crimes and other social problems. Such assessments help shape more effective policies and also avoid wasting resources. Unfortunately, these assessments have not been conducted in the realm of hate crimes, perhaps in part because it is a relatively recent domain.

Why Do Research?

In this section, and throughout this book, I have repeatedly called for more research on hate crimes. A practical question arises, however: What is the point? Certainly scholarship has value in and of itself, but is the research likely to have any impact on real-life problems?

The answer is, I do not know.

Even though hate crime research itself is scanty and still in its infancy, there is surely some well-established literature in areas that are relevant to hate crimes. Studies on prejudice reduction and group influence, for example, have been around for many decades. And yet they have played little or no part in the formation of hate crime prevention policies. It is unclear whether additional research would fare any better.

Why do policymakers not pay more attention to research? There are several possible answers. First, they are probably often unaware of its existence. For the most part, for example, legislators do not read academic journal articles, monographs, and textbooks. To the extent that these individuals are familiar with the literature on hate crimes, it is most likely the legal literature that they know, and those sources tend to focus primarily on constitutional and related issues.

A second, more cynical explanation is that policymakers (legislators especially) are not interested in what the research says might work—they are interested in what will make them look good in the public eye. More and harsher punishments seem to meet many people's approval, and so that is the approach that is often taken.[4] Similarly, antihate public relations campaigns and commissions composed of citizens and representatives from multiple agencies seem like a good idea and make a good impression. For a police department, it is relatively easy to form a bias crime unit, perhaps give the members of that unit some nominal training, and then say the department has done its part to fight hate. As Walker and Katz (1995) found, however, the gulf between what is declared and what is reality can sometimes be quite wide.

Another explanation for the lack of attention to research is that research tends to suggest that problems are complex, so their solutions must be as well. Complex solutions, however, are difficult to implement in real life; if the prejudices that infuse so much of our culture and society are a major stimulus of hate crimes, they cannot be reduced in any simple manner. For instance, the jigsaw classroom studies of Aronson, Blaney, Stephan, Sikes, and Snapp (1978), as well as the work of other social psychologists, indicate that one way to foster tolerance is to provide members of different groups the opportunity to work together to reach a common goal and to ensure that certain rather specific conditions apply (see Chapter 7). This model can be achieved in the classroom sometimes and, to a limited extent, among small numbers of individuals in other specialized settings. But how can this technique be used to foster tolerance among the public at large?

These problems are not insurmountable, however. Policymakers can be informed about the results of research. They can be persuaded to attempt programs that will be not merely popular but also effective. They can be assisted to craft strategies that take a multifaceted, holistic approach rather than a simplistic one. Perhaps grassroots and advocacy organizations are the logical mediums through which to create a nexus between researchers and policymakers. This plan would take some effort on the parts of scholars. If research is to have a practical impact, researchers must take care not only to preach to the converted—other researchers—but also to reach lawmakers and others who control policy as well.

BETTER DATA COLLECTION

Related to the need for more research on hate crimes is the need for better data collection by government agencies. A step in the right direction was taken more than a decade ago when Congress enacted the Hate Crime Statistics Act. However, as discussed in Chapter 3 of this book, the data collected pursuant to this Act have a number of serious flaws. Furthermore, although the number of participating agencies has increased considerably since the law first came into effect, there has been no indication in the past few years that reporting under the Act will improve any further. However, there are a number of steps that could be taken to enhance official data.

One suggestion raised in the previous section is to include hate crimes in the National Crime Victimization Survey. Other crimes are generally measured both through the NCVS and through the Uniform Crime Reports (UCR);[5] why not do the same with hate crimes? Because the implementation of the NCVS is already in place, it would be relatively simple to add a question or two related to hate crime, as was done in the United Kingdom. Surely the results would usefully augment the UCR.

A second way to improve reporting would be for all states to mandate that local law enforcement agencies participate. Currently, some states do require this, and some, such as California, even require data collection by the state government. Not all states do, however. And although federal law requires the U.S. Department of Justice to collect and report hate crime statistics, participation by local agencies is voluntary. In 2000, about 17,000 agencies contributed hate crime data to the FBI, and those agencies represented about 84% of the U.S. population (FBI, 2001c). Mandatory reporting would not only increase this coverage, but it would also, possibly, send a message to local agencies that hate crime reporting is important.

Even if every police agency in the country were to contribute hate crime statistics to the FBI, the FBI report would still remain inadequate. As we have discussed, only a small percentage of these crimes ever get reported to the police in the first place. And even when victims or others do tell the police, a variety of factors influence whether the officers involved will record the offense as a hate crime. Thus, any meaningful effort to make official reporting more accurate must focus on victim and police responses to hate crimes and on the interaction between law enforcement and the community.

LAW ENFORCEMENT RESPONSES

Police

Police departments play a very important role in the treatment of hate crimes. They are responsible not only for collecting statistics but also for implementing hate crime legislation. They act as the primary liaison between victims and the legal system as well as communities and the legal system. They have the opportunity to take actions to prevent such crimes from happening at all. And they also, through their policies and procedures, as well as the attitudes and behaviors of their officers, influence the way people in general (members of minority groups especially) are perceived and treated by the state and by the public. Therefore, the police response to hate crimes is a significant issue.

Research both in the United States and abroad indicates that, by and large, the police response to hate crimes is insufficient at best, harmful at worst. Lack of trust of the police is cited as one of the most common reasons that victims fail to report hate crimes. This lack of trust is probably not very surprising, given the long history of discord between law enforcement and the groups that are frequently targeted for hate crimes.

Many police departments appear to be well aware of this issue and have taken some steps to address it. However, it is unlikely that a few hours of diversity or hate crime training, or the nominal designation of a bias crime unit, will do much good. A few hours of training will not adequately cover the complex subject of hate crimes,[6] nor will it change ingrained patterns of bias. And bias crime units often exist more in name than in substance.[7] Even where they are a meaningful reality, Bell (2002b) points out that such units may have little actual power and may be unpopular, be disrespected, or inspire jealousy among other officers, which has direct effects on the officers involved:

> Officers who spent time in the unit cited personal and sometimes professional costs associated with enforcing law with which their fellow officers did not agree. . . . Officers in the unit who had once had cordial relations with other police officers lost friends after they were transferred to the unit. The constant criticism distressed some of the detectives. One White detective who had spent several years in the unit, fighting the community and bureaucracy, began to cry as he related how he had been personally affected by the department and the community's dislike of him. As tears welled up, he said, "There's a lot of pain here. It's very painful." (Bell, 2002b, p. 120)

Furthermore, just because an officer is a member of a bias crime unit, he or she is not necessarily a paragon of tolerance and respect. Bell relates several examples of bias crime officers who themselves expressed bigoted ideas (sometimes to the victims) or behaviors. Others tended to minimize the seriousness of the offenses, dismissing them as "name calling" or "kid stuff" (Bell, 2002b, p. 125).

Most important, however, bias crime units, hate crime data collection, minimal training, and the like will do little or nothing to improve relations between police and their communities or to reduce institutionalized bias within the agencies themselves. Both of these issues are extremely difficult problems that clearly cannot be fixed with simple solutions. Meaningful police responses to hate crimes must be comprehensive and must focus on all aspects of prejudice: offenders, victims, officers, and the community. This treatment might be costly, would probably meet with great resistance from many fronts, and would certainly be a complex undertaking. Thus far, it does not appear that any police department has attempted it. On the other hand, if it were successful, not only would it result in an effective implementation of hate crime laws, but it would improve police-community relations and, potentially, reduce societal prejudice in general.

Prosecutors

The police are not the only component of law enforcement; prosecutors are important as well. Some of the problems related to prosecutors are very similar to those concerning the police: lack of training, lack of resources, and, in many cases, bias within the institutions themselves. Moreover, even the most dedicated and well-educated prosecutor is faced with the difficulties inherent in proving offenders' motives, as well as the political pressures to obtain high conviction rates. On the other hand, the prosecutorial mission concerning hate crimes is considerably less complex than that of the police, and district attorneys' offices are considerably more insulated from community relations difficulties than the police are. Furthermore, although victims and witnesses may often be reluctant to cooperate with prosecutors in hate crime cases, at least they have usually taken the initial step of reporting the crime, so they are generally interested in seeking some official response. Nevertheless, as discussed in Chapter 3, hate crime prosecutions are very rare, and convictions are even rarer.

A few prosecutors' offices have dealt with this situation in much the same ways that police departments have, such as the creation of special prosecutorial units. Some also participate in interagency task forces or commissions. The efficacy of these methods is unknown. Even if they are effective, they have been adopted by very few district attorneys' offices; most government-based approaches have focused exclusively on the police. Existing policies need to be evaluated, and perhaps prosecutors in more jurisdictions should consider creating special hate crime programs.

LEGISLATION

Less than three decades ago, special hate crime laws did not exist. Since their creation, they have continued to evolve in response to public pressure and judicial decisions. More laws have been enacted, more groups have been added to the statutes, and definitions have been made more precise. It is likely that this evolution will continue, at least for the near future.

Some scholars have raised serious doubts concerning hate crime laws' effectiveness in actually reducing bias-motivated offenses. Certainly, there is not much evidence that the laws have any direct positive effect. Their potential indirect effects, such as the symbolic value of protecting certain groups or of focusing public attention on the problem, are nearly impossible to measure. Their passage often also creates difficult court challenges. For example, if a federal hate crime law is ever enacted, it will certainly be challenged as exceeding Congress's constitutional authority, and such a challenge might very well be successful.

Despite these limitations, however, there is virtually no question that the push to enact more hate crime laws will continue. Currently, several advocacy organizations, including the Anti-Defamation League (ADL), the Southern Poverty Law Center (SPLC), and the Simon Wiesenthal Center (SWC), strongly support the enactment of a federal hate crime law, as well as expansions of existing state laws.

It is not always easy to get hate crime laws passed. Often resistance is because of opposition to the inclusion of certain protected categories, especially sexual orientation. This provision has been the major stumbling block to the federal law being passed. Sometimes, people oppose hate crime laws because they misunderstand them. They believe (erroneously) that the laws create special protections for minorities only,[8] or, like then-Texas Governor George W. Bush, they announce that *all* crimes are hate crimes ("Governor of Texas," 2001). Some oppose the laws because they are concerned that they may infringe on certain rights, especially those protected by the First Amendment. A small number of people oppose the laws because they believe they do not work and might even be counterproductive.

Despite all of this opposition, hate crime laws never really die. Potential federal legislation (most recently titled the Law Enforcement Enhancement Act) has been reintroduced annually. The Nebraska legislature killed a hate crime law at least twice before it was resurrected once more in 1997 and, finally, passed. Texas's hate crime law, which was rejected in 1999, was passed in 2001 (and named after prominent hate crime victim James Byrd, Jr.). In Wyoming, proposals for hate crime laws have arisen several times, especially after the death of Matthew Shepard. Although the state still has not passed such a law, the city of Laramie, where Shepard attended college, did enact a hate crime ordinance in 2000.

If hate crime laws have no demonstrated positive effects, and if they are subject to so much resistance, why do they keep being proposed (and enacted)? In part, this situation may be due to the lobbying efforts of advocacy organizations as well as media pressure every time a high-profile hate crime occurs. The laws tend to meet with public acceptance, largely because Americans are used to thinking of harsher criminal sanctions as appropriate solutions to complex social ills.[9] Furthermore, many legislators embrace the laws as a (simple) method of demonstrating their dual commitments to promote tolerance and crack down on crime.

Almost all states have some form of hate crime law, and a federal law is probably eventually inevitable. If advocacy groups and others wish to continue to focus their efforts on encouraging legislation, there are three primary areas on which they could concentrate.

Groups Protected

First, attention should be paid to which groups are protected by existing laws. As discussed in Chapter 3, 16 states currently do not include sexual orientation within their laws. This fact is significant because not only are gays and lesbians among the most frequent victims of hate crimes, but these crimes are motivated by the same types of biases that lie behind attacks on people of color and religious minorities. Hate crime laws may or may not actually reduce hate crimes, but *not* including sexual orientation within the laws sends a message that homophobia and its consequences are acceptable. Thus, the existence of hate crime laws that lack this category may actually add to a culture of intolerance and may indirectly sanction violence.

Civil Actions

A second area of emphasis might be civil actions. The ADL's model law includes a provision allowing victims of hate crimes to sue their attackers. Several states have enacted similar provisions, some of which allow successful plaintiffs to obtain extra damages. No statistics are available on how common these lawsuits are, and it is unclear whether many people are aware of these laws or have the desire to take advantage of them.

Because they are young, most hate crime offenders likely have little in the way of assets, so most victims probably would not receive great amounts of money. However, civil lawsuits, like criminal prosecutions, can have symbolic value. Moreover, civil suits are initiated by the victim, unlike criminal cases, which require action first by the police and then by prosecutors. Thus, victims might choose to pursue civil suits in cases where government authorities will not take action. Civil cases also require a lighter standard of proof than criminal cases—preponderance of the evidence instead of beyond a reasonable doubt—so a civil suit might be successful in instances in which a criminal conviction is unlikely or impossible.

Hate crime opponents could lobby more states to enact civil hate crime provisions. They could also publicize existing laws to a greater extent and offer legal assistance (as does the SPLC) to those who wish to pursue civil suits. Again, there is no evidence that any of this would actually decrease the incidence of hate crimes, but it is about as likely as criminal laws to have such an effect, and at least it has the possibility of offering victims some direct benefits.

Legislation Concerning Law Enforcement

The final area in which more efforts could be made is legislation concerning law enforcement. Only a few states mandate police officer training on hate crimes, and even those requirements are generally fairly minimal. Some advocacy organizations, such as the SPLC, provide hate crime training for police officers, and the Department of Justice has prepared a model training curriculum, but it is unclear how many officers actually receive such training. Of course, the impact of the training is also not known, but it is certainly one area worth pursuing.

GLOBALIZED HATE

A topic that is certain to receive more attention in the future, and deservedly so, is the globalization of hate. As discussed in Chapter 8, there are indications that extremists in the United States are currently expanding their ties to extremists in other countries. Certainly, American hate groups have established a large presence for themselves on the Internet, which has facilitated communication with sympathizers abroad.

The potential consequences of these alliances are quite serious. As I write this, the U.S. government (as well as local governments) is quite busy attempting to establish ways of avoiding terrorist attacks from foreign sources such as Al Qaeda. Among other efforts, intelligence efforts abroad have been increased, as have border searches and immigration restrictions. As we should have learned from Oklahoma City, however, not all threats to domestic security come from foreigners. Imagine the possibilities for death and damage within our country if organizations such as Al Qaeda were to form close working relationships with domestic extremists.

Another issue that must be addressed is the regulation of international hate, especially over the Internet. As discussed earlier, the United States' tolerance of hate speech has resulted in our becoming, to some extent, a virtual haven for those who hate. This in turn has created tension with other countries as well as thorny legal problems. It is unclear how these issues can be addressed without damage to our important and constitutionally protected freedoms.

In fact, technology is evolving so quickly that it is possible that any solutions that are found to these problems will only be short-lived. Two decades ago, very few people would have imagined that it would be possible for a person in nearly any part of the world to get on a computer and instantly communicate with like-minded individuals around the globe, or, in the space of a few minutes, download libraries' worth of text, music, videos, or games.[10] The possibilities of electronic media seem endless for all of us; unfortunately, that includes extremists as well.

FIGHTING HATE

If hate crimes are truly to be reduced, piecemeal and one-dimensional approaches are not satisfactory. Instead, government and private agencies and individuals must craft cooperative, multifaceted efforts. Those efforts should include several components.

Focus on Young People

A great deal of the efforts should be aimed at young people, in part because they are the ones who commit most hate crimes. Antihate or pro-diversity curricula such as those that have already been developed may be incorporated, but the efforts must not end there. It is very unlikely that a few hours of lessons on tolerance will overcome a lifetime of exposure to prejudice, nor are they apt to surmount the peer pressures that are so strong at this stage in life.

Successful programs would encourage young people to interact cooperatively with those who are of different groups and would foster an atmosphere in which intergroup friendships can form. A culture of more than political correctness or even mere tolerance would be created: People must learn to view the differences between them as unimportant. They must also be given the opportunity and incentive to overcome the cognitive and attitudinal effects of stereotypes. Furthermore, all of these efforts to promote tolerance must be supported not only by teachers, parents, and other authority figures, but also by peers.

Some attention might also be paid to limiting the potential attractiveness of extremist groups. Censoring those groups or limiting access to their messages is probably not an appropriate way to go about this: Not only would doing so raise significant constitutional questions, but it might also simply add to those messages the allure of forbidden fruit. Moreover, experience has shown that attempts to silence extremist groups will almost certainly fail. Instead, those who might be attracted to extremist groups could be provided with alternative social outlets that have greater benefits for them. For those who are already involved in extremist groups, ways could be found to create social pushes from the group and social pulls toward other activities (Aho, 1988).

Focus on Bias in State Institutions

A second area in which hate crime prevention efforts must concentrate is the state. It is both inaccurate and dangerous to think of hate crimes as deviant acts committed by fanatics. In truth, both in the past and in the present, hate has been sanctioned, encouraged, and even acted on by agents of the state. It may be comforting to think that some of the most egregious forms of this hate, such as enslavement of African Americans and genocide of Native Americans, are no longer practiced. But discrimination still occurs. For example, it was recently reported that nine linguists were discharged from the Army because they are gay (Mason, 2002).[11] What message does it send when people wish to provide an urgently needed service to their country, and that country refuses to take advantage of their skills merely because of their sexual orientation? Other examples of bias within the government may be less obvious than the "Don't ask, don't tell" policy, but that makes them no less insidious.

It is easy to single out police departments as the government agencies most in need of reform when it comes to hate, and certainly there is a great need there. However, institutionalized bias is much more widespread than that. Some government bias occurs at the national or statewide level, and some occurs locally. Some occurs as a result of official or unwritten policy, whereas some is manifested by small groups or single government agents. Occasionally, such bias is publicized in the media, such as when New York City police officers sexually assaulted Haitian immigrant Abner Louima in 1997, but much more often it goes unreported. It is unrealistic to believe that while the government talks out of both sides of its mouth, prohibiting hate crimes here while engaging in bias there, any real reform will occur.

Focus on Prejudices in Society

Last, a meaningful attempt to reduce hate crimes must find a way to decrease prejudices within society at large. Hate crimes do not occur in a vacuum. Research from the United States and other countries strongly suggests that hate crimes are the result of a complex combination of individual and social factors, and general societal prejudice certainly plays an important role.

When skinheads attacked African and Asian immigrants in Hoyerswerda, Germany, local residents cheered them on. When nationalists and white supremacists such as David Duke and Jean-Marie Le Pen have run for office in the United States and abroad, it is not only the neo-Nazis or the Klansmen who have cast votes for them. Retired Nebraska Judge Albert Walsh belonged to no hate group when he publicly opposed including sexual orientation within Nebraska's hate crime law because "Homosexuals and lesbians are not victims. They are an affluent, powerful class" (Heinzl, 1997, p. 13SF). In other words, the prejudices that manifest themselves in hate crimes are only a reflection of the prejudices that exist in the public at large. Unless bigotry is reduced within the culture, hate crimes will probably never be prevented.

CONCLUSION

Of course, it is a simple matter for me to sit here and make these suggestions; it is quite another thing to think of practical ways to undertake them. Any comprehensive effort to fight hate crimes will clearly be a massive task requiring creative planning and the cooperation of many people. All of this will be extremely difficult. What is more, no matter how good the programs, we can never hope to completely eliminate either hate crimes or prejudice itself. As we learn more, however, and if groups and individuals are willing to make the appropriate efforts, it is possible that we can find successful ways to control hate crimes.

DISCUSSION QUESTIONS

1. This chapter outlines a number of areas related to hate crimes that are most in need of more research. Are there other topics you would add to the discussion? If you had the opportunity to study hate crimes, what would you focus on and why? How would you go about doing your research? Can you think of ways to encourage more research in this field?

2. Most of the empirical research on hate crimes has been conducted by sociologists, psychologists, and criminologists. Are there other disciplines that you feel could make a contribution to this literature?

3. As this chapter discusses, all the research in the world will be of little use if policymakers choose to ignore it. Why do you believe that this seems to happen so often? Can you suggest ways to make policymakers more aware of hate crime research and to encourage them to use that research to shape policy? What changes would you expect to see if policy were more heavily influenced by research on hate crimes?

4. What do you see as the biggest impediments to the creation of effective hate crime approaches within law enforcement agencies? How would you suggest overcoming them?

5. One major problem with effective police responses to hate crimes is the police culture, which is often ridden with bias. Why do you think bias is such a problem within police departments? What are some ways of decreasing it? Assess the likelihood that police departments would adopt prejudice-reduction programs.

6. What do you predict will be the future of hate crime laws? Do you agree that federal legislation in inevitable? If you headed an advocacy organization, how much would your organization concentrate on hate crime laws, and in which specific areas?

7. How can policymakers effectively address the problems associated with the globalization of hate? What are some of the impediments to effective policy? How real a threat do you believe this globalization presents?

8. Imagine that you are a school administrator, police chief, or community leader. Describe a plan you would implement to decrease hate crimes. What would be the components of that plan? Whom would it involve, and how would you encourage involvement? How would your plan interact with those created by other groups or agencies? How would you assess its success? What would be the costs involved?

9. What will you take with you as the major lessons or themes of this book? In what ways, if any, do you think that these will affect your personal, academic, or professional life? Are there aspects of hate crimes that you believe were not adequately covered here or about which you would like to learn more?

INTERNET EXERCISES

1. Randy Blazak, a professor at Portland State University, directs the Hate Crimes Research Network, which is dedicated to linking academic research on hate crimes. The Network's Web site, www.hatecrime.net, contains abstracts of articles in several disciplines related to hate crimes, links to many hate crimes researchers, and other useful information.

2. The students in my hate crime class have created a Web site on hate crimes. You can view it at cjwww.csustan.edu/hatecrimes/. You might want to consider creating your own. If you have never made a Web page, here are a few resources to get you started:
   • http://wp.netscape.com/browsers/createsites/index.html
   • www.cln.org/themes/webpages_intro.html
   • www.learningspace.org/tech/FrontPage/
   • http://teams.lacoe.edu/documentation/internet/webpages.html

NOTES

1. The same argument could be made, incidentally, about a large number of criminal justice and social issues. Those issues, however, are beyond the scope of this book.

2. For a recent review of hate crime research, as well as recommendations for the future, see Green, McFalls, & Smith (2001).

3. The FBI data indicate a rate of 3.3 hate crimes per 100,000 residents, whereas the U.K. data indicate 468.4 per 100,000.

4. Especially in the United States, although not exclusively so. British Prime Minister Tony Blair recently announced, for example, that his government would be instituting stricter sentences, as well as eliminating provisions such as double jeopardy prohibitions, in an attempt to reduce crime (Hughes, 2002).

5. The Uniform Crime Reports (UCR) is issued annually by the Department of Justice and includes counts of most major crimes. It is based on reports from nearly all local law enforcement agencies within the United States. Hate crime statistics are now a part of the UCR, although the FBI also issues a separate hate crime report each year. Both the UCR and the hate crime reports for the past several years can be viewed at www.fbi.gov/ucr/ucr.htm. NCVS information and data are available from the Bureau of Justice Statistics (www.ojp. usdoj.gov/bjs/pubalp2.htm) and the National Archive of Criminal Justice Data (www.icpsr. umich.edu/NACJD/NCVS/).

6. I teach a semester-long hate crimes class, in which students spend nearly 40 hours in the classroom, and probably several times that amount of time reading, studying, and writing on the topic. Still, there are many aspects that we must touch on only briefly or skip altogether.

7. Police in a great many jurisdictions—probably the majority within the United States—lack even these minimal things (bias crime units, special training, and so on).

8. Extremist groups promote this misconception quite heavily.

9. Americans seem to like simplistic approaches as well, such as the "Just Say No!" antidrug campaign.

10. Last year, a colleague spent a month in Ghana and was able to e-mail us about her adventures as she went. Just this week, I've communicated via e-mail with people on five continents.

11. This discharge occurred despite the fact that the military and intelligence agencies are facing a severe shortage of language specialists, and six of the nine discharged men were trained in Arabic, a language for which there is a particularly urgent need.

APPENDIX A

Selected International
Hate Crime Laws

GERMANY

The German Penal Code

Section 86a Use of Symbols of Unconstitutional Organizations

(1) Whoever:

1. domestically distributes or publicly uses, in a meeting or in writings (Section 11 subsection (3)) disseminated by him, symbols of one of the parties or organizations indicated in Section 86 subsection (1), nos. 1, 2 and 4; or

2. produces, stocks, imports or exports objects which depict or contain such symbols for distribution or use domestically or abroad, in the manner indicated in number 1, shall be punished with imprisonment for not more than three years or a fine.

(2) Symbols, within the meaning of subsection (1), shall be, in particular, flags, insignia, uniforms, slogans and forms of greeting. Symbols which are so similar as to be mistaken for those named in sentence 1 shall be deemed to be equivalent thereto.

Section 130 Agitation of the People

(1) Whoever, in a manner that is capable of disturbing the public peace:

1. incites hatred against segments of the population or calls for violent or arbitrary measures against them; or

2. assaults the human dignity of others by insulting, maliciously maligning, or defaming segments of the population,

shall be punished with imprisonment from three months to five years.

(2) Whoever:

1. with respect to writings (Section 11 subsection (3)), which incite hatred against segments of the population or a national, racial or religious group, or one characterized by its folk customs, which call for violent or arbitrary measures against them, or which assault the human dignity of others by insulting, maliciously maligning or defaming segments of the population or a previously indicated group:

a) disseminates them;

b) publicly displays, posts, presents, or otherwise makes them accessible;

c) offers, gives or makes accessible to a person under eighteen years; or

d) produces, obtains, supplies, stocks, offers, announces, commends, undertakes to import or export them, in order to use them or copies obtained from them within the meaning of numbers a through c or facilitate such use by another; or

2. disseminates a presentation of the content indicated in number 1 by radio,

shall be punished with imprisonment for not more than three years or a fine.

(3) Whoever publicly or in a meeting approves of, denies or renders harmless an act committed under the rule of National Socialism of the type indicated in Section 220a subsection (1), in a manner capable of disturbing the public peace shall be punished with imprisonment for not more than five years or a fine.

(4) Subsection (2) shall also apply to writings (Section 11 subsection (3)) with content such as is indicated in subsection (3).

(5) In cases under subsection (2), also in conjunction with subsection (4), and in cases of subsection (3), Section 86 subsection (3), shall apply correspondingly.

Section 130a Instructions for Crimes

(1) Whoever disseminates, publicly displays, posts, presents, or otherwise makes accessible a writing (Section 11 subsection (3)) which is capable of serving as instructions for an unlawful act named in Section 126 subsection (1), and is intended by its content to encourage or awaken the readiness of others to commit such an act, shall be punished with imprisonment for not more than three years or a fine.

(2) Whoever:

1. disseminates, publicly displays, posts, presents, or otherwise makes accessible a writing (Section 11 subsection (3)) which is capable of serving as instructions for an unlawful act named in Section 126 subsection (1); or

2. gives instructions for an unlawful act named in Section 126 subsection (1), publicly or in a meeting, in order to encourage or awaken the readiness of others to commit such an act,

shall be similarly punished.

(3) Section 86 subsection (3), shall apply correspondingly.

Section 131 Representation of Violence

(1) Whoever, in relation to writings (Section 11 subsection (3)), which describe cruel or otherwise inhuman acts of violence against human beings in a manner which expresses a glorification or rendering harmless of such acts of violence or which represents the cruel or inhuman aspects of the event in a manner which injures human dignity:

1. disseminates them;

2. publicly displays, posts, presents, or otherwise makes them accessible;

3. offers, gives or makes them accessible to a person under eighteen years; or

4. produces, obtains, supplies, stocks, offers, announces, commends, undertakes to import or export them, in order to use them or copies obtained from them within the meaning of numbers 1 through 3 or facilitate such use by another,

shall be punished with imprisonment for not more than one year or a fine.

(2) Whoever disseminates a presentation of the content indicated in subsection (1) by radio, shall be similarly punished.

(3) Subsections (1) and (2) shall not apply if the act serves as reporting about current or historical events.

(4) Subsection (1), number 3 shall not be applicable if the person authorized to care for the person acts.

Section 185 Insult

Insult shall be punished with imprisonment for not more than one year or a fine and, if the insult is committed by means of violence, with imprisonment for not more than two years or a fine.

Section 189 Disparagement of the Memory of Deceased Persons

Whoever disparages the memory of a deceased person shall be punished with imprisonment for not more than two years or a fine.

GREAT BRITAIN

The Crime and Disorder Act

Meaning of "racially aggravated"

28. (1) An offense is racially aggravated for the purposes of sections 29 to 32 below if—

(a) at the time of committing the offense, or immediately before or after doing so, the offender demonstrates toward the victim of the offence hostility based on the victim's membership (or presumed membership) of a racial group; or

(b) the offence is motivated (wholly or partly) by hostility towards member of a racial group based on their membership of that group.

(2) In subsection (1)(a) above—

"membership," in relation to a racial group, includes association with members of that group;

"presumed" means presumed by the offender.

(3) It is immaterial for the purposes of paragraph (a) or (b) of subsection (1) above whether or not the offender's hostility is also based, to any extent, on—

(a) the fact or presumption that any person or group of persons belongs to any religious group; or

(b) any other factor not mentioned in that paragraph.

(4) In this section "racial group" means a group of persons defined by reference to race, colour, nationality (including citizenship) or ethnic or national origins.

Racially aggravated assaults

29. (1) A person is guilty of an offence under this section if he commits—

(a) an offence under section 20 of the Offences Against the Person Act 1861 (malicious wounding or grievous bodily harm);

(b) an offence under section 47 of that Act (actual bodily harm); or

(c) common assault,

which is racially aggravated for the purposes of this section.

(2) A person guilty of an offence falling within subsection (1)(a) or (b) above shall be liable—

(a) on summary conviction, to imprisonment for a term not exceeding six months or to a fine not exceeding the statutory maximum, or to both;

(b) on conviction on indictment, to imprisonment for a term not exceeding seven years or to a fine, or to both

(3) A person guilty of an offence falling within subsection (1)(c) above shall be liable—

(a) on summary conviction, to imprisonment for a term not exceeding six months or to a fine not exceeding the statutory maximum, or to both;

(b) on conviction on indictment, to imprisonment for a term not exceeding two years or to a fine, or to both.

[Other provisions of the law concern criminal damage, public order offenses, and harassment.]

AUSTRALIA

Federal Racial Anti-Discrimination Act

Section 18C

(1) It is unlawful for a person to do an act, otherwise than in private, if:

(a) the act is reasonably likely, in all the circumstances, to offend, insult, humiliate or intimidate another person or a group of people; and

(b) the act is done because of the race, colour or national or ethnic origin of the other person or of some or all of the people in the group.

(2) For the purposes of subsection (1), an act is taken not to be done in private if it:

(a) causes words, sounds, images or writing to be communicated to the public; or

(b) is done in a public place; or

(c) is done in the sight or hearing of people who are in a public place.

(3) In this section:

public place includes any place to which the public have access as of right or by invitation, whether express or implied and whether or not a charge is made for admission to the place.

Section 18D

Section 18C does not render unlawful anything said or done reasonably and in good faith:

(a) in the performance, exhibition or distribution of an artistic work; or

(b) in the course of any statement, publication, discussion or debate made or held for any genuine academic, artistic or scientific purpose or any other genuine purpose in the public interest; or

(c) in making or publishing:

(i) a fair and accurate report of any event or matter of public interest; or

(ii) a fair comment on any event or matter of public interest if the comment is an expression of a genuine belief held by the person making the comment.

New South Wales Anti-Discrimination Act

Section 20C

(1) It is unlawful for a person, by a public act, to incite hatred towards, serious contempt for, or severe ridicule of, a person or group of persons on the ground of the race of the person or members of the group.

(2) Nothing in this section renders unlawful:

(a) a fair report of a public act referred to in subsection (1), or

(b) a communication or the distribution or dissemination of any matter comprising a publication referred to in Division 3 of Part 3 of the Defamation Act 1974 or which is otherwise subject to a defence of absolute privilege in proceedings for defamation, or

(c) a public act, done reasonably and in good faith, for academic, artistic, scientific or research purposes or for other purposes in the public interest, including discussion or debate about and expositions of any act or matter.

Section 20D

(1) A person shall not, by a public act, incite hatred towards, serious contempt for, or severe ridicule of, a person or group of persons on the ground of the race of the person or members of the group by means which include:

threatening physical harm towards, or towards any property of, the person or group of persons, or

inciting others to threaten physical harm towards, or towards any property of, the person or group of persons.

Maximum penalty:

In the case of an individual—50 penalty units or imprisonment for 6 months, or both.

In the case of a corporation—100 penalty units.

(2) A person shall not be prosecuted for an offence under this section unless the Attorney General has consented to the prosecution.

CANADA

Criminal Code Sections 318 through 320

318. (1) Every one who advocates or promotes genocide is guilty of an indictable offence and liable to imprisonment for a term not exceeding five years. . . .

319. (1) Every one who, by communicating statements in any public place, incites hatred against any identifiable group where such incitement is likely to lead to a breach of the peace is guilty of

an indictable offence and is liable to imprisonment for a term not exceeding two years; or

an offence punishable on summary conviction

(2) Every one who, by communicating statements, other than in private conversation, willfully promotes hatred against an identifiable group is guilty of

an indictable offence and is liable to imprisonment for a term not exceeding two years; or

an offence punishable on summary conviction

(3) No person shall be convicted of an offence under subsection (2)

if he establishes that the statements communicated were true;

if, in good faith, he expressed or attempted to establish by opinion an argument on a religious subject;

if the statements were relevant to any matter of public interest, the discussion of which was for the public benefit, and if on reasonable grounds he believed them to be true; or

if, in good faith, he intended to point out, for the purposes of removal, matters producing or tending to produce feelings of hatred toward an identifiable group in Canada. . . .

320. (1) A judge who is satisfied by information on oath that there are reasonable grounds for believing that any publication, copies of which are kept for sale or distribution within the jurisdiction of the court, is hate propaganda shall issue a warrant under his hand authorizing seizure of the copies. . . .

(8) In this section, "hate propaganda" means any writing, sign or visible representation that advocates or promotes genocide or the communication of which by any person would constitute an offence under section 319.

Criminal Code 718.2

A court that imposes a sentence shall also take into consideration the following principles:

(*a*) a sentence should be increased or reduced to account for any relevant aggravating or mitigating circumstances relating to the offence or the offender, and, without limiting the generality of the foregoing,

(i) evidence that the offence was motivated by bias, prejudice or hate based on race, national or ethnic origin, language, colour, religion, sex, age, mental or physical disability, sexual orientation, or any other similar factor. . . .

Canadian Human Rights Act 13(1)

It is a discriminatory practice for a person or a group of persons acting in concert to communicate telephonically or to cause to be so communicated, repeatedly, in whole or in part by means of the facilities of a telecommunication undertaking within the legislative authority of Parliament, any matter that is likely to expose a person or persons to hatred or contempt by reason of the fact that that person or those persons are identifiable on the basis of a prohibited ground of discrimination.

APPENDIX B

Selected Resources on Hate Crimes

T he following is a list of some of the major resources for additional information on hate crimes. Many other sources of information are available as well, but these provide a good starting place.

American-Arab Anti-Discrimination Committee (ADC)

4201 Connecticut Ave.
Washington, DC 20008
www.adc.org

Anti-Defamation League (ADL)

823 United Nations Plaza
New York, NY 10017
www.adl.org

Center for Democratic Renewal (CDR)

P.O. Box 50469
Atlanta, GA 30302
www.publiceye.org/cdr/cdr.html

FBI Hate Crime Statistics

www.fbi.gov/ucr/01hate.pdf

Hate Crime Web Site at
California State University, Stanislaus

cjwww.csustan.edu/hatecrimes/

Hate Crimes Research Network

www.hatecrime.net

The Hate Directory: Hate Groups on the Internet

www.bcpl.net/~rfrankli/hatedir.htm

National Asian Pacific American Legal Consortium (NAPALC)

1140 Connecticut Ave. N.W., Suite 1200
Washington, DC 20036
www.napalc.org

National Center for Hate Crime Prevention (NCHCP)

www.edc.org/HHD/hatecrime/id1_homepage.htm

National Criminal Justice Reference Service (NCJRS)

Hate Crime Resources

www.ncjrs.org/hate_crimes/summary.html

National Gay and Lesbian Task Force (NGLTF)

1325 Massachusetts Ave. N.W., Suite 600
Washington, DC 20005
www.ngltf.org

Not in Our Town

www.pbs.org/niot/

Partners Against Hate (PAH)

www.partnersagainsthate.org

Political Research Associates (PRA)

1310 Broadway, Suite 201
Somerville, MA 02144
www.publiceye.org

Sexual Orientation: Science, Education, and Policy

http://psychology.ucdavis.edu/rainbow/index.html

Simon Wiesenthal Center (SWC)

1399 South Roxbury Drive
Los Angeles, CA 90035
www.wiesenthal.com

Southern Poverty Law Center (SPLC)

400 Washington Avenue
Montgomery, AL 36104
www.splcenter.org

REFERENCES

Abood v. Detroit Board of Education, 431 U.S. 239 (1977).

Adams, H. E., Wright, L. E., & Lohr, B. A. (1996). Is homophobia associated with homosexual arousal? *Journal of Abnormal Psychology, 105,* 440-445.

Adorno, T. W., Frenkel-Brunswick, E., Levinson, D. J., & Sanford, R. N. (1950). *The authoritarian personality.* New York: Harper & Row.

Aho, J. A. (1988). Out of hate: A sociology of defection from neo-Nazism. *Current Research on Peace and Violence, 11,* 159-168.

Aho, J. A. (1990). *The politics of righteousness: Idaho Christian patriotism.* Seattle: University of Washington Press.

Aho, J. A. (1994). *This thing of darkness: A sociology of the enemy.* Seattle: University of Washington Press.

Allen, W. R. (1991). Klan, cloth, and constitution: Anti-mask laws and the first amendment. *Georgia Law Review, 25,* 819-860.

Allport, G. W. (1954). *The nature of prejudice.* Reading, MA: Addison-Wesley.

Almaguer, T. (1994). *Racial fault lines: The historical origins of white supremacy in California.* Berkeley: University of California Press.

Altemeyer, B. (1981). *Right-wing authoritarianism.* Winnipeg: University of Manitoba Press.

Altemeyer, B. (1998). The other "Authoritarian Personality." *Advances in Experimental Social Psychology, 30,* 47-92.

American-Arab Anti-Discrimination Committee. (2002). ADC fact sheet: The condition of Arab Americans post-9/11. Retrieved March 13, 2003, from www.adc.org/terror_attack/9-11 aftermath.pdf

American Psychiatric Association. (1998). *Psychiatric treatment and sexual orientation: Position statement.* Retrieved February 25, 2003, from www.psych.org/archives/980020.pdf

American Psychological Association. (1997). *Resolution on appropriate therapeutic responses to sexual orientation.* Retrieved February 24, 2003, from www.apa.org/pi/lgbpolicy/orient.html

Anderson, J. F., Mangels, N. J., & Dyson, L. (2001). A gang by any other name is just a gang: Towards an expanded definition of gangs. *Journal of Gang Research, 8*(4), 19-34.

Anderson, T. (2001, March 27). White supremacist gets life. *Daily News of Los Angeles,* p. N1.

Anti-Defamation League. (1988). *Hate crimes: Policies and procedures for law enforcement agencies.* New York: Author.

Anti-Defamation League. (1994). *Hate crime laws: A comprehensive guide.* New York: Author.

Anti-Defamation League. (1997). *Vigilante justice: Militias and "common law courts" wage war against the government.* New York: Author.

Anti-Defamation League. (1998). *Feminism perverted: Extremist women on the World Wide Web.* Retrieved February 19, 2003, from www.adl.org/special_reports/extremist_women_on_web/print.asp

Anti-Defamation League. (2000). *Audit of anti-Semitic incidents.* New York: Author.

Anti-Defamation League. (2001). *Hate crimes laws.* Retrieved February 19, 2003, from www.adl.org/99hatecrime/intro.asp.

Anti-Defamation League. (2002a). *Anti-Semitism in America, 2002.* Retrieved February 25, 2003, from www.adl.org/anti_semitism/2002/as_survey.pdf

Anti-Defamation League. (2002b). *Anti-Semitism in Russia: 2002 update.* Retrieved February 19, 2003, from www.adl.org/Anti_semitism/Russia_AntiSem_Rise.asp

Anti-Defamation League. (2002c). *Bigots who rock: An ADL list of hate music groups.* Retrieved February 19, 2003, from www.adl.org/extremism/bands/default.asp

Anti-Defamation League. (2002d). *Extremism in America: Ernst Zundel.* Retrieved March 7, 2003, from www.adl.org/learn/Ext_US/zundel.asp?xpicked=2&item=zundel

Anti-Defamation League. (2002e). *The new global Anti-Semitism.* Retrieved February 19, 2003, from www.adl.org/Anti_semitism/anti-semitism_global.asp

Anti-Defamation League. (2002f). *2001 audit of Anti-Semitic incidents.* Retrieved February 25, 2003, from www.adl.org/2001audit/adlaudit2001.pdf

Anti-Defamation League. (2002g). *World Church of the Creator.* Retrieved February 19, 2003, from www.adl.org/learn/Ext_US/WCOTC.asp?xpicked=3&item=17

Apprendi v. New Jersey, 530 U.S. 466 (2000).

Aronowitz, A. (1994). Germany's xenophobic violence: Criminal justice and social responses. In M. S. Hamm (Ed.), *Hate crime: International perspectives on causes and control* (pp. 37-69). Cincinnati: Anderson.

Aronson, E. (1999). *The social animal* (8th ed.). New York: W. H. Freeman & Co.

Aronson, E. (2000). *Jigsaw Classroom: Jigsaw in 10 Easy Steps.* Retrieved February 24, 2003, from www.jigsaw.org/steps.htm

Aronson, E., Blaney, N., Stephan, C., Sikes, J., & Snapp, M. (1978). *The jigsaw classroom.* Beverly Hills, CA: Sage.

Asch, S. (1956). Studies of independence and conformity: A minority of one against a unanimous majority. *Psychological Monographs, 70,* No. 9, Whole No. 416.

Associated Press. (2002, August 16). Robbers targeting old men. *Modesto Bee,* p. A-3.

Bailey, E. (2001, Sept. 8). Temple arson pleas clear the way for murder trial. *Los Angeles Times,* part 2, p. 10.

Bailey, J. M., & Pillard, R. (1991). A genetic study of male sexual orientation. *Archives of General Psychiatry, 48,* 1089-1096.

Baldus, D. C., & Woodworth, G. (1998). Racial discrimination and the death penalty: An empirical and legal overview. In J. R. Acker, R. M. Bohm, and C. S. Lanier (Eds.), *America's experiment with capital punishment: Reflections on the past, present and future of the ultimate penal sanction* (pp. 385-417). Durham, NC: Carolina Academic Press.

Barclay v. Florida, 463 U.S. 939 (1983).

Barkun, M. (1994). *Religion and the racist right: The origins of the Christian Identity movement.* Chapel Hill: University of North Carolina Press.

Barnes, A., & Ephross, P. H. (1994). The impact of hate violence on victims: Emotional and behavioral responses to attacks. *Social Work, 39,* 247-251.

Bartsch, R. A., Burnett, T., Diller, T. R., & Rankin-Williams, E. (2000).Gender representation in television commercials: Updating an update. *Sex Roles, 43,* 735-743.

Beale, S. S. (2000). Federalizing hate crimes: Symbolic politics, expressive law, or tool for criminal enforcement? *Boston University Law Review, 80,* 1227-1281.

Bell, J. (1997). Policing hatred: Police bias units and the construction of a hate crime. *Michigan Journal of Race and Law, 2,* 421-460.

Bell, J. (2002a). Deciding when hate is a crime: The first amendment, police detectives, and the identification of hate crime. *Rutgers Race and Law Review, 4,* 33-76.

Bell, J. (2002b). *Policing hatred: Law enforcement, civil rights, and hate crime.* New York: New York University Press.

Bendick, M., Jr., Jackson, C. W., & Reinoso, V. A. (1999). Measuring employment discrimination through controlled experiments. In F. L. Pincus & H. J. Ehrlich (Eds.), *Race and ethnic conflict* (pp. 140-151). Boulder: Westview Press

Bennett, D. H. (1995). *The party of fear: The American far right from nativism to the militia movement* (Rev. ed.). New York: Vintage Books.

Bergmann, W. (1997). Antisemitism and xenophobia in Germany since unification. In H. Kurthen, W. Bergmann, & R. Erb (Eds.), *Antisemitism and xenophobia in Germany after unification* (pp. 21-38). New York: Oxford University Press.

Berlet, C., & Lyons, M. N. (2000). *Right-wing populism in American: Too close for comfort.* New York: Guilford Press.

Berman, E. G. (2001). The international commission of inquiry (Rwanda): Lessons and observations from the field. *American Behavioral Scientist, 45,* 616-625.

Berrill, K. T. (1992). Anti-gay violence and victimization in the United States: An overview. In G. M. Herek & K. T. Berrill (Eds.), *Hate crimes: Confronting violence against lesbians and gay men* (pp. 19-45). Newbury Park, CA: Sage.

Berry, J. W., & Kalin, R. (1995). Multicultural and ethnic attitudes in Canada: An overview of the 1991 national survey. *Canadian Journal of Behavioural Science, 27,* 301-320.

Bjørgo, T. (1998). Entry, bridge-burning, and exit options: What happens to young people who join racist groups—and want to leave. In J. Kaplan & T. Bjørgo (Eds.), *Nation and race: The developing Euro-American racist subculture* (pp. 231-258). Boston: Northeastern University Press.

Blazak, R. (2001). White boys to terrorist men: Target recruitment of Nazi skinheads. *American Behavioral Scientist, 44,* 982-1000.

Blee, K. M. (1991). *Women of the Klan: Racism and gender in the 1920s.* Berkeley: University of California Press.

Blee, K. M. (2002). *Inside organized racism: Women in the hate movement.* Berkeley: University of California Press.

Bobak, L. (2002, September 8). Muslim: We are not evil. *Toronto Sun,* p. 36.

Bohlen, A., Rafferty, K., & Ridgeway, J. (Directors). (1991). *Blood in the face.* United States: Right Thinking.

Bowling, B. (1994). Racial harassment in East London. In M. S. Hamm (Ed.), *Hate crime: International perspectives on causes and control* (pp. 1-36). Cincinnati: Anderson.

Boyd, E. A, Berk, R. A., & Hamner, K. M. (1996). "Motivated by hatred or prejudice": Categorization of hate-motivated crimes in two police divisions. *Law & Society Review, 30,* 819-850.

Boyle, K. (2001). Hate speech—The United States versus the rest of the world? *Maine Law Review, 53,* 487-502.

Brearly, M. (2001). The persecution of Gypsies in Europe. *American Behavioral Scientist, 45,* 588-599.

Brenner, C. (1992). Survivor's story: Eight bullets. In G. M. Herek & K. T. Berrill (Eds.), *Hate crimes: Confronting violence against lesbians and gay men* (pp. 11-13). Newbury Park, CA: Sage.

Brewer, M. B. (1988). A dual process model of impression formation. In T. K. Skrull & R. S. Wyer (Eds.), *Advances in social cognition* (Vol. 1, pp. 1-36). Hillsdale, NJ: Earlbaum.

Brigham, J. C. (1971). Ethnic stereotypes. *Psychological Bulletin, 76,* 15-38.

Broadus, T. (2000). Vote no if you believe in marriage: Lessons from the No On Knight/No On Proposition 22 campaign. *Berkeley Women's Law Journal, 15,* 1-13.

Brooks, T. D. (1994). First Amendment—Penalty enhancement for hate crimes: Content regulation, questionable state interests and non-traditional sentencing. *Journal of Criminal Law & Criminology, 84,* 703-742.

Brown, R. (1995). *Prejudice: Its social psychology.* Cambridge, MA: Blackwell.

Brown v. Board of Education of Topeka, Kansas, 347 U.S. 483 (1954).

Bruner, J. S. (1957). On perceptual readiness. *Psychological Review, 64,* 123-151.

Bruner, T. K. (2002, May 1). "Anti-Muslim" incidents on rise in U.S., group says. *Atlanta Journal and Constitution,* p. 14A.

Buchanan, P. J. (2001). *The death of the West: How dying populations and immigrant invasions imperil our country and civilization.* New York: Thomas Dunne Books.

Burbach, C. (2001, January 13). Brandon case back in court. *Omaha World-Herald,* p. 13.

Bureau of Justice Assistance. (1997). *A policy maker's guide to hate crimes.* Washington, DC: Author.

Bureau of Justice Assistance. (2001). *Hate crimes on campus: The problem and efforts to confront it.* Washington, DC: Author. Retrieved February 25, 2003, from www.ncjrs.org/pdffiles1/bja/187249.pdf

Bushart, H. L., Craig, J. R., & Barnes, M. (1998). *Soldiers of God: White supremacists and their holy war for America.* New York: Pinnacle Books.

Byers, B. D., & Crider, B. W. (2002). Hate crimes against the Amish: A qualitative analysis of bias motivation using routine activities theory. *Deviant Behavior, 23,* 115-148.

Byers, B. D., Crider, B. W., & Biggers, G. K. (1999). Bias crime motivation: A study of hate crime and offender neutralization techniques used against the Amish. *Journal of Contemporary Criminal Justice, 15,* 78-96.

Byrd, W. M., & Clayton, L. A. (2001). *An American health dilemma: Race, medicine, and health care in the United States* (Vol. 2). New York: Routledge.

California Attorney General. (2001). *Reporting hate crimes.* Retrieved February 16, 2003, from http://caag.state.ca.us

California Attorney General. (2002). *Hate crime in California: 2001.* Retrieved February 25, 2003, from http://caag.state.ca.us/cjsc/publications/hatecrimes/hc01/preface.pdf

California Department of Justice. (2001a). *Crime and delinquency in California: 2000.* Retrieved February 25, 2003, from http://caag.state.ca.us/cjsc/publications/candd/cd00/ar1.pdf

California Department of Justice. (2001b). *Hate crime in California: 2000.* Retrieved February 25, 2003, from http://caag.state.ca.us/cjsc/publications/hatecrimes/hc2000/preface.pdf

California Department of Justice. (2002). *Crime and delinquency in California: 2001.* Retrieved February 25, 2003, from http://ag.ca.gov/cjsc/publications/candd/cd01/cr1.pdf

California Penal Code § 190.2 (2003). Retrieved March 13, 2003, from www.leginfo.ca.gov/cgi-bin/displaycode?section=pen&group=00001-01000&file=187-199

California v. Joshua H., 17 Cal. Rptr. 2d 291 (Cal. App. 6 Dist. 1993).

Cambanis, T. (2002, May 17). UPS sued over alleged 9/11 hate crime. *Boston Globe,* p. B2.

Campbell, D. T. (1965). Ethnocentric and other altruistic motives. In D. Levine (Ed.), *Nebraska symposium on motivation* (pp. 283-311). Lincoln: University of Nebraska Press.

Candeub, A. (1994). Motive crimes and other minds. *University of Pennsylvania Law Review, 142,* 2071-2123.

Caplan, R., & Feffer, J. (Eds.). (1996). *Europe's new nationalism: States and minorities in conflict.* New York: Oxford University Press.

Carlsmith, J. M., Collins, B. E., & Helmreich, R. L. (1966). Studies in forced compliance: I. The effect of pressure for compliance on attitude change produced by face-to-face role playing and anonymous essay writing. *Journal of Personality and Social Psychology, 4,* 1-13.

Cart, J. (1999, December 31). Killer of gay student is spared death penalty. *Los Angeles Times,* p. A1.

Carter, C. S. (1999). Church burning in African American communities: Implications for empowerment practice. *Social Work, 44,* 62-68.

Chalmers, D. L. (1965). *Hooded Americanism: The first century of the Ku Klux Klan, 1865-1965.* Garden City, NY: Doubleday.

Chandler, M. (1998). *Forgotten fires* [Documentary film]. United States: UC Extension Media.

Chang, D. (1994). Beyond uncompromising positions: Hate crimes legislation and the common ground between conservative Republicans and gay rights advocates. *Fordham Urban Law Journal, 21,* 1097-1105.

Chen, K. (1997). Including gender in bias crime statutes: Feminist and evolutionary perspectives. *William & Mary Journal of Women & the Law, 3,* 277-328.

Chen, T. (2000). Hate violence as border patrol: An Asian American theory of hate violence. *Asian Law Journal, 7,* 69-101.

Chorba, C. (2001). The danger of federalizing hate crimes: Congressional misconceptions and the unintended consequences of the Hate Crimes Prevention Act. *Virginia Law Review, 87,* 319-379.

Civil Rights Acts, 18 U.S.C. §§ 241-242 (1948). Retrieved March 13, 2003, from www4.law.cornell. edu/uscode/18/241.html

Civil Rights Acts, 42 U.S.C. 1983 (1871). Retrieved March 13, 2003, from www4.law.cornell.edu/uscode/42/1983.html

Clancy, A., Hough, M., Aust, R., & Kershaw, C. (2001). *Crime, policing and justice: The experience of ethnic minorities* [Findings from the 2000 British Crime Survey; Home Office Research Study 223]. Retrieved February 25, 2003, from www.homeoffice.gov.uk/rds/pdfs/hors223.pdf

Clark, I., & Moody, S. (2002). *Racist crime and victimisation in Scotland.* Edinburgh: Scottish Executive Central Research Unit.

Clarkson, F. (1998). Anti-abortion extremism: Anti-abortion extremists, "patriots" and racists converge. *Intelligence Report, 91,* 8-12.

Cleary, E. J. (1994). *Beyond the burning cross: The First Amendment and the landmark R. A. V. case.* New York: Random House.

Coates, J. (1987). *Armed and dangerous: The rise of the survivalist right.* New York: Hill and Wang.

Cohan, S. (2002). The world was silent. *Teaching Tolerance, 22*(Fall), 50-56.

Cohen, A. (1964). *Attitude change and social influence.* New York: Basic Books.

Coliver, S. (Ed.). (1992). *Striking a balance: Hate speech, freedom of expression and non-discrimination.* London: Article 19.

Colorado Bureau of Investigation. (2002). *2001 hate crime report.* Retrieved February 19, 2003, from http://cbi.state.co.us/dr/cic2k1/supplemental_reports/hate_crime.htm

Comstock, G. D. (1989). Victims of anti-gay/lesbian violence. *Journal of Interpersonal Violence, 4,* 101-106.

Cook, W., Jr., & Kelly, R. J. (1999). The dispossessed: Domestic terror and political extremism in the American heartland. *International Journal of Comparative & Applied Criminal Justice, 23,* 246-256.

Cooperman, A. (2002, January 20). September 11 backlash murders and the state of "hate." *Washington Post,* p. A03.

Corcoran, J. (1990). *Bitter harvest: Gordon Kahl and the Posse Comitatus: Murder in the heartland.* New York: Penguin.

Craig, K. M. (1999). Retaliation, fear, or rage: An investigation of African American and white reactions to racist hate crimes. *Journal of Interpersonal Violence, 14,* 138-151.

Craig, K. M. (2002). Examining hate-motivated aggression: A review of the social psychological literature on hate crimes as a distinct form of aggression. *Aggression and Violent Behavior, 7,* 85-101.

Craig, K. M., & Waldo, C. R. (1996). "So what's a hate crime anyway?" Young adults' perceptions of hate crimes, victims and perpetrators. *Law & Human Behavior, 20,* 113-129.

Crosland, A. (Director). (1927). *The jazz singer* [Motion picture]. United States: Warner Brothers.

Crowder, C. (1998, October 11). Clinton saddened by Wyoming "hate" assault. *Denver Rocky Mountain News,* p. 5A.

Crowe, D. M. (1996). *A history of the Gypsies of Eastern Europe and Russia.* New York: St. Martin's Press.

Cunneen, C. (1997). Hysteria and hate: The vilification of Aboriginal and Torres Strait islander people. In C. Cunneen, D. Fraser, & S. Tomsen (Eds.), *Faces of hate: Hate crime in Australia* (pp. 137-161). Leichardt, NSW: Hawkins Press.

Cunneen, C., & de Rome, L. (1993). Monitoring hate crimes: A report on a pilot project in New South Wales. *Current Issues in Criminal Justice, 5*(2), 160-172.

Cunneen, C., Fraser, D., & Tomsen, S. (1997). Introduction: Defining the issues. In C. Cunneen, D. Fraser, & S. Tomsen (Eds.), *Faces of hate: Hate crime in Australia* (pp. 1-14). Leichardt, NSW: Hawkins Press.

Cutler, J. E. (1905). *Lynch-law: An investigation into the history of lynching in the United States.* New York: Longmans, Green.

Daniels, J. (1997). *White lies: Race, class, gender, and sexuality in white supremacist discourse.* New York: Routledge.

Davidson, O. G. (1996). *The best of enemies: Race and redemption in the New South.* New York: Scribner.

Dawson v. Delaware, 503 U.S. 159 (1992).

Dean, L., Wu, S., & Martin, J. L. (1992). Trends in violence and discrimination against gay men in New York City: 1984 to 1990. In G. M. Herek & K. T. Berrill (Eds.), *Hate crimes: Confronting violence against lesbians and gay men* (pp. 46-64). Newbury Park, CA: Sage.

Deci, E. (1975). *Intrinsic motivation.* New York: Plenum Press.

Dees, M., & Fiffer, S. (1993). *Hate on trial: The case against America's most dangerous neo-Nazi.* New York: Villard Books.

Degan, M. S. (1993). Adding the First Amendment to the fire: Cross burning and hate crime laws. *Creighton Law Review, 26,* 1109-1153.

Delgado, R. (1993). Words that wound: A tort action for racial insults, epithets, and name calling. In M. J. Matsuda, C. R. Lawrence III, R. Delgado, & K. W. Crenshaw (Eds.), *Words that wound: Critical race theory, assaultive speech, and the First Amendment* (pp. 89-110). Boulder, CO: Westview Press.

Department of Justice. (2001a). *Federal criminal case processing, 2000.* Retrieved February 25, 2003, from www.ojp.usdoj.gov/bjs/pub/pdf/fccp00.pdf

Department of Justice. (2001b). *State court prosecutors in large districts, 2001.* Retrieved March 13, 2003, from www.ojp.usdoj.gov/bjs/pub/pdf/scpld01.pdf

Department of Justice. (2002a). *Criminal victimization in the United States, 2000.* Retrieved February 25, 2003, from www.ojp.usdoj.gov/bjs/pub/pdf/cvus0005.pdf

Department of Justice. (2002b, April 10). *News conference with USA John Brownlee: Indictment of Darrell David Rice.* DOJ Conference Center, Washington, DC. Retrieved March 14, 2003, from www.usdoj.gov/ag/speeches/2002/041002newsconferenceindictment.htm

Devine, P. G. (1989). Stereotypes and prejudice: Their automatic and controlled components. *Journal of Personality and Social Psychology, 56,* 5-18.

Dharmapala, D., & Garoupa, N. (2001). *Penalty enhancement for hate crimes: An economic analysis.* Retrieved February 19, 2003, from http://papers.ssrn.com/paper.taf?abstract_id=268644

Dharmapala, D., & McAdams, R. (2001). *Words that kill: An economic perspective on hate speech and hate crimes.* Retrieved February 19, 2003, from http://papers.ssrn.com/abstract=300695

Diagnostic and statistical manual of mental disorders (4th ed., text revision). (2000). Arlington, VA: American Psychiatric Publishing.

Dickerson, D. (1997). *Clifford's Tower: Massacre at York (1190).* Retrieved February 24, 2003, from http://ddickerson.igc.org/cliffords-tower.html

Dillof, A. M. (1997). Punishing bias: An examination of the theoretical foundations of bias crime statutes. *Northwestern University Law Review, 91,* 1015-1081.

Dinnerstein, L. (1994). *Antisemitism in America.* New York: Oxford University Press.

Dion, K. L. (2001). Immigrants' perceptions of housing discrimination in Toronto: The Housing New Canadians Project. *Journal of Social Issues, 57,* 523-539.

DiPersio, V., & Guttentag, W. (2000). *Hate.com: Extremists on the Internet* [Documentary film]. United States: HBO.

Dixon, T., Jr. (1905). *The clansman.* New York: Grossett and Dunlap.

Dixon, T. L., & Linz, D. (2000). Race and the misrepresentation of victimization on local television news. *Communication Research, 27,* 547-573.

Dobbins v. Florida, 605 So.2d 922 (Fla. App. 5th 1992).

Dobratz, B. A., & Shanks-Meile, S. (1995). Conflict in the white supremacist/racialist movement in the United States. *International Journal of Group Tensions, 25,* 57-75.

Dobratz, B. A., & Shanks-Meile, S. (1997). *"White power, white pride!": The white separatist movement in the United States.* New York: Twayne Publishers.

Donahue, P. (1999, February 8). No parole in slay at Howard Beach. *New York Daily News,* p. 10.

Dorfman, L., & Schiraldi, V. (2001). *Off balance: Youth, race and crime in the news: Executive summary and full report.* Washington, DC: Building Blocks for Youth, Youth Law Center.

Dovidio, J. F., & Gaertner, S. L. (1991). Changes in the expression of racial prejudice. In H. Knopke, J. Norell, & R. Rogers (Eds.), *Opening doors: An appraisal of race relations in contemporary America* (pp. 201-241). Tuscaloosa: University of Alabama Press.

Downey, J. P., & Stage, F. K. (1999). Hate crimes and violence on college and university campuses. *Journal of College Student Development, 40,* 3-9.

Downs, D. A. (1985). Skokie revisited: Hate group speech and the first amendment. *Notre Dame Law Review, 60,* 629-685.

Dral, C. H., & Phillips, J. J. (2001). Commerce by another name: The impact of United States v. Lopez and United States v. Morrison. *Tennessee Law Review, 68,* 605-633.

Duke, D. (2001, September 17). *Will anyone dare to ask why?* Retrieved February 24, 2003, from www.davidduke.com/writings/09-17-01.shtml

Dunbar, E. (1999). Defending the indefensible: A critique and analysis of psycholegal defense arguments of hate crime perpetrators. *Journal of Contemporary Criminal Justice, 15,* 64-77.

Dunbar, E. (in press). Symbolic, relational, and ideological signifiers of bias motivated offenders: Toward a strategy of assessment. *American Journal of Orthopsychiatry.*

Edds, K. (2002, September 26). L.A. gay community unnerved by attacks. *Washington Post,* p. A04.

Esses, V. M., Dovidio, J. F., Jackson, L. M., & Armstrong, T. L. (2001). The immigration dilemma: The role of perceived group competition, ethnic prejudice, and national identity. *Journal of Social Issues, 57,* 389-412.

Ezekiel, R. S. (1995). *The racist mind: Portraits of American neo-Nazis and Klansmen.* New York: Viking.

Ezekiel, R. S. (2002). An ethnographer looks at neo-Nazi and Klan groups: *The racist mind* revisited. *American Behavioral Scientist, 46,* 51-71.

Faludi, S. (1992). *Backlash: The undeclared war against American women.* New York: Anchor.

Fangen, K. (1998). Living out our ethnic instincts: Ideological beliefs among right-wing activists in Norway. In J. Kaplan & T. Bjørgo (Eds.), *Nation and race: The developing Euro-American racist subculture* (pp. 202-230). Boston: Northeastern University Press.

Fazio, R. H., Jackson, J. R., Dunton, B. C., & Williams, C. J. (1995). Variability in automatic activation as an unobtrusive measure of racial attitudes: A bona fide pipeline? *Journal of Personality and Social Psychology, 69,* 1013-1027.

Federal Bureau of Investigation. (2001a). *Crime in the United States, 2000.* Retrieved February 25, 2003, from www.fbi.gov/ucr/cius_00/contents.pdf

Federal Bureau of Investigation. (2001b). *Hate crimes statistics, 1999.* Retrieved February 25, 2003, from www.fbi.gov/ucr/99hate.pdf

Federal Bureau of Investigation. (2001c). *Hate crime statistics, 2000.* Retrieved February 25, 2003, from www.fbi.gov/ucr/cius_00/hate00.pdf

Federal Bureau of Investigation. (2002). *Uniform Crime Reports: Crime trends, 2001 preliminary figures.* Retrieved February 25, 2003, from www.fbi.gov/ucr/01prelim.pdf

Ferber, A. L. (1998). *White man falling: Race, gender, and white supremacy.* Lanham, MD: Rowman & Littlefield.

Ferrante, J., & Brown, P., Jr. (1999). Classifying people by race. In F. L. Pincus & H. J. Ehrlich (Eds.), *Race and ethnic conflict* (pp. 14-23). Boulder, CO: Westview Press.

Festinger, L., & Carlsmith, J. M. (1959). Cognitive consequences of forced compliance. *Journal of Abnormal and Social Psychology, 58,* 203-210.

Fewster, D. (1999). *Racist violence, structural racism and the impact of legislative measures to confront racism.* Retrieved February 25, 2003, from *E-Valuate, 1*(1), www.law.ecel.uwa.edu.au/ elawjournal/Volume%201/Articles%20Vol_1/racistviolence.pdf

Fijalkowski, J. (1996). Aggressive nationalism and immigration in Germany. In R. Caplan & J. Feffer (Eds.), *Europe's new nationalism: States and minorities in conflict* (pp. 138-150). New York: Oxford University Press.

Fiske, S. T. (1993). Social cognition and social perception. *Annual Review of Psychology, 44,* 155-194.

Fiske, S. T., & Neuberg, S. L. (1990). A continuum of impression formation, from category-based to individuating processes: Influences of information and motivation on attention and interpretation. In M. P. Zanna (Ed.), *Advances in experimental and social psychology* (Vol. 23, pp. 1-74). New York: Academic Press.

Fleischauer, M. (1990). Teeth for a paper tiger: A proposal to add enforceability to Florida's hate crimes act. *Florida State University Law Review, 17,* 675-711.

Fleisher, M. (1994). Down the passage which we should not go: The folly of hate crime legislation. *Journal of Law and Policy, 2,* 1-50.

Florida v. Stalder, 630 So. 2d 1072 (1994).

Ford, P. (2002, May 15). Across Europe, the far right rises. *Christian Science Monitor.* Retrieved February 19, 2003, from www.csmonitor.com/2002/0515/p01s03-woeu.html

Franklin, K. (2000). Antigay behaviors among young adults: Prevalence, patterns, and motivators in a noncriminal population. *Journal of Interpersonal Violence, 15,* 339-362.

Franklin, K. (2002). Good intentions: The enforcement of hate crime penalty-enhancement statutes. *American Behavioral Scientist, 46,* 154-172.

Franklin, R. A. (2001, November 15). *The hate directory.* Retrieved February 19, 2003, from www.bcpl.net/~rfrankli/hatedir.htm

Frazier, D. (2002, June 4). Killer of teen gets 40 years. *Rocky Mountain News,* p. 16A.

Fredella, H. F., Carroll, M. R., Chamberlain, E., & Melendez, R. A. (2002). Sexual orientation, justice, and higher education: Student attitudes toward gay civil rights and hate crimes. *Law and Sexuality, 11,* 11-51.

Freeman, G. (1998, November 3). Hate crime laws are necessary to send clear message. *St. Louis Post-Dispatch,* p. B1.

Friebert, S. (1999, July 26). [Letter to the editor]. *Milwaukie Journal Sentinel,* p. 9.

Gaertner, S. L., & Dovidio, J. F. (1986). The aversive form of racism. In J. F. Dovidio & S. L. Gaertner (Eds.), *Prejudice, discrimination, and racism* (pp. 61-89). New York: Academic Press.

Gaertner, S. L., Dovidio, J. F., Anastasio, P. A., Bachman, B. A., & Rust, M. C. (1993). The Common Ingroup Identity Model: Recategorization and the reduction of intergroup bias. In W. Stroebe & M. Hewstone (Eds.), *European review of social psychology* (Vol. 4, pp. 1-26). London: Wiley.

Gaertner, S. L., Dovidio, J. F., Nier, J. A, Ward, C. M., & Banker, B. S. (1999). Across cultural divides: The value of a superordinate identity. In D. A. Prentice & D. T. Miller (Eds.), *Cultural divides: Understanding and overcoming group conflict* (pp. 173-212). New York: Russell Sage Foundation.

Gale, L. R., Heath, W. C., & Ressler, R. W. (1999). *An economic interpretation of hate crime.* Unpublished manuscript, University of Southwestern Louisiana, Lafayette.

Gardner, M. R. (1993). The mens rea enigma: Observations on the role of motive in the criminal law, past and present. *Utah Law Review, 1993,* 635-750.

Garland, J. A. (2001). The low road to violence: Governmental discrimination as a catalyst for pandemic hate crime. *Law and Sexuality: A Review of Gay, Bisexual, and Transgender Legal Issues, 10,* 1-91.

Garofalo, J. (1997). Hate crime victimization in the United States. In R. C. Davis, A. J. Lurigio, & W. G. Skogan (Eds.), *Victims of crime* (2d ed., pp. 134-145). Thousand Oaks, CA: Sage.

Garofalo, J., & Martin, S. E. (1993). The law enforcement response to bias-motivated crimes. In R. J. Kelley (Ed.), *Bias crime* (pp. 64-80). Chicago: Office of International Criminal Justice.

Gaumer, C. P. (1994). Punishment for prejudice: A commentary on the constitutionality and utility of state statutory responses to the problem of hate crimes. *South Dakota Law Review, 39,* 1-48.

Gelber, K. (2000). Hate crimes: Public policy implications of the inclusion of gender. *Australian Journal of Political Science, 35*(2), 275-289.

Gellman, S. (1991). Sticks and stones can put you in jail, but can words increase your sentence? Constitutional and policy dilemmas of ethnic intimidation laws. *UCLA Law Review, 39,* 333-396.

Gellman, S. (1992). "Brother, you can't go to jail for what you're thinking": Motives, effects, and "hate crime" laws. *Criminal Justice Ethics, 11,* 24-29.

George, J., & Wilcox, L. (1996). *American extremists: Militias, supremacists, Klansmen, communists, and others.* Amherst, NY: Prometheus Books.

German Embassy. (2001). *Background paper: Freedom of speech and recent legal controversies in Germany.* Retrieved February 19, 2003, from www.germany-info.org/relaunch/info/archives/background/speech.html

German Ministry of the Interior. (2001). *First periodical report on crime and crime control in Germany* (Abridged edition). Berlin: Federal Ministry of the Interior.

Gerstenfeld, P. B. (1992). Smile when you call me that! The problems with punishing hate motivated behavior. *Behavioral Sciences and the Law, 10,* 259-285.

Gerstenfeld, P. B. (1998). Reported hate crimes in America. *CSU Stanislaus Journal of Research, 2*(1), 35-43.

Gerstenfeld, P. B. (2002). A time to hate: Situational antecedents of intergroup bias. *Analyses of Social Issues and Public Policy, 2002,* pp. 61-67.

Gerstenfeld, P. B. (in press). Juror decision making in hate crime cases. *Criminal Justice Policy Review.*

Gerstenfeld, P. B., Grant, D. R., & Chiang, C. (2003). Hate online: A content analysis of extremist internet sites. *Analyses of Social Issues and Public Policy, 3*(1), 29-44.

Gey, S. G. (1997). What if Wisconsin v. Mitchell had involved Martin Luther King, Jr.? The constitutional flaws of hate crime enhancement statutes. *George Washington Law Review, 65,* 1014-1070.

Gibson, R. J. (2002). *The Negro Holocaust: Lynching and race riots in the United States, 1880-1950.* Retrieved February 19, 2003, from www.yale.edu/ynhti/curriculum/units/1979/2/79.02.04.x.html

Gilbert, A. M., & Marchand, E. D. (1999). Splitting the atom or splitting hairs—The Hate Crimes Prevention Act of 1999. *St. Mary's Law Journal, 30,* 931-986.

Gilbert, G. M. (1951). Stereotype persistence and change among college students. *Journal of Abnormal and Social Psychology, 46,* 245-254.

Gilmour, G. A. (1994). *Hate-motivated violence* [Funded by Canadian Department of Justice]. Retrieved February 19, 2003, from http://canada.justice.gc.ca/en/dept/pub/hmv/

Glass, D. (1964). Changes in liking as a means of reducing cognitive discrepancies between self-esteem and aggression. *Journal of Personality, 32,* 531-549.

Goldscheid, J., & Kaufman, R. E. (2001). Seeking redress for gender-based bias crimes: Charting new ground in familiar legal territory. *Michigan Journal of Race & Law, 6,* 265-283.

Goldsmith, A. (2002, June 7). Bigotry by the Bay. *Jerusalem Post,* p. 6B.

Goodman, M. (1964). *Race awareness in young children* (2nd ed.). New York: Crowell-Collier.

Gordon, R. (2002, September 19). Feds sue S.F. public housing authority. *San Francisco Chronicle,* p. A1.

Governor of Texas signs hate crimes bill. (2001, May 12). *New York Times,* p. A8.

Grannis, E. J. (1993). Fighting words and fighting freestyle: The constitutionality of penalty enhancement for bias crimes. *Columbia Law Review, 93,* 178-230.

Grattet, R., & Jenness, V. (2001a). The birth and maturation of hate crime policy in the United States. *American Behavioral Scientist, 45,* 668-696.

Grattet, R., & Jenness, V. (2001b). Examining the boundaries of hate crime law: Disabilities and the "dilemma of difference." *Journal of Criminal Law & Criminology, 91,* 653-697.

Grattet, R., Jenness, V., & Curry, T. R. (1998). The homogenization and differentiation of hate crime law in the United States, 1978-1995: Innovation and diffusion in the criminalization of bigotry. *American Sociological Review, 63,* 286-307.

Greason, D. (1997). Australia's racist far-right. In C. Cunneen, D. Fraser, & S. Tomsen (Eds.), *Faces of hate: Hate crime in Australia* (pp. 188-213). Leichardt, NSW: Hawkins Press.

Green, C. (1998, October 11). Outlaw hate in Wyoming. *Denver Post,* p. B-01.

Green, D. P., Abelson, R. P., & Garnett, M. (1999). The distinctive political views of hate-crime per-petrators and White supremacists. In D. A. Prentice & D. Miller (Eds.), *Cultural divides: Under standing and overcoming group conflict* (pp. 429-464). New York: Russell Sage Foundation.

Green, D. P., Glaser, J., & Rich, A. (1998). From lynching to gay bashing: The elusive connection between economic conditions and hate crime. *Journal of Personality and Social Psychology, 75,* 82-92.

Green, D. P., McFalls, L. H., & Smith, J. K. (2001). Hate crime: An emergent research agenda. *Annual Review of Sociology, 27,* 479-504.

Green, D. P., Strolovitch, D., & Wong, J. (1998). Defended neighborhoods, integration, and racially motivated crime. *American Journal of Sociology, 104,* 372-403.

Green, D. P., Strolovitch, D. Z., Wong, J. S., & Bailey, R. S. (2001). Measuring gay populations and antigay hate crime. *Social Science Quarterly, 82*(2), 281-296.

Greenawalt, K. (1995). *Fighting words: Individuals, communities, and liberties of speech.* Princeton, NJ: Princeton University Press.

Greenberg, S. J. (1993). *The evolution of state hate crime statutes: Social change through criminal law.* Unpublished manuscript.

Gremillion v. NAACP, 366 U.S. 293 (1961).

Griffith, D. W. (Producer/director). (1915). *The birth of a nation* [Motion picture]. United States: David W. Griffith Corp.

Grobman, G. M. (1990). *Classical and Christian anti-Semitism.* Retrieved February 24, 2003, from http://remember.org/History.root.classical.html

Haiman, F. S. (1993). *"Speech acts" and the First Amendment.* Carbondale: Southern Illinois University Press.

Hajela, D. (1999, August 25). Father of Yusuf Hawkins, Sharpton gather at site of youth's murder. Associated Press.

Hamm, M. (1993). *American skinheads: The criminology and control of hate crime.* Westport, CT: Praeger.

Hamm, M. S. (Ed.). (1994). *Hate crime: International perspectives on causes and control.* Cincinnati: Anderson.

Hamm, M. S. (1998). Terrorism, hate crime, and antigovernment violence: A review of the research. In H. Kushner (Ed.), *The future of terrorism: Violence in the new millennium* (pp. 59-96). Thousand Oaks, CA: Sage.

Harnishmacher, R., & Kelly, R. J. (1998). The neo-Nazis and skinheads of Germany: Purveyors of hate. In R. J. Kelly & J. Maghan (Eds.), *Hate crime: The global politics of polarization* (pp. 37-50). Carbondale: Southern Illinois University Press.

Harrison, J. V. (2000). Peeping through the closet keyhole: Sodomy, homosexuality, and the amorphous right of privacy. *St. John's Law Review, 74,* 1087-1138.

Hasenstab, D. (2001). Is hate a form of commerce? The questionable constitutionality of federal "hate crime" legislation. *St. Louis University Law Journal, 45,* 973-1017.

Heart of Atlanta v. United States, 379 U.S. 241 (1964).

Hegarty, S. (2002, August 30). 9/11 fuels hate crimes in Florida. *St. Petersburg Times,* p. 1A.

Heinzl, T. (1997, February 14). State hate-crimes law urged. *Omaha World Herald,* p. 13SF.

Herek, G. M. (1992). The social context of hate crimes: Notes on cultural heterosexism. In G. M. Herek & K. T. Berrill (Eds.), *Hate crimes: Confronting violence against lesbians and gay men* (pp. 89-104). Newbury Park, CA: Sage.

Herek, G. (1999). "Reparative therapy" and other attempts to alter sexual orientation: A background paper. Accessed February 25, 2003, from http://psychology.ucdavis.edu/rainbow/html/reptherapy.pdf

Herek, G. M. (2000a). The psychology of sexual prejudice. *Current Directions in Psychological Science, 9*(1), 19-22.

Herek, G. M. (2000b). Sexual prejudice and gender: Do heterosexuals' attitudes towards lesbians and gay men differ? *Journal of Social Issues, 56*(2), 252-266.

Herek, G. M. (2002a). Gender gaps in public opinion about lesbians and gay men. *Public Opinion Quarterly, 66*(1), 40-66.

Herek, G. M. (2002b). *Sexual prejudice: Prevalence.* Retrieved February 19, 2003, from http://psychology. ucdavis.edu/rainbow/html/prej_prev.html

Herek, G. M., & Berrill, K. T. (1992a). Documenting the victimization of lesbians and gay men: Methodological issues. In G. M. Herek & K. T. Berrill (Eds.), *Hate crimes: Confronting violence against lesbians and gay men* (pp. 270-286). Newbury Park, CA: Sage.

Herek, G. M., & Berrill, K. T. (Eds.). (1992b). *Hate crimes: Confronting violence against lesbians and gay men.* Newbury Park, CA: Sage.

Herek, G. M., & Capitanio, J. P. (1999). AIDS stigma and sexual prejudice. *American Behavioral Scientist, 42,* 1126-1143.

Herek, G. M., Capitanio, J. P., & Widaman, K. F. (2002). HIV-related stigma and knowledge in the United States: Prevalence and trends, 1991-1999. *American Journal of Public Health, 92*(3), 371-377.

Herek, G. M., Cogan, J. C., & Gillis, J. R. (2002). Victim experiences in hate crimes based on sexual orientation. *Journal of Social Issues, 58,* 319-339.

Herek, G. M., Gillis, J. R., & Cogan, J. C. (1999). Psychological sequelae of hate-crime victimization among lesbian, gay, and bisexual adults. *Journal of Consulting and Clinical Psychology, 67,* 945-951.

Herek, G. M., Gillis, J. R., Cogan, J. C., & Glunt, E. K. (1997). Hate crime victimization among lesbian, gay, and bisexual adults: Prevalence, psychological correlates, and methodological issues. *Journal of Interpersonal Violence, 12,* 195-215.

Herman, S. (2001). *The USA Patriot Act and the US Department of Justice: Losing our balances?* Retrieved March 8, 2003, from http://jurist.law.pitt.edu/forum/forumnew40.htm

Hernandez, T. K. (1990). Bias crimes: Unconscious racism in the prosecution of "racially motivated violence." *Yale Law Journal, 99,* 845-864.

Hershberger, S. L., & D'Augelli, A. R. (1995). The impact of victimization on the mental health and suicidality of lesbian, gay, and bisexual youth. *Developmental Psychology, 31,* 65-74.

Heumann, M., & Church, T. W. (Eds.). (1997). *Hate speech on campus: Cases, case studies, and commentary.* Boston: Northeastern University Press.

Hewstone, M., & Jaspars, J. M. F. (1984). Social dimensions of attribution. In H. Tajfel (Ed.), *The social dimension: European developments in social psychology* (pp. 379-404). Cambridge: Cambridge University Press.

Hockenos, P. (1993). *Free to hate: The rise of the right in post-communist Eastern Europe.* New York: Routledge.

Holzer, H. M. (Ed.). (1994). *Speaking freely: The case against speech codes.* Studio City, CA: Second Thoughts Books.

Home Secretary. (1999). *Stephen Lawrence inquiry: Home Secretary's action plan.* Retrieved February 25, 2003, at www.homeoffice.gov.uk/ppd/oppu/slpages.pdf

Hovland, C. I., & Sears, R. (1940). Minor studies in aggression: VI. Correlations of lynchings with economic indices. *Journal of Psychology, 9,* 301-310.

Howard, J. W., & Rothbart, M. (1980). Social categorization and memory for in-group and out-group behavior. *Journal of Personality and Social Psychology, 38,* 301-310.

Hughes, D. (2002, November 14). Tough on crime, but is it just talk. *London Daily Mail,* pp. 18-19.

Hughes, P. R. (2002, June 15). Big rise in hate crimes. *Houston Chronicle,* p. A-38.

Huigens, K. (2002). Solving the Apprendi puzzle. *Georgetown Law Journal, 90,* 387-459.

Hunter, J. (1992). Violence against lesbian and gay male youths. In G. M. Herek & K. T. Berrill (Eds.), *Hate crimes: Confronting violence against lesbians and gay men* (pp. 76-82). Newbury Park, CA: Sage.

Hurtz, W., & Durkin, K. (1997). Gender role stereotyping in Australian radio commercials. *Sex Roles, 36,* 103-114.

Hwang, V. M. (2001). The interrelationship between anti-Asian violence and Asian America. In P. W. Hall & V. M. Hwang (Eds.), *Anti-Asian violence in North America: Asian American and Asian Canadian reflections on hate, healing, and resistance* (pp. 43-66). Walnut Creek, CA: Rowman & Littlefield.

Ibish, H. (2001a, September 19). *ADC shocked by Congressman Cooksey's remarks* [on American Arab Anti-Discrimination Committee Web site]. Retrieved November 15, 2001, from www.adc.org/press/2001/19september2001.htm

Ibish, H. (2001b, September 21). *Anti-Arab hate crimes, discrimination continue* [on American Arab Anti-Discrimination Committee Web site]. Retrieved November 15, 2001, from www.adc.org/press/2001/21september2001.htm

Iganski, P. (1999a). Legislating against hate: Outlawing racism and anti-Semitism in Britain. *Critical Social Policy, 58,* 129-141.

Iganski, P. (1999b). Why make "hate" a crime? *Critical Social Policy, 60,* 386-395.

Iganski, P. (2001). Hate crimes hurt more. *American Behavioral Scientist, 45,* 626-638.

Illinois State Police. (2002). *Crime in Illinois 2001.* Retrieved March 13, 2003, from www.isp.state.il.us/crime/cii2002.htm

Illinois v. Nitz, 674 N.E.2d 802 (1996).

In protest of KKK rally, Schumer launches "project lemonade" to raise money for anti-bias. (1999, October 22). Retieved March 7, 2003, from http://schumer.senate.gov/SchumerWebsite/pressroom/press_releases/PR00063.html

Jackson, J. S., Brown, K. T., & Kirby, D. C. (1998). International perspectives on prejudice and racism. In J. L. Eberhardt & S. T. Fiske (Eds.), *Confronting racism: The problem and the response* (pp. 101-135). Thousand Oaks, CA: Sage.

Jacobs, J. B., & Potter, K. (1997). Hate crimes: A critical perspective. *Crime & Justice, 22,* 1-42.

Jacobs, J. B., & Potter, K. (1998). *Hate crimes: Criminal law and identity politics.* New York: Oxford University Press.

Janis, I. L. (1972). *Victims of groupthink: A psychological study of foreign-policy decisions and fiascoes.* Boston: Houghton Mifflin.

Janis, I. L. (1989). *Crucial decisions: Leadership in policymaking and crisis management.* New York: Free Press.

Jeansonne, G. (1996). *Women of the far right: The Mothers' Movement and World War II*. Chicago: University of Chicago Press.

Jenness, V., & Broad, K. (1997). *Hate crimes: New social movements and the politics of violence*. New York: Aldine de Gruyter.

Jenness, V., & Grattet, R. (1996). The criminalization of hate: A comparison of structural and polity influences on the passage of "bias-crime" legislation in the United States. *Sociological Perspectives, 39,* 129-154.

Jenness, V., & Grattet, R. (2001). *Making hate a crime: From social movement to law enforcement*. New York: Russell Sage Foundation.

Johnstone, G. (2002). *Restorative justice: Ideals, values, debates*. London: Willan Publishing.

Jolly-Ryan, J. (1999). Strengthening hate crime laws in Kentucky. *Kentucky Law Journal, 88,* 63-86.

Jones, J. M. (1999). The changing nature of prejudice. In F. L. Pincus & H. J. Ehrlich (Eds.), *Race and ethnic conflict* (pp. 65-76). Boulder, CO: Westview Press.

Jones, M. (1997). The legal response: Dealing with hatred—a user's guide. In C. Cunneen, D. Fraser, & S. Tomsen (Eds.), *Faces of hate: Hate crime in Australia* (pp. 214-241). Leichardt, NSW: Hawkins Press.

Jones, R. A. (1982). Perceiving other people: Stereotyping as a process of social cognition. In A. Miller (Ed.), *In the eye of the beholder: Contemporary issues in stereotyping* (pp. 41-91). New York: Praeger.

Judge doubles sentence. (1992, July 18). *Portland Oregonian*. Retrieved February 22, 2003, from www.oregonlive.com/search/oregonian/

Jury to get complicated assault case. (1992, June 16). *Portland Oregonian*. Retrieved February 22, 2003, from www.oregonlive.com/search/oregonian/

Kalin, R., & Berry, J. W. (1996). Interethnic attitudes in Canada: Ethnocentrism, consensual hierarchy and reciprocity. *Canadian Journal of Behavioural Science, 28,* 253-261.

Karlins, M., Coffman, T. L., & Walters, G. (1969). On the fading of social stereotypes: Studies in three generations of college students. *Journal of Personality and Social Psychology, 13,* 1-16.

Katz, D., & Braly, K. (1933). Racial stereotypes of one hundred college students. *Journal of Abnormal and Social Psychology, 28,* 282-290.

Katz, P. A. (1976). The acquisition of racial attitudes in children. In P. A. Katz (Ed.), *Towards the elimination of racism* (pp. 125-156). New York: Pergamon.

Katz, P. A. (1983). Developmental foundations of gender and racial attitudes. In R. L. Leahy (Ed.), *The child's construction of social inequality* (pp. 41-78). New York: Academic Press.

Katzenbach v. McClung, 379 U.S. 294 (1964).

Kaye, T. (Director). (1998). *American History X* [Motion picture]. United States: New Line Cinema.

Keane, F. (1995). *A season of blood: A Rwandan journey*. New York: Penguin.

Keeton, W. P., Dobbs, D. B., Keeton, R. E., & Owen, D. G. (1984). *Prosser and Keeton on torts* (5th ed.). St. Paul, MN: West.

Kellman, L. (2002, April 11). [Untitled Associated Press story]. Retrieved March 14, 2003, from www.aldha.org/arrest02.htm

Kelly, R. J., & Maghan, J. (Eds.). (1998). *Hate crime: The global politics of polarization*. Carbondale: Southern Illinois University Press.

Kelly, S., & Robinson, M. (2002, September 11). Muslims target of hate crimes: "Environment permanently changed." *Denver Post*, p. B-02.

Kinder, D. R., & Sears, D. O. (1981). Prejudice and politics: Symbolic racism versus racial threats to the good life. *Journal of Personality and Social Psychology, 40,* 414-431.

King, J. (2002). *Hate crime: The story of a dragging in Jasper, Texas*. New York: Pantheon.

King, L. L. (1969). *Confessions of a white racist*. New York: Viking Press.

Kinsella, W. (1994). *Web of hate: Inside Canada's far right network*. Toronto: Harper Collins.

Kite, M. E., & Whitley, B. E. (1998). Do heterosexual women and men differ in their attitudes towards homosexuality? A conceptual and methodological analysis. In G. M. Herek (Ed.), *Stigma and sexual orientation: Understanding prejudice against lesbians, gay men, and bisexuals* (pp. 39-61). Thousand Oaks, CA: Sage.

Kitschelt, H., & McGann, A. J. (1995). *The radical right in Western Europe: A comparative analysis*. Ann Arbor: University of Michigan Press.

Kleg, M. (1993). *Hate, prejudice and racism.* Albany: SUNY Press.

Knights Party. (1991). *The Knights' Party Platform.* Retrieved March 14, 2003, from www.kukluxklan. org/program.htm

Korem, D. (1995). *The art of profiling: Reading people right the first time.* Richardson, TX: International Focus Press.

Kühnel, W. (1998). Hitler's grandchildren? The reemergence of a right-wing social movement in Germany. In J. Kaplan & T. Bjørgo (Eds.), *Nation and race: The developing Euro-American racist subculture* (pp. 148-174). Boston: Northeastern University Press.

Kürti, L. (1998). The emergence of postcommunist youth identities in Eastern Europe: From communist youth, to skinheads, to national Socialists and beyond. In J. Kaplan & T. Bjørgo (Eds.), *Nation and race: The developing Euro-American racist subculture* (pp. 175-201). Boston: Northeastern University Press.

Landis, J. (Director). (1980). *The Blues Brothers* [Motion picture]. United States: Universal Pictures.

Lane, G. (1999, November 4). Jury convicts McKinney. *Denver Post,* p. A1.

Lawrence, C. R., III. (1992). If he hollers, let him go: Regulating racist speech on campus. M. J. Matsuda, C. R., Lawrence III, R. Delgado, & K. W. Crenshaw (Eds.), *Words that wound: Critical race theory, assaultive speech, and the First Amendment* (pp. 53-88). Boulder, CO: Westview Press.

Lawrence, F. M. (1994). The punishment of hate: Toward a normative theory of bias-motivated crimes. *Michigan Law Review, 93,* 320-381.

Lawrence, F. M. (1999). *Punishing hate: Bias crimes under American law.* Cambridge, MA: Harvard University Press.

Lawrence, F. M. (2000a, March 15). *Bias crime in a multi-cultural society.* Paper presented at the International Human Rights Conference, Eugene, Oregon. Retrieved March 13, 2003, from http://papers.ssrn.com/paper.taf?abstract_id=239320

Lawrence, F. M. (2000b). Federal bias crime law symposium. *Boston University Law Review, 80,* 1437-1449.

League for Human Rights of B'nai B'rith Canada. (1993). *Victim impact of racially motivated crime: A study conducted for the Commission on Systemic Racism in the Ontario Criminal Justice System.* Downsview, Ontario: Author.

League for Human Rights of B'nai B'rith Canada. (2002). *2002 interim audit of antisemitic incidents.* Downsview, Ontario: Author.

Leavitt, M. F. (2001). Keenan v. Aryan Nations: Making hate groups liable for the torts of their neighbors. *Idaho Law Review, 37,* 603-639.

Lee, M. A. (1997). *The beast reawakens: Fascism's resurgence from Hitler's spymasters to today's neo-Nazi groups and right-wing extremists.* New York: Routledge.

Lee, M. A. (2002). The swastika and the crescent. *Intelligence Report, 105,* 18-26.

Lee, M. K. (2001). Hate crime on the internet: The University of California, Irvine case. In P. W. Hall & V. M. Hwang (Eds.), *Anti-Asian violence in North America: Asian American and Asian Canadian reflections on hate, healing, and resistance* (pp. 67-76). Walnut Creek, CA: Rowman & Littlefield.

Leets, L. (2001). Responses to internet hate sites: Is speech too free in cyberspace? *Communication Law and Policy, 6,* 287-317.

Lefkowitz, B. (1998). *Our guys: The Glen Ridge rape and the secret life of the perfect suburb.* New York: Vintage Books.

Lemley, E. C. (2001). Designing restorative justice policy: An analytical perspective. *Criminal Justice Policy Review, 12,* 43-65.

Lepper, M. R., & Greene, D. (1975). Turning play into work: Effects of adult surveillance and extrinsic rewards on children's intrinsic motivation. *Journal of Personality and Social Psychology, 31,* 479-486.

LeVay, S. (1991). A difference in hypothalamic structure between heterosexual and homosexual men. *Science, 253,* 1034-1037.

Levenson, L. (1994). The future of state and federal civil rights prosecutions: The lessons of the Rodney King trial. *UCLA Law Review, 41,* 509-608.

Levin, B. (1993). A dream deferred: The social and legal implications of hate crimes in the 1990s. *The Journal of Intergroup Relations, 20*(3), 3-27.

Levin, B. (1998). The patriot movement: Past, present, and future. In H. Kushner (Ed.), *The future of terrorism: Violence in the new millennium* (pp. 97-131). Thousand Oaks, CA: Sage.

Levin, B. (1999). Hate crimes: Worse by definition. *Journal of Contemporary Criminal Justice, 15,* 1-21.

Levin, B. (2001a). History as a weapon: How extremists deny the Holocaust in North America. *American Behavioral Scientist, 44,* 1001-1031.

Levin, B. (2001b). The vindication of hate violence victims via criminal and civil adjudications. *Journal of Hate Studies, 1,* 133-165.

Levin, B. (2002). Cyberhate: A legal and historical analysis of extremists' use of computer networks in America. *American Behavioral Scientist, 45,* 958-988.

Levin, J. (2002). *The violence of hate: Confronting racism, anti-Semitism, and other forms of bigotry.* Boston: Allyn & Bacon.

Levin, J., & McDevitt, J. (1993). *Hate crimes: The rising tide of bigotry and bloodshed.* New York: Plenum.

Li, P. S. (2001). The racial subtext in Canada's immigration discourse. *International Journal of Migration and Integration, 2*(1), 77-97.

Linville, P. W., & Jones, E. E. (1980). Polarized appraisals of out-group members. *Journal of Personality and Social Psychology, 38,* 689-703.

Linville, P. W., Salovey, P., & Fischer, G. W. (1986). Stereotyping and perceived distributions of social characteristics: An application to ingroup-outgroup perception. In J. Dovidio & S. L. Gaertner (Eds.), *Prejudice, discrimination, and racism* (pp. 165-208). New York: Academic Press.

Luzadder, D. (1999, January 15). Hate crime bill passes hurdle. *Rocky Mountain News,* p. 7A.

MacGinty, R. (2001). Ethno-national conflict and hate crime. *American Behavioral Scientist, 45,* 639-653

Mackie, D. M., & Worth, L. T. (1990). Differential recall of subcategory information about in-group and out-group members. *Personality and Social Psychology Bulletin, 15,* 401-413.

Macpherson Report. (1999, February 24). Retrieved March 5, 2003, from www.archive.official-documents.co.uk/document/cm42/4262/sli-00.htm

Macrae, C. N., Stangor, C., & Milne, A. B. (1994). Activating social stereotypes: A functional analysis. *Journal of Experimental Social Psychology, 30,* 370-389.

Madani, A. O. (2000). *Depiction of Arabs and Muslims in the United States news media.* Unpublished doctoral dissertation, California School of Professional Psychology, Alameda, CA.

Maddison, D. S. (n.d.). *A brief chronology of anti-Semitism.* Retrieved February 24, 2003, from www.geocities.com/Athens/Cyprus/8815/chrono.html

Mallourides, E., & Turner, G. (2002). Control of immigration: Statistics 2001 [U.K. Home Office]. Retrieved February 25, 2003, from www.homeoffice.gov.uk/rds/pdfs2/hosb1102.pdf

Marcus-Newhall, A., Blake, L. P., & Baumann, J. (2002). Perceptions of hate crime perpetrators and victims as influenced by race, political orientation, and peer group. *American Behavioral Scientist, 46,* 108-135.

Maroney, T. A. (1998). The struggle against hate crime: Movement at a crossroads. *New York University Law Review, 73,* 564-620.

Martin, S. E. (1995). "A cross-burning is not just an arson": Police social construction of hate crimes in Baltimore County. *Criminology, 33*(3), 303-326.

Mason, M. (2002, November 15). Army tosses gay foreign language students. *San Francisco Chronicle,* p. A21.

Massachusetts General Laws Chapter 265, § 39(b). Retrieved March 13, 2003, from www.state.ma.us/legis/laws/mgl/265%2D39.htm

Mastro, D. E., & Robinson, A. L. (2000). Cops and crooks: Images of minorities on primetime television. *Journal of Criminal Justice, 28,* 385-396.

Matsuda, M. J. (1993). Public response to racist speech: Considering the victim's story. In M. J. Matsuda, C. R. Lawrence III, R. Delgado, & K. W. Crenshaw (Eds.), *Words that wound: Critical race theory, assaultive speech, and the First Amendment* (pp. 17-52). Boulder, CO: Westview Press.

Matsuda, M. J., Lawrence, C. R., III, Delgado, R., & Crenshaw, K. W. (Eds.). (1993). *Words that wound: Critical race theory, assaultive speech, and the First Amendment.* Boulder, CO: Westview Press.

Mazur-Hart, H. L. (1982). Racial and religious intimidation: An analysis of Oregon's 1981 law. *Willamette Law Review, 18,* 197-218.

McConahay, J. B. (1986). Modern racism, ambivalence, and the Modern Racism Scale. In J. F. Dovidio & S. L. Gaertner (Eds.), *Prejudice, discrimination, and racism* (pp. 91-125). New York: Academic Press.

McCullen, K. (1998, October 10). Attack on student in Laramie. *Denver Rocky Mountain News,* p. 7A.

McCurrie, T. F. (1998). White supremacist gang members: A behavioral profile. *Journal of Gang Research, 5*(2), 51-60.

McDevitt, J. (1990, November). *The study of the character of civil rights crimes in Massachusetts, 1983-1989.* Paper presented at the meeting of the American Society of Criminology, Reno, Nevada.

McDevitt, J., Balboni, J., Garcia, L., & Gu, J. (2001). Consequences for victims: A comparison of bias- and non-bias-motivated assaults. *American Behavioral Scientist, 45,* 697-713.

McDevitt, J., Levin, J., & Bennett, S. (2002). Hate crime offenders: An expanded typology. *Journal of Social Issues, 58,* 303-317.

McGlade, H. (1997). The international prohibition of racist organisations: An Australian perspective. *Murdoch University Electronic Journal of Law, 7*(1), www.murdoch.edu.au/elaw/issues/ v7n1/mcglade71_text.html

McLaughlin, K. A., Brilliant, K., & Lang, C. (1995). *National bias crimes training for law enforcement and victim assistance professionals.* Washington, DC: Office for Victims of Crime.

McPhail, B. A. (2002). Gender-bias hate crimes: A review. *Trauma, Violence, and Abuse, 3*(2), 125-143.

Medoff, M. H. (1999). Allocation of time and behavior: A theoretical and positive analysis of hate and hate crimes. *American Journal of Economics and Sociology, 58,* 959-973.

Melton, G. B., & Saks, M. J. (1986). The law as an instrument of socialization and social structure. In G. B. Melton (Ed.), *Nebraska Symposium on Motivation: Vol. 33. The Law as a Behavioral Instrument* (pp. 235-277). Lincoln: University of Nebraska Press.

Merton, R. K. (1957). *Social theory and social structure.* Glencoe, IL: Free Press.

Metzger, T. (n.d.). [Response to a question about when parents should teach their children about racism]. Retrieved February 15, 2003, from www.resist.com/updates/2002updates/8.25.02aryanupdate.htm

Micacci, R. (1995). Wisconsin v. Mitchell: Punishable conduct v. protected thought. *New England Journal of Criminal and Civil Confinement, 21,* 131ff.

Milgram, S. (1974). *Obedience to authority.* New York: Harper and Row.

Milner, D. (1975). *Children and race.* London: Penguin.

Miniclier, K. (1998, October 10). Intimidation complaints neglected, lawmakers say. *Denver Post,* p. A16.

Moore, J. B. (1993). *Skinheads shaved for battle: A cultural history of American skinheads.* Bowling Green, OH: Bowling Green State University Press.

Moran, K. (2002, February 23). Santa Fe ISD settled lawsuit that charges anti-Semitism. *Houston Chronicle,* p. A39.

Morris, E. (Director). (1999). *Mr. Death: The rise and fall of Fred A. Leuchter, Jr.* [Documentary film]. United States: Lions Gate.

Morris, J. R. (1991). Racial attitudes of undergraduates in Greek housing. *College Student Journal, 25,* 501-505.

Morton, M. L. (2001). Hitler's ghosts: The interplay between international organizations and their member states in response to the rise of neo-nazism in society and government. *Georgia Journal of International & Comparative Law, 30,* 71-100.

Mouzos, J., & Thompson, S. (2000). Gay-hate related homicides: An overview of findings in New South Wales. *Trends and Issues in Crime and Criminal Justice, 155,* 1-6.

Mukherjee, S. (1999). *Ethnicity and crime: An Australian research study* [Australian Institute of Criminology]. Retrieved February 25, 2003, from www.aic.gov.au/publications/ethnicity-crime/ ethnic.pdf

NAACP v. Alabama, 357 U.S. 449 (1958).

NAACP v. Alabama, 377 U.S. 288 (1964).

NAACP v. Button, 371 U.S. 415 (1963).

National Alliance Web site. (n.d.). *General principles: The law of inequality*. Retrieved February 18, 2003, from www.natall.com/what-is-na/na1.html#law

National Asian Pacific American Legal Consortium. (2001). *Executive Summary of 2000 audit of violence against Asian Pacific Americans*. Retrieved March 8, 2003, from www.napalc.org/literature/annual_report/2000.htm

National Asian Pacific American Legal Consortium. (2002). *Backlash: When America turned on its own*. Retrieved March 13, 2003, from www.napalc.org/literature/annual_report/9-11_report.htm

National Church Arson Task Force. (2000, September 15). *National Church Arson Task Force issues fourth report*. Retrieved February 24, 2003, from www.atf.treas.gov/press/fy00press/091500ncatf4th.htm

National Coalition of Anti-Violence Programs. (2002). *Anti-lesbian, gay, bisexual and transgender violence in 2001*. Retrieved February 25, 2003, from www.lambda.org/2001ncavpbiasrpt.pdf

National Gay & Lesbian Task Force. (2001). *Hate crime laws in the U.S.—June 2001*. Retrieved February 25, 2003, from www.ngltf.org/downloads/hatemap0601.pdf

National Institute Against Prejudice and Violence. (1986). *Striking back at bigotry: Remedies under federal and state law for violence motivated by racial, religious, and ethnic violence*. Baltimore: Author.

National Institute Against Prejudice and Violence. (1989). *National victimization survey*. Baltimore: Author.

Nelson, J. (1993). *Terror in the night: The Klan's campaign against the Jews*. Jackson: University Press of Mississippi.

Nelson, T. D. (2002). *The psychology of prejudice*. Boston: Allyn & Bacon.

Nemes, I. (1997). Antisemitic hostility. In C. Cunneen, D. Fraser, & S. Tomsen (Eds.), *Faces of hate: Hate crime in Australia* (pp. 44-74). Leichardt, NSW: Hawkins Press.

Newton, M., & Newton, J. A. (1991). *Racial and religious violence in America: A chronology*. New York: Garland.

Nickerson, S., Mayo, C., & Smith, A. (1986). Racism in the courtroom. In J. P. Dovidio & S. L. Gaertner (Eds.), *Prejudice, discrimination, and racism*. San Diego: Academic Press.

Noelle, M. (2002). The ripple effect of the Matthew Shepard murder: Impact on the assumptive worlds of members of the targeted group. *American Behavioral Scientist, 46,* 27-50.

Nolan, J. J., & Akiyama, Y. (1999). An analysis of factors that affect law enforcement participation in hate crime reporting. *Journal of Contemporary Criminal Justice, 15,* 111-127.

Nolan, J. J., Akiyama, Y., & Berhanu, S. (2002). The Hate Crime Statistics Act of 1990: Developing a method for measuring the occurrence of hate violence. *American Behavioral Scientist, 46,* 136-153.

Novick, M. (1995). *White lies, white power: The fight against white supremacy and reactionary violence*. Monroe, ME: Common Courage Press.

Official Georgia Code Annotated, § 16-11-38 (2003). Retrieved March 13, 2003, from www.legis.state.ga.us/cgi-bin/gl_codes_detail.pl?code=16-11-38

Ohio v. Wyant, 597 N.E.2d 450 (1992).

Oliver, M. B. (1994). Portrayals of crime, race, and aggression in "reality-based" police shows: A content analysis. *Journal of Broadcasting and Electronic Media,* 179-192.

Olivero, J. M., & Murataya, R. (2001). Homophobia and university law enforcement students. *Journal of Criminal Justice Sciences, 12,* 271-281.

O'Neill, P., & Miller, R. (Producers). (1995). *Not in our town* [Documentary film broadcast on PBS]. United States: The Working Group.

One People's Project. (2001, November 14). *Ann Coulter angers students with university visit*. Retrieved March 7, 2003, from www.onepeoplesproject.com/coulter2.htm

Operario, D., & Fiske, S. T. (1998). Racism equals power plus prejudice: A social psychological equation for racial oppression. In J. L. Eberhardt & S. T. Fiske (Eds.), *Confronting racism: The problem and responses* (pp. 33-53). Thousand Oaks, CA: Sage.

Orange County Human Relations Commission. (2002). *Hate crimes and incidents in Orange County in the year 2001*. Retrieved February 25, 2003, from www.oc.ca.gov/csa/hrc/programs/HC.pdf

Oregon v. Beebe, 680 P.2d 11 (1984).

Oregon v. Plowman, 838 P.2d 558 (1992).

Ostendorf, D. (2001). Christian Identity: An American heresy. *Journal of Hate Studies, 1,* 24-55.

Owyoung, B. H. (1999). *The psychological effects and treatment of hate crime victimization on Chinese-Americans and Asian-Americans residing in Castro Valley, California.* Unpublished doctoral dissertation, California School of Professional Psychology, Alameda, California.

Peek, G. S. (2001). Where are we going with federal hate crimes legislation? Congress and the politics of sexual orientation. *Marquette Law Review, 85,* 537-577.

Pendo, E. A. (1993). Recognizing violence against women: Gender and the Hate Crime Statistics Act. *Harvard Women's Law Journal, 17,* 157-183.

Peritz, I. (1994, November 26). Hate-crimes legislation calls for harsher penalties. *Montreal Gazette,* p. A4.

Perry, B. (2000). "Button-down terror": The metamorphosis of the hate movement. *Sociological Focus, 33,* 113-131.

Perry, B. (2001). *In the name of hate: Understanding hate crimes.* New York: Routledge.

Perry, B. (2002). Defending the color line: Racially and ethnically motivated hate crime. *American Behavioral Scientist, 46,* 72-92.

Petrosino, C. (1999). Connecting the past to the future: Hate crime in America. *Journal of Contemporary Criminal Justice, 15,* 22-47.

Pfeifer, J. E., & Ogloff, J. R. (1991). Ambiguity and guilt determinations: A modern racism perspective. *Journal of Applied Social Psychology, 21,* 1713-1725.

Pierce, K. (Director). (1999). *Boys don't cry* [Motion picture]. United States: Fox Searchlight Pictures.

Pierce, W. (1978). *The Turner diaries.* Fort Lee, NJ: Barricade Books.

Pilkington, N. W., & D'Augelli, A. R. (1995). Victimization of lesbian, gay, and bisexual youth in community settings. *Journal of Community Psychology, 23,* 33-56.

Pincus, F. L., & Ehrlich, H. J. (Eds.). (1999). *Race and ethnic conflict.* Boulder, CO: Westview Press.

Pitcavage, M. (2001). Camouflage and conspiracy: The militia movement from Ruby Ridge to Y2K. *American Behavioral Scientist, 44,* 957-981.

Poll: Anti-Semitism rising in Germany. (2002, September 7). *Modesto Bee,* p. A-15.

Prager, D., & Telushkin, J. (1983). *Why the Jews? The reason for anti-Semitism.* New York: Simon & Schuster.

Quarles, C. L. (1999). *The Ku Klux Klan and related American racialist and anti-Semitic organizations: A history and analysis.* Jefferson, NC: McFarland & Co.

R. A. V. v. City of St. Paul, 505 U.S. 377 (1992).

Rahman, A. U. (1998). Hate crimes in India: A historical perspective. In R. J. Kelly & J. Maghan (Eds.), *Hate crime: The global politics of polarization* (pp. 111-134). Carbondale: Southern Illinois University Press.

Ramet, S. P. (1999). Defining the radical right: The values and behaviors of organized intolerance. In S. P. Ramet (Ed.), *The radical right in Central and Eastern Europe since 1989* (pp. 3-28). University Park: Pennsylvania State University Press.

Redish, M. H. (1992). Freedom of thought as freedom of expression: Hate crime sentencing enhancement and First Amendment theory. *Criminal Justice Ethics,* 29-42.

Rennison, C. M. (2000). *Criminal victimization 1999: Changes 1998-99 with trends 1993-99.* (NCJ No. 182734). Washington, DC: Bureau of Justice Statistics. Retrieved March 13, 2003, from www.ojp.usdoj.gov/bjs/pub/pdf/cv99.pdf

Repeat cemetery vandal gets 5-year sentence. (2002, November 17). *Bergen County Record,* p. A06.

Rey, A. M., & Gibson, P. R. (1997). Beyond high school: Heterosexuals' self-reported anti-gay/lesbian behaviors and attitudes. *Journal of Gay & Lesbian Social Services, 7*(4), 65-84.

Reynolds, K. J., Turner, J. C., Haslam, S. A., & Ryan, M. K. (2001). The role of personality and group factors in explaining prejudice. *Journal of Experimental Social Psychology, 37,* 427-434.

Richards v. Florida, 608 So. 2d 917 (Fla. App. 1992).

Ridgeway, J. (1995). *Blood in the face: The Ku Klux Klan, Aryan Nations, Nazi skinheads, and the rise of a new white culture.* New York: Thunder's Mouth Press.

Roberts, J. V. (1995). *Disproportionate harm: Hate crime in Canada.* Ottawa: Canadian Department of Justice.

Roberts, J. V., & Hastings, A. J. A. (2001). Sentencing in cases of hate-motivated crime: An analysis of subparagraph 718.2(a)(i) of the Criminal Code. *Queen's Law Journal, 27,* 93-126.

Rohan, M. J., & Zanna, M. P. (1996). Value transmission in families. In C. Seligman, J. M. Olson, & M. P. Zanna (Eds.), *The psychology of values: The Ontario symposium* (vol. 8, pp. 253-276). Mahwah, NJ: Erlbaum.

Rohter, L. (1991, August 25). Without smiling, to call Floridian a 'cracker' may be a crime. *New York Times,* sec. 1, p. 1.

Rokeach, M. (1956). Political and religious dogmatism: An alternative to the authoritarian personality. *Psychological Monographs, 70*(18), whole issue.

Ronson, J. (2002). *Them: Adventures with extremists.* New York: Simon & Schuster.

Rosenthal, S. (1998, October 23). Hate-crime laws send society a message. *Rocky Mountain News,* p. 69A.

Ross, J. I. (1994). Hate crime in Canada: Growing pains with new legislation. In M. S. Hamm (Ed.), *Hate crime: International perspectives on causes and control* (pp. 151-172). Cincinnati: Anderson.

Rossi, R. (1992, April 28). Hate case jury sends message. *Chicago Sun-Times,* p. 4.

Rothbart, M., Evans, M., & Fulero, S. (1979). Recall for confirming events: Memory processes and the maintenance of social stereotypes. *Journal of Experimental Social Psychology, 15,* 343-355.

Rushton, J. P. (2000). *Race, evolution, and behavior: A life history perspective.* Port Huron, MI: Charles Darwin Research Institute.

Samolinski, C. J. (2001, December 8). Slurs won't keep family from home. *Tampa Tribune,* p. 1.

Santiago, R. (1999, December 21). Race-slay victim honored. *New York Daily News,* p. 1.

Schmaltz, W. H. (1999). *Hate: George Lincoln Rockwell and the American Nazi Party.* Washington: Brassey's.

Schope, R. D., & Eliason, M. J. (2000). Thinking versus acting: Assessing the relationship between heterosexual attitudes and behaviors toward homosexuals. *Journal of Gay & Lesbian Social Services, 11*(4), 69-92.

Sears, D. O. (1998). Racism and politics in the United States. In J. L. Eberhardt & S. T. Fiske (Eds.), *Confronting racism: The problem and responses* (pp. 76-100). Thousand Oaks, CA: Sage.

Selbin, J. D. (1993). Bashers beware: The continuing constitutionality of hate crimes after R. A. V. *Oregon Law Review, 72,* 157ff.

Shapiro, A. (Producer). (1999). *Teen files: The truth about hate* [Documentary film]. United States: Arnold Shapiro and Allison Grodner Productions.

Shenk, A. H. (2001). Victim-offender mediation: The road to repairing hate crime injustice. *Ohio State Journal on Dispute Resolution, 17,* 185-217.

Sherif, M., Harvey, O. J., White, B. J., Hood, W. R., & Sherif, C. W. (1961). *Intergroup conflict and cooperation: The Robber's Cave experiment.* Norman: University of Oklahoma, Institute of Group Relations.

Sibbitt, R. (1997). *The perpetrators of racial harassment and racial violence* [Home Office Research Study 176]. Retrieved February 25, 2003, from www.homeoffice.gov.uk/rds/pdfs/hors176.pdf

Siegel, M. L. (1999). Hate speech, civil rights, and the internet: The jurisdictional and human rights nightmare. *Albany Law Journal of Science and Technology, 9,* 375-398.

Simms, P. (1996). *The Klan.* Lexington, KY: University Press of Kentucky.

Simon Wiesenthal Center. (2002a). *Digital Hate 2002* [CD-ROM]. Los Angeles: Author.

Simon Wiesenthal Center. (2002b). Europe: An intolerable climate of hate. *Response,* 1-4. Retrieved March 13, 2003, from www.wiesenthal.com/social/pdf/index.cfm?ItemID=6491

Simon Wiesenthal Center. (2002c). *The making of a skinhead: Interactive Q & A.* Retrieved March 8, 2003, from www.wiesenthal.com/taskforce/qanda2.cfm

Sloan, L. M., King, L., & Sheppard, S. (1998). Hate crimes motivated by sexual orientation: Police reporting and training. *Journal of Gay & Lesbian Social Services, 8*(3), 25-39.

Sloane, R. L. (2002). Barbarian at the gates: Revisiting the case of Matthew F. Hale to reaffirm that character and fitness evaluations appropriately preclude racists from the practice of law. *Georgetown Journal of Legal Ethics, 15,* 397-439.

Smith, B. L., & Damphousse, K. R. (1998). Two decades of terror: Characteristics, trends, and prospects for the future of American terrorism. In H. Kushner (Ed.), *The future of terrorism: Violence in the new millennium* (pp. 132-156). Thousand Oaks, CA: Sage.

Smith, E. W. (2001). Apprendi v. New Jersey: The United States Supreme Court restricts judicial discretion and raises troubling constitutional questions concerning sentencing statutes and reforms nationwide. *Arkansas Law Review, 54,* 649-702.

Smith, T. W. (1996). Anti-Semitism in contemporary America: A review. *Research in Micropolitics, 5,* 125-178.

Smolla, R. A. (1992). *Free speech in an open society.* New York: Vintage Books.

Sommers, S. R., & Ellsworth, P. C. (2000). Race in the courtroom: Perceptions of guilt and dispositional attributions. *Personality and Social Psychology Bulletin, 14,* 694-708.

Soule, S. A., & Earle, J. (2001). The enactment of state-level hate crime law in the United States: Intrastate and interstate factors. *Sociological Perspectives, 44,* 281-305.

Soule, S. A., & Van Dyke, N. (1999). Black church arson in the United States, 1989-1996. *Ethnic and Racial Studies, 22,* 724-742.

South Asian American Leaders of Tomorrow. (2001). *American backlash: Terrorists bring war home in more ways than one.* Retrieved February 25, 2003, from www.saalt.org/biasreport.pdf

Southern Poverty Law Center. (2001, fall). Ties that bind. *Intelligence Report, #103.* Retrieved March 7, 2003, from www.splcenter.org/intelligenceproject/ip-4s3.html

Southern Poverty Law Center. (2002a). *Active "patriot" groups in the United States in 2001.* Retrieved March 7, 2003, from www.splcenter.org/intelligenceproject/ip-index.html

Southern Poverty Law Center. (2002b, summer). "Patriot" free fall. *Intelligence Report, #106,* 30-31. Retrieved March 15, 2003, from www.splcenter.org/intelligenceproject/ip-index.html

Southern Poverty Law Center. (2002c, summer). The puppeteer. *Intelligence Report, #106,* 44-51. Retrieved March 15, 2003, from www.splcenter.org/intelligenceproject/ip-index.html

Southern Poverty Law Center. (2002d). *10 ways to fight hate.* Retrieved March 7, 2003, from www.tolerance.org/10_ways/unite/05.html

Southern Poverty Law Center. (2002e). *U.S. map of hate groups.* Retrieved March 10, 2003, from www.tolerance.org/maps/hate/index.html

Stanton, B. (1992). *Klanwatch: Bringing the Ku Klux Klan to justice.* New York: Mentor Books.

Stern, K. B. (2001). Hate and the internet. *Journal of Hate Studies, 1,* 55-108.

Stern, S. (2001). *Jews in Germany today: Dramatic change, dramatic growth.* Retrieved March 8, 2003, from www.germany-info.org/relaunch/info/archives/background/jews.html

Stern-Larosa, C., & Bettmann, E. H. (2000). *Hate hurts: How children learn and unlearn prejudice.* New York: Scholastic.

Strom, K. J. (2001). *Hate crimes reported in NIBRS, 1997-99.* (NCJ No. 186765). Washington, DC: Bureau of Justice Statistics.

Strossen, N. (1990). Regulating racist speech on campus: A modest proposal? *Duke law Journal, 1990,* 484-573.

Sutherland, E., & Cressey, E. (1970). *Principles of criminology.* New York: J. B. Lippincott.

Sykes, G., & Matza, D. (1957). Techniques of neutralization. *American Sociological Review, 22,* 664-670.

Szaniszlo, M. (2002, September 25). Study: 9/11 fuels anti-Arab crime. *Boston Herald,* p. 003.

Tajfel, H. (1969). Cognitive aspects of prejudice. *Journal of Social Issues, 25*(4), 79-97.

Tang, E., Ho, M., Thompson, J. P., Kim, U., & Ganz, M. (2001). Divide and conquer: The challenges of multiracial politics: Communities organizing against anti-Asian violence. *New York University Review of Law and Social Change, 27,* 31-62.

Tang, S. H., & Hall, V. C. (1995). The overjustification effect: A meta-analysis. *Applied Cognitive Psychology, 9,* 365-404.

Taslitz, A. E. (1999). Condemning the racist personality: Why the critics of hate crimes legislation are wrong. *Boston College Law Review, 40,* 739-785.

Temple-Raston, D. (2002). *A death in Texas: A story of race, murder, and a small town's struggle for redemption.* New York: Henry Holt & Co.

Terkel, S. (1992). *Race: How blacks and whites think and feel about the American obsession.* New York: Doubleday.

Texas Department of Public Safety. (2001). *2000 Crime in Texas.* Retrieved February 25, 2003, from www.txdps.state.tx.us/crimereports/Crime%20in%20Texas%20TOC.pdf

Texas v. Johnson, 491 U.S. 397 (1989).

Thompson, S. K. (1975). Gender labels and early sex-role development. *Child Development, 46,* 339-347.

Tiby, E. (2001). Victimization and fear among lesbians and gay men in Stockholm. *International Review of Victimology, 8,* 217-243.

Tolnay, S. T., & Beck, E. M. (1995). *A festival of violence: An analysis of Southern lynchings, 1882-1930.* Urbana: University of Illinois Press.

Tomsen, S. (2001, June). Hate crimes and masculinity: New crimes, new responses and some familiar patterns. Paper presented at the 4th National Outlook Symposium on Crime, Canberra, Australia.

Torres, S. (1999). Hate crimes against African Americans: The extent of the problem. *Journal of Contemporary Criminal Justice, 15,* 48-63.

Totten, S., Parsons, W. S., & Charny, I. W. (1997). *Century of genocide: Eyewitness accounts and critical views.* New York: Garland.

Tough hate-crime laws urged after beating. (1998, October 11). *Milwaukie Journal Sentinel,* p. 28.

Tsesis, A. (2000). The empirical shortcomings of first amendment jurisprudence: A historical perspective on the power of hate speech. *Santa Clara Law Review, 40,* 729-780.

Turner, J., Stanton, B., Vahala, M., & Williams, R. (1982). *The Ku Klux Klan: A history of racism and violence.* Montgomery, AL: Southern Poverty Law Center.

Turpin-Petrosino, C. (2002). Hateful sirens . . . who hears their song? An Examination of student attitudes toward hate groups and affiliation potential. *Journal of Social Issues, 58,* 281-301.

Ugwuegbu, D. C. E. (1979). Racial and evidential factors in juror attribution of legal responsibility. *Journal of Experimental Social Psychology, 15,* 133-146.

Uhrich, C. L. (1999). Hate crime legislation: A policy analysis. *Houston Law Review, 36,* 1467-1529.

Umbreit, M. S., Coates, R. B., & Vos, B. (2001). The impact of victim-offender mediation: Two decades of research. *Federal Probation, 65*(3), 29-35.

United States Census (2001). *2000 statistical abstract of the United States.* Retrieved February 25, 2003, from www.census.gov/prod/2001pubs/statab/sec01.pdf

United States v. Lopez, 514 U.S. 549 (1995).

United States v. Morrison, 529 U.S. 598 (2000).

United States v. Russell, 471 U.S. 858 (1985).

United States v. Schwimmer, 279 U.S. 644 (1929).

Van Ness, D. W., & Strong, K. H. (2001). *Restoring justice* (2nd ed.). Cincinnati: Anderson.

Vermont Statutes Annotated, Title 13 § 1456 (2003). Retrieved March 13, 2003, from www.leg.state. vt.us/statutes/sections.cfm?Title=13&Chapter=031

von Schulthess, B. (1992). Violence in the streets: Anti-lesbian assault and harassment in San Francisco. In G. M. Herek & K. T. Berrill (Eds.), *Hate crimes: Confronting violence against lesbians and gay men* (pp. 65-75). Newbury Park, CA: Sage.

Walker, N. (1980). *Punishment, danger, and stigma: The morality of criminal justice.* Lanham, MD: Rowman & Littlefield.

Walker, S. (1994). *Hate speech: The history of an American controversy.* Lincoln: University of Nebraska Press.

Walker, S., Delone, M., & Spohn, C. S. (1999). *The color of justice: Race, ethnicity, and crime in America.* Belmont, CA: Wadsworth.

Walker, S., & Katz, C. M. (1995). Less than meets the eye: Police department bias-crime units. *American Journal of Police, 14,* 29-48.

Waller, J. (2001). Perpetrators of genocide: An explanatory model of extraordinary human evil. *Journal of Hate Studies, 1,* 5-22.

Wang, L. (1997). The transforming power of "hate": Social cognition theory and the harms of bias-related crime. *Southern California Law Review, 71,* 47-135.

Wang, L. (2002). Hate crime and everyday discrimination: Influences of and on the social context. *Rutgers Race & Law Review, 4,* 1-31.

Warmbir, S. (2001a, June 1). Lawsuit accuses cops of abusing gay man. *Chicago Sun-Times,* 14.

Warmbir, S. (2001b, January 12). Man sues city, says cops beat him. *Chicago Sun-Times,* p. 14.

Wassmuth, B., & Bryant, M. J. (2001). Not in our world: A perspective of community organizing against hate. *Journal of Hate Studies, 1,* 109-131.

Watts, M. W. (2001). Aggressive youth cultures and hate crime: Skinheads and xenophobic youth in Germany. *American Behavioral Scientist, 45,* 600-615.

Weinberg, L. (1998). An overview of right-wing extremism in the Western world: A study of convergence, linkage, and identity. In J. Kaplan & T. Bjørgo (Eds.), *Nation and race* (pp. 3-33). Boston: Northeastern University Press.

Weine, S. M. (1999). *When history is a nightmare: Lives and memories of ethnic cleansing in Bosnia-Herzegovina.* New Brunswick, NJ: Rutgers University Press.

Weisburd, S. B., & Levin, B. (1994). "On the basis of sex": Recognizing gender-based bias crimes. *Stanford Law and Policy Review, 5,* 21-43.

Weiss, A., & Chermak, S. M. (1998). The news value of African-American victims: An examination of the media's presentation of homicide. *Journal of Crime and Justice, 21*(2), 71-88.

Weissman, E. (1992). Kids who attack gays. In G. M. Herek & K. T. Berrill (Eds.), *Hate crimes: Confronting violence against lesbians and gay men* (pp. 170-178). Newbury Park, CA: Sage.

Weller, P., Feldman, A., & Purdam, K. (2001). *Religious discrimination in England and Wales* [Home Office Research Study 220]. Retrieved February 25, 2003, from www.homeoffice.gov.uk/rds/pdfs/hors220.pdf

Wetzel, J. (1997). Antisemitism among right-wing extremist groups, organizations, and parties in postunification Germany. In H. Kurthen, W. Bergmann, & R. Erb (Eds.), *Antisemitism and xenophobia in Germany after unification* (pp. 159-173). New York: Oxford University Press.

White, R., & Perrone, S. (2001). Racism, ethnicity and hate crime. *Communal/Plural, 9*(2), 161-181.

White Aryan Resistance. (2002). *Aryan Update August 25, 2002.* Retrieved February 18, 2003, from www.resist.com/updates/2002updates/8.25.02aryanupdate.htm

Wilson, S., & Greider-Durango, J. (1998). Social cleansing in Colombia: The war on street children. In R. J. Kelly & J. Maghan (Eds.), *Hate crime: The global politics of polarization* (pp. 135-149). Carbondale: Southern Illinois University Press.

Winkel, F. W. (1997). Hate crime and anti-racism campaigning: Testing the Ø approach of portraying stereotypical information-processing. *Issues in Criminological and Legal Psychology, 29,* 14-19.

Wisconsin Statutes §§ 939.50, 939.645, 940.19, 943.012. (2003). Retrieved March 13, 2003, from www.legis.state.wi.us/rsb/stats.html

Wisconsin v. Mitchell, 485 N.W.2d 807 (1992).

Wisconsin v. Mitchell, 508 U.S. 476 (1993).

Yang, A. S. (1998). *From wrongs to rights: Public opinion on gay and lesbian Americans moves toward equality.* Retrieved March 13, 2003, from www.ngltf.org/downloads/yang.pdf

Yi, M. (2001, July 28). East Bay dairy sued by workers. *San Francisco Chronicle,* p. A16.

Yinger, J. M. (1964). *Anti-Semitism: A case study in prejudice and discrimination.* New York: Freedom Books.

Zimbardo, P. G. (1969). *The cognitive control of motivation.* Glencoe, IL: Scott, Foresman.

Zimbardo, P. G., Haney, C., Banks, W. C., & Jaffe, D. (1974). The psychology of imprisonment: Privation, power and pathology. In Z. Rubin (Ed.), *Doing unto others: Explorations in social behavior* (pp. 61-73). Englewood Cliffs, NJ: Prentice-Hall.

Zittrain, J., & Edelman, B. (2002). *Localized Google search result exclusions.* Retrieved March 10, 2003, from http://cyber.law.harvard.edu/filtering/google/

Zollo, F., & Colesberry, R. F. (Producers), & Parker, A. (Director). (1988). Mississippi burning [Motion picture]. United States: Orion Pictures.

INDEX

All page numbers appearing in bold type refer to **boxed text**, pages that contain *tables* or *figures* appear in italic type.

ABOUT THE AUTHOR

Phyllis B. Gerstenfeld has a B.A. in Psychology from Reed College, an M.A. and a Ph.D. in Psychology from the University of Nebraska–Lincoln, and a J.D. from the University of Nebraska. She is an Associate Professor of Criminal Justice at California State University, Stanislaus, where she has taught a class on hate crimes for the past 8 years. In addition to hate crimes, her research interests include juvenile justice, capital punishment, and psychology and law.